Methods in Molecular Biology

Series Editor
John M. Walker
School of Life and Medical Sciences
University of Hertfordshire
Hatfield, Hertfordshire, UK

For further volumes:
http://www.springer.com/series/7651

For over 35 years, biological scientists have come to rely on the research protocols and methodologies in the critically acclaimed *Methods in Molecular Biology* series. The series was the first to introduce the step-by-step protocols approach that has become the standard in all biomedical protocol publishing. Each protocol is provided in readily-reproducible step-by step fashion, opening with an introductory overview, a list of the materials and reagents needed to complete the experiment, and followed by a detailed procedure that is supported with a helpful notes section offering tips and tricks of the trade as well as troubleshooting advice. These hallmark features were introduced by series editor Dr. John Walker and constitute the key ingredient in each and every volume of the *Methods in Molecular Biology* series. Tested and trusted, comprehensive and reliable, all protocols from the series are indexed in PubMed.

Recombinant Glycoproteins

Methods and Protocols

Edited by

Steven B. Bradfute

Center for Global Health, University of New Mexico Health Sciences Center, Albuquerque, NM, USA

Humana Press

Editor
Steven B. Bradfute
Center for Global Health
University of New Mexico Health Sciences Center
Albuquerque, NM, USA

ISSN 1064-3745 ISSN 1940-6029 (electronic)
Methods in Molecular Biology
ISBN 978-1-0716-3665-7 ISBN 978-1-0716-3666-4 (eBook)
https://doi.org/10.1007/978-1-0716-3666-4

This Humana imprint is published by the registered company Springer Science+Business Media, LLC, part of Springer Nature.
The registered company address is: 1 New York Plaza, New York, NY 10004, U.S.A.

Paper in this product is recyclable.

Preface

Glycoproteins are simply defined as proteins that have attached carbohydrate groups. However, glycoprotein biology is extraordinarily complex. Glycoproteins play a vital role in many biological processes, and modifications in the type, placement, and amount of glycosylation can greatly affect protein structure and function. There are different types of glycosylation, such as attachment to asparagine (N-linked) or serine or threonine (O-linked) residues. There is a vast array of glycans that can be attached to proteins, and glycan composition varies widely in different organisms and cell types. Therefore, it is important to consider glycosylation when studying glycoproteins.

Producing and purifying glycoproteins for therapeutic use or basic research requires an in-depth understanding of the various cellular and acellular systems available for these purposes, their advantages and disadvantages, and considerations for choosing the most appropriate system for the desired application. Recent advances in glycoprotein production have greatly expanded the tools available for studying these proteins. This book centers on the production and purification of glycoproteins, including viral-encoded glycoproteins (Part I) and those found in other pathogens (Part II). Production of mammalian glycoproteins is covered by Part III. Great strides have been made in analyzing glycoprotein content, as is described in Part IV. Finally, Part V covers cell-free synthesis of glycoproteins and considerations for production and purification of glycoproteins. I hope this book provides a wide range of guidelines for studying these vitally important proteins.

Albuquerque, NM, USA *Steven B. Bradfute*

Contents

Contributors

MOYINOLUWA ADENIYI • *Department of Chemistry and Biochemistry, Texas Tech University, Lubbock, TX, USA*

PARISA AHMADI • *Department of Chemistry and Biochemistry, Texas Tech University, Lubbock, TX, USA*

M. JAVAD AMAN • *AbVacc, Rockville, MD, USA*

ANKUR ANNAPAREDDY • *Department of Molecular BioSciences, University of Texas at Austin, Austin, TX, USA*

MOJGAN ATASHI • *Department of Chemistry and Biochemistry, Texas Tech University, Lubbock, TX, USA*

ROCHELLE AW • *Department of Chemical and Biological Engineering, Northwestern University, Evanston, IL, USA*

JASON BAARDSNES • *Quality Attributes and Characterization, Human Health Therapeutics, National Research Council Canada, Montréal, QC, Canada*

CHANDRA KANTH BANDI • *Robert F. Smith School of Chemical and Biomolecular Engineering, Cornell University, Ithaca, NY, USA*

ANDREW I. BENNETT • *Department of Chemistry and Biochemistry, Texas Tech University, Lubbock, TX, USA*

MD MOSTOFA AL AMIN BHUIYAN • *Department of Chemistry and Biochemistry, Texas Tech University, Lubbock, TX, USA*

ERIK J. BIDSTRUP • *Robert F. Smith School of Chemical and Biomolecular Engineering, Cornell University, Ithaca, NY, USA*

LOUIS BISSON • *Human Health Therapeutics Research Centre, National Research Council of Canada, Montreal, QC, Canada*

PAUL J. BRETT • *Department of Microbiology and Immunology, University of Nevada, Reno School of Medicine, Reno, NV, USA; Department of Microbiology and Immunology, Faculty of Tropical Medicine, Mahidol University, Bangkok, Thailand*

ALINA BURLACU • *Human Health Therapeutics Research Centre, National Research Council of Canada, Montreal, QC, Canada*

MARY N. BURTNICK • *Department of Microbiology and Immunology, University of Nevada, Reno School of Medicine, Reno, NV, USA; Department of Microbiology and Immunology, Faculty of Tropical Medicine, Mahidol University, Bangkok, Thailand*

BRIAN CASS • *Human Health Therapeutics Research Centre, National Research Council of Canada, Montreal, QC, Canada*

ELIZABETH C. CLARKE • *Center for Global Health, Division of Infectious Diseases, Department of Internal Medicine, University of New Mexico, Albuquerque, NM, USA*

JOHN COOPER • *Department of Molecular BioSciences, University of Texas at Austin, Austin, TX, USA*

LOCHLAIN CORLISS • *Department of Microbiology, University of Alabama at Birmingham, Birmingham, AL, USA*

GREGORY DE CRESCENZO • *Department of Chemical Engineering, Polytechnique Montreal, Montreal, QC, Canada*

OLUWATOSIN DARAMOLA • *Department of Chemistry and Biochemistry, Texas Tech University, Lubbock, TX, USA*

LAURENCE DELAFOSSE • *Human Health Therapeutics Research Centre, National Research Council of Canada, Montreal, QC, Canada*

MATTHEW P. DELISA • *Robert F. Smith School of Chemical and Biomolecular Engineering, Cornell University, Ithaca, NY, USA; Cornell Institute of Biotechnology, Cornell University, Ithaca, NY, USA*

YVES DUROCHER • *Human Health Therapeutics Research Centre, National Research Council of Canada, Montreal, QC, Canada; Department of Biochemistry and Molecular Medicine, University of Montreal, Montreal, QC, Canada*

OSHADI EDIRISINGHE • *Department of Chemistry and Biochemistry, University of Arkansas, Fayetteville, AR, USA*

ILYA J. FINKELSTEIN • *Department of Molecular BioSciences, University of Texas at Austin, Austin, TX, USA; Center for Systems and Synthetic Biology, University of Texas at Austin, Austin, TX, USA*

CATHERINE FOREST-NAULT • *Department of Chemical Engineering, Polytechnique Montreal, Montreal, QC, Canada*

MOJIBOLA FOWOWE • *Department of Chemistry and Biochemistry, Texas Tech University, Lubbock, TX, USA*

BENJAMIN FRANCIS • *Department of Microbiology, Icahn School of Medicine at Mount Sinai, New York, NY, USA; Center for Vaccine Research and Pandemic Preparedness (C-VaRPP), Icahn School of Medicine at Mount Sinai, New York, United States*

JIMMY GAUDREAULT • *Department of Chemical Engineering, Polytechnique Montreal, Montreal, QC, Canada*

MONA GOLI • *Department of Chemistry and Biochemistry, Texas Tech University, Lubbock, TX, USA*

PABLO GUARDADO-CALVO • *G5 Structural Biology of Infectious Diseases, Institut Pasteur, Université Paris Cité, Paris, France*

CRISTIAN D. GUTIERREZ REYES • *Department of Chemistry and Biochemistry, Texas Tech University, Lubbock, TX, USA*

MD ABDUL HAKIM • *Department of Chemistry and Biochemistry, Texas Tech University, Lubbock, TX, USA*

LISA HAUEIS • *Fraunhofer Institute for Cell Therapy and Immunology (IZI), Branch Bioanalytics and Bioprocesses (IZI-BB), Potsdam, Germany; Institute of Biochemistry and Biology, University of Potsdam, Potsdam, Germany*

OLIVIER HENRY • *Department of Chemical Engineering, Polytechnique Montreal, Montreal, QC, Canada*

KAMYAB JAVANMARDI • *Department of Molecular BioSciences, University of Texas at Austin, Austin, TX, USA*

MICHAEL C. JEWETT • *Department of Chemical and Biological Engineering, Northwestern University, Evanston, IL, USA; Department of Bioengineering, Stanford University, Stanford, CA, USA*

PEILIN JIANG • *Department of Chemistry and Biochemistry, Texas Tech University, Lubbock, TX, USA*

SHWETA KAILASAN • *AbVacc, Rockville, MD, USA*

KEEHUN KIM • *Robert F. Smith School of Chemical and Biomolecular Engineering, Cornell University, Ithaca, NY, USA*

IZEL KOYUTURK • *Human Health Therapeutics Research Centre, National Research Council of Canada, Montreal, QC, Canada*

FLORIAN KRAMMER • *Department of Microbiology, Icahn School of Medicine at Mount Sinai, New York, NY, USA; Center for Vaccine Research and Pandemic Preparedness (C-VaRPP), Icahn School of Medicine at Mount Sinai, New York, NY, USA; Department of Pathology, Molecular and Cell Based Medicine, Icahn School of Medicine at Mount Sinai, New York, NY, USA*

STEFAN KUBICK • *Fraunhofer Institute for Cell Therapy and Immunology (IZI), Branch Bioanalytics and Bioprocesses (IZI-BB), Potsdam, Germany; Institute of Chemistry and Biochemistry, Freie Universität Berlin, Berlin, Germany; Faculty of Health Sciences, Joint Faculty of the Brandenburg University of Technology Cottbus–Senftenberg, The Brandenburg Medical School Theodor Fontane and the University of Potsdam, Potsdam, Germany*

THALLAPURANAM KRISHNASWAMY SURESH KUMAR • *Department of Chemistry and Biochemistry, University of Arkansas, Fayetteville, AR, USA*

YONG HYUN KWON • *Robert F. Smith School of Chemical and Biomolecular Engineering, Cornell University, Ithaca, NY, USA*

DENIS L'ABBÉ • *Human Health Therapeutics Research Centre, National Research Council of Canada, Montreal, QC, Canada*

NICHOLAS J. LENNEMANN • *Department of Microbiology, University of Alabama at Birmingham, Birmingham, AL, USA*

MADHUMATHI LOGANATHAN • *Department of Microbiology, Icahn School of Medicine at Mount Sinai, New York, NY, USA; Center for Vaccine Research and Pandemic Preparedness (C-VaRPP), Icahn School of Medicine at Mount Sinai, New York, United States*

WENDY MAURY • *Department of Microbiology and Immunology, University of Iowa, Iowa City, IA, USA*

YEHIA MECHREF • *Department of Chemistry and Biochemistry, Texas Tech University, Lubbock, TX, USA*

ANNALISA MEOLA • *G5 Structural Biology of Infectious Diseases, Institut Pasteur, Université Paris Cité, Paris, France*

DAMIR MOGUT • *Department of Food Biochemistry, Faculty of Food Science, University of Warmia and Mazury in Olsztyn, Olsztyn, Poland*

MOHAMMAD NAGHIZADEH • *Department for Congenital Disorders, Statens Serum Institut, Copenhagen, Denmark; Centre for Medical Parasitology at Department of Immunology and Microbiology, University of Copenhagen, Copenhagen, Denmark*

HECTOR R. NAJERA GONZALEZ • *Institute of Genomics for Crop Abiotic Stress Tolerance, Department of Plant and Soil Science, Texas Tech University, Lubbock, TX, USA*

MADELEINE NOONAN-SHUEH • *AbVacc, Rockville, MD, USA*

JUDITH NWAIWU • *Department of Chemistry and Biochemistry, Texas Tech University, Lubbock, TX, USA*

EBENEZER ADDO OFORI • *Department for Congenital Disorders, Statens Serum Institut, Copenhagen, Denmark; Centre for Medical Parasitology at Department of Immunology and Microbiology, University of Copenhagen, Copenhagen, Denmark*

CARSTEN OHLHOFF • *Fraunhofer Institute for Cell Therapy and Immunology (IZI), Branch Bioanalytics and Bioprocesses (IZI-BB), Potsdam, Germany*

PATIENCE SALVALINA OKOTO • *Department of Chemistry and Biochemistry, University of Arkansas, Fayetteville, AR, USA*

SHERIFDEEN ONIGBINDE • *Department of Chemistry and Biochemistry, Texas Tech University, Lubbock, TX, USA*

CAITLYN E. ORNE • *Department of Microbiology and Immunology, University of Nevada, Reno School of Medicine, Reno, NV, USA*
KATHARINA PASCHINGER • *Institut für Biochemie, Department für Chemie, Universität für Bodenkultur, Vienna, Austria*
BÉATRICE PAUL-ROC • *Quality Attributes and Characterization, Human Health Therapeutics, National Research Council Canada, Montréal, QC, Canada*
ALEX PELLETIER • *Human Health Therapeutics Research Centre, National Research Council of Canada, Montreal, QC, Canada*
PHUC PHAN • *Department of Chemistry and Biochemistry, University of Arkansas, Fayetteville, AR, USA*
JORDAN PLIESKATT • *Department for Congenital Disorders, Statens Serum Institut, Copenhagen, Denmark*
WAZIHA PURBA • *Department of Chemistry and Biochemistry, Texas Tech University, Lubbock, TX, USA*
ANDREA QUEZADA • *Department of Molecular BioSciences, University of Texas at Austin, Austin, TX, USA*
VISHAL SANDILYA • *Department of Chemistry and Biochemistry, Texas Tech University, Lubbock, TX, USA*
AKEEM SANNI • *Department of Chemistry and Biochemistry, Texas Tech University, Lubbock, TX, USA*
SUSHEEL K. SINGH • *Biotherapeutic and Vaccine Research Division, ICMR-Regional Medical Research Centre, Bhubaneswar, Odisha, India*
SHIVAKUMAR SONNAILA • *Department of Chemistry and Biochemistry, University of Arkansas, Fayetteville, AR, USA*
MARLITT STECH • *Fraunhofer Institute for Cell Therapy and Immunology (IZI), Branch Bioanalytics and Bioprocesses (IZI-BB), Potsdam, Germany*
LAUREN STUART • *Mapp Biopharmaceutical, San Diego, CA, USA*
MATTHEW STUIBLE • *Human Health Therapeutics Research Centre, National Research Council of Canada, Montreal, QC, Canada*
GAETANE TERNIER • *Department of Chemistry and Biochemistry, University of Arkansas, Fayetteville, AR, USA*
MICHAEL THEISEN • *Department for Congenital Disorders, Statens Serum Institut, Copenhagen, Denmark; Centre for Medical Parasitology at Department of Immunology and Microbiology, University of Copenhagen, Copenhagen, Denmark*
JESSICA ULLRICH • *Fraunhofer Institute for Cell Therapy and Immunology (IZI), Branch Bioanalytics and Bioprocesses (IZI-BB), Potsdam, Germany; Institute of Biotechnology, Technische Universität Berlin, Berlin, Germany*
FEDERICO URBANO-MUNOZ • *Department of Microbiology and Immunology, University of Nevada, Reno School of Medicine, Reno, NV, USA*
JORICK VANBESELAERE • *Institut für Biochemie, Department für Chemie, Universität für Bodenkultur, Vienna, Austria*
JUNYAO WANG • *Department of Chemistry and Biochemistry, Texas Tech University, Lubbock, TX, USA*
IAIN B. H. WILSON • *Institut für Biochemie, Department für Chemie, Universität für Bodenkultur, Vienna, Austria*
ANNE ZEMELLA • *Fraunhofer Institute for Cell Therapy and Immunology (IZI), Branch Bioanalytics and Bioprocesses (IZI-BB), Potsdam, Germany*

Part I

Viral Surface Glycoproteins

Chapter 1

Production and Purification of Hantavirus Glycoproteins in *Drosophila melanogaster* S2 Cells

Annalisa Meola and Pablo Guardado-Calvo

Abstract

Hantaviruses, are rodent-borne viruses found worldwide that are transmitted to humans through inhalation of contaminated excreta. They can cause a renal or a pulmonary syndrome, depending on the virus, and no effective treatment is currently available for either of these diseases. Hantaviral particles are covered by a protein lattice composed of two glycoproteins (Gn and Gc) that mediate adsorption to target cells and fusion with endosomal membranes, making them prime targets for neutralizing antibodies. Here we present the methodology to produce soluble recombinant glycoproteins in different conformations, either alone or as a stabilized Gn/Gc complex, using stably transfected Drosophila S2 cells.

Key words Hantavirus, Bunyavirus, Glycoproteins, Protein expression, S2, Insect cells

1 Introduction

1.1 Hantavirus Glycoproteins

Hantaviruses are rodent-borne viruses present worldwide and transmitted to humans by breathing contaminated aerosols. New-World hantaviruses, such as Andes (ANDV) and Sin Nombre (SNV) viruses, produce a respiratory disease termed hantavirus cardiopulmonary syndrome (HCPS) with up to 40% fatality rates [1]. Old-World hantaviruses, which include Puumala (PUUV) or Hantaan (HNTV) virus, produce a renal syndrome denominated hemorrhagic fever with renal syndrome (HFRS), less deadly but with a higher incidence. Neither of the two diseases has any approved treatments or vaccines, and patients only receive supportive care to ease their symptoms.

The surface of hantaviruses is covered by a protein lattice consisting of two glycoproteins termed Gn and Gc [2]. These proteins mediate virion assembly [3, 4], adsorption to the plasma membrane of susceptible cells [5], and fusion in the acidic lumen of endosomes [6]. They are the only targets of neutralizing antibodies produced during infection [7–9]. Gc is a class-II fusion protein [6, 10] with

Steven B. Bradfute (ed.), *Recombinant Glycoproteins: Methods and Protocols*, Methods in Molecular Biology, vol. 2762, https://doi.org/10.1007/978-1-0716-3666-4_1,

three beta-structured domains and Gn is a companion protein with two globular regions denominated head (Gn^H) and base (Gn^B) [11]. This organization is strictly conserved throughout the family and the boundaries of Gn^H, Gn^B, and Gc are easily distinguishable in the amino acidic sequence (Figs. 1a and 2) [12]. During biogenesis, Gn and Gc fold together to form a metastable Gn/Gc heterodimer [13, 14], which further associates to form the spikes that make up the outer lattice mentioned above (Fig. 1b). Structural analysis of the recombinant glycoproteins revealed three different conformations of Gc: the pre-fusion conformation, which is acquired in complex with Gn on the viral surface [13, 15]; a transient intermediate conformation [6], which is formed after dissociation of the heterodimer at acidic pH; and a stable post-fusion conformation [6, 10], which is a homotrimer formed after the fusion reaction (Fig. 1c). In isolation, Gn^H is monomeric at neutral pH [13, 16] and tetrameric at acidic pH [17]. Gn^B forms a tetramer [13]. Several studies have identified and characterized neutralizing antibodies targeting Gn^H [8, 18–20], Gc [7, 15], and the Gn/Gc interface [7]. So far, no antibodies recognizing Gn^B have been reported.

1.2 Production of Hantavirus Glycoproteins

Hantavirus glycoproteins have been used as immunogens [19] or as tools to isolate and characterize monoclonal antibodies [7, 8, 15, 18]. To this end, the glycoprotein domains mentioned above have been produced recombinantly using a variety of expression systems. Gn^H from PUUV [17, 21] and HNTV [13] have been expressed using mammalian and Drosophila S2 cells, respectively. Gn^B from ANDV has been produced using S2 cells [13]. The complete ectodomain of Gc from PUUV has been obtained from insect cells infected with a baculovirus [10] and using mammalian cells [15] and Gc from HNTV purified from the supernatants of S2 cells [13]. The stabilized heterodimers from ANDV and Maporal virus (MPRV) [13] have been purified from stable transfected S2 cells.

In this chapter, we will detail how to express all the glycoprotein domains of hantaviruses by using Drosophila melanogaster S2 cells. Although other expression systems can also produce these proteins, we prioritize the S2 cell system due to its cost-effectiveness and glycosylation properties. S2 cells can grow at room temperature without CO_2 and do not require a dedicated incubator. They are semi-adherent and can be grown in suspension at high densities, making scaling up culture easy using spinners flasks or Erlenmeyer. Generating stable S2 cell lines is a slow process, but they are highly stable, allowing for long-term storage and no need for re-transfection. S2 cells secrete and accumulate glycoproteins in the supernatant without cell lysis, which reduces the

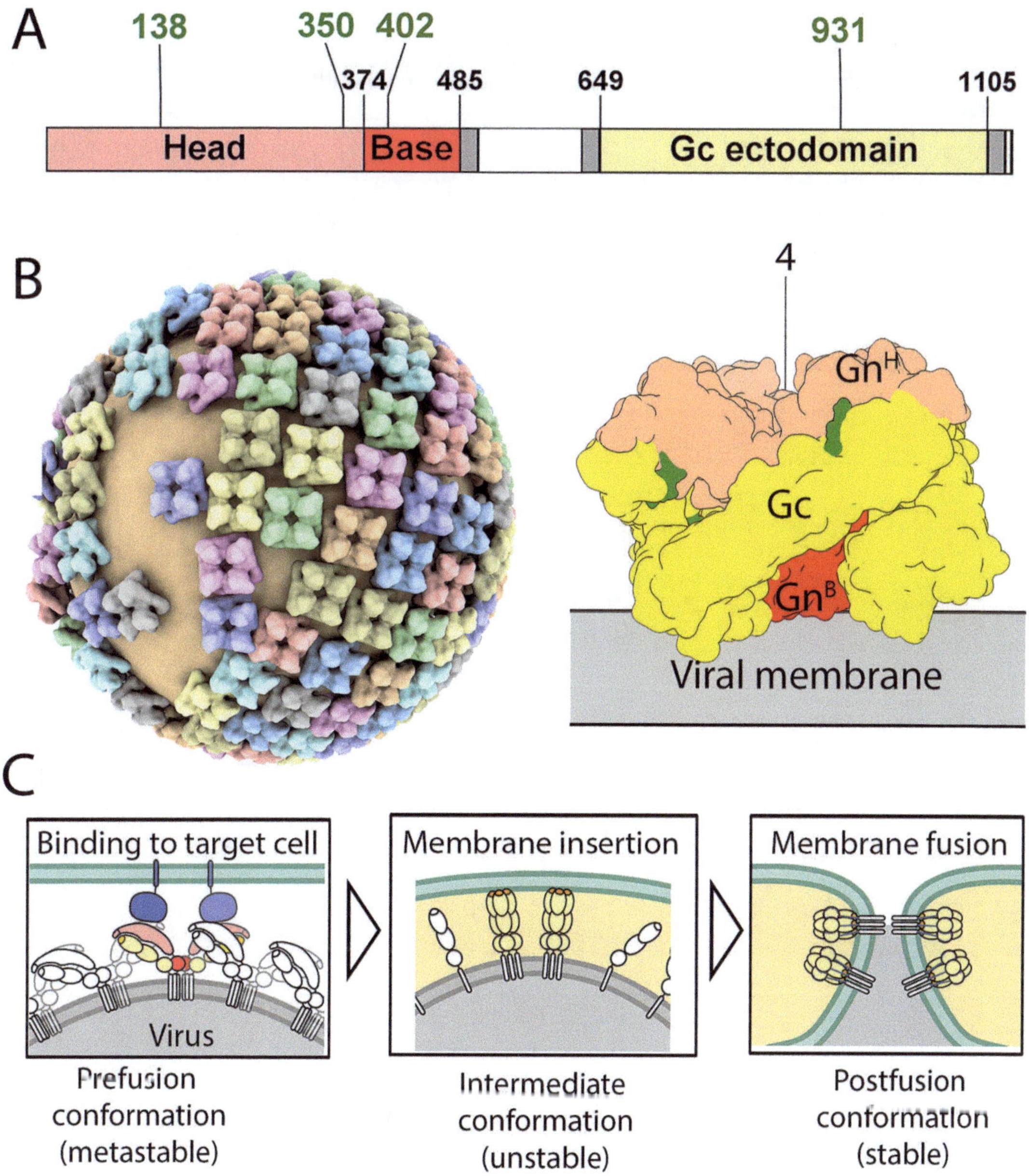

Fig. 1 The organization of the hantavirus glycoproteins. (**a**) Organization of the glycoprotein precursor. The different regions of Gn (Gn^{H}, Gn^{B}) are colored in light and dark red and Gc in yellow with the boundaries indicated on the diagram in black numbers. The TM segments are shown in gray and the conserved glycosylation sites are annotated above in green numbers. The numbering in this diagram belongs to the Andes virus sequence (ANDV, NP_604472.1). (**b**) The left panel is a reconstruction of the glycoprotein outer lattice on the virion surface. For clarity, each tetrameric spike is colored differently and the viral membrane is colored in wheat. The right panel shows the organization of a single spike in the side view colored as in **a**. The approximate position of the viral membrane is indicated. (**c**) Scheme showing the different conformations of Gc during viral entry. Gn^{H}, Gn^{B}, and Gc are colored as in **a**

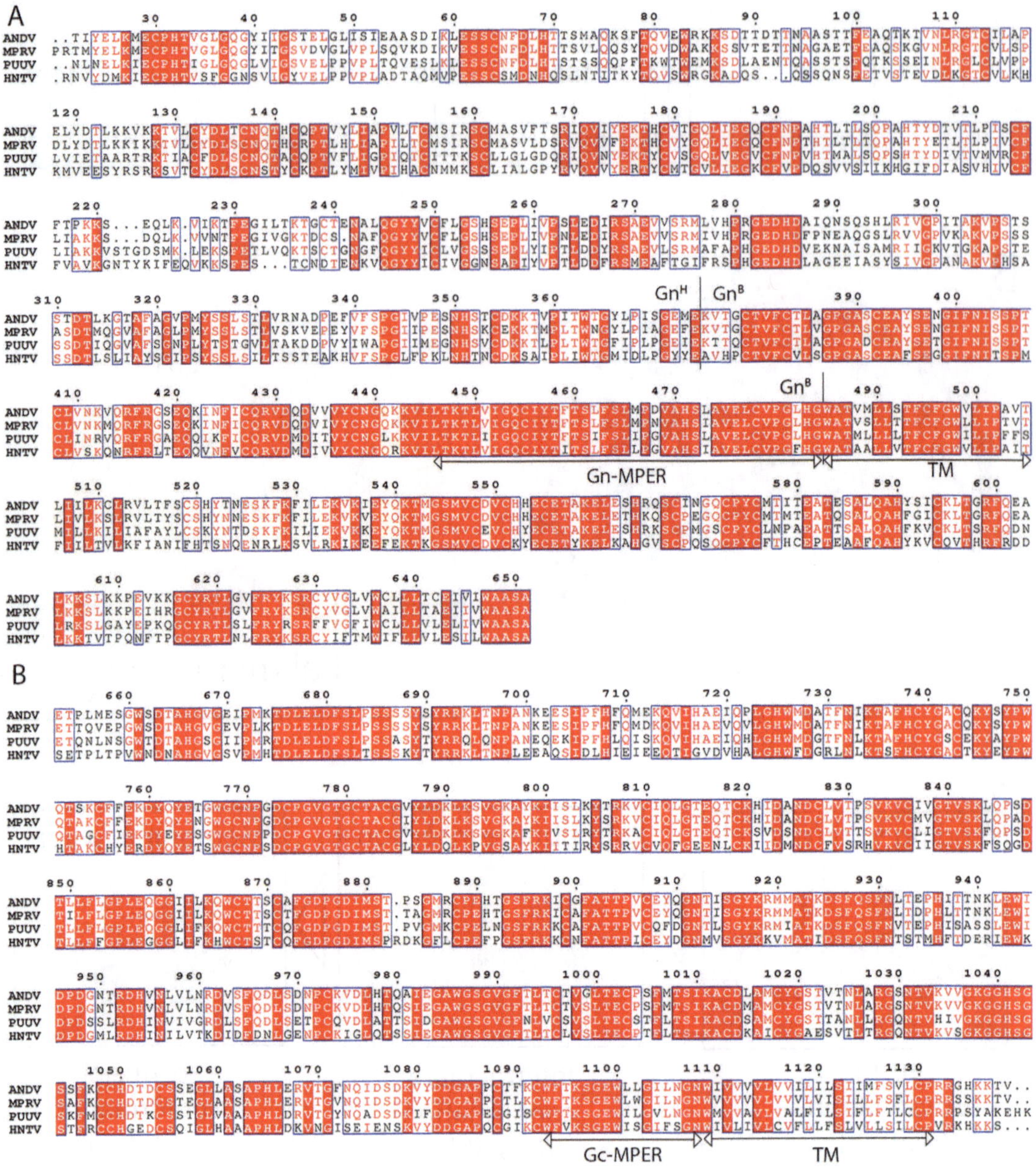

Fig. 2 Sequence analysis. Multiple sequence alignment of Gn (**a**) and Gc (**b**) from four representative hantaviruses: Andes virus (ANDV, NP_604472.1), Maporal virus (MPRV, YP_009362281), Puumala virus (PUUV, CAB43026.1), and Hantaan virus (HNTV, CAA68456.1). Residues are colored according to conservation: strictly conserved residues are colored in white on a red background, partially conserved in red on a white background and the non-conserved residues in black on a white background. Gn^{H} and Gn^{B} boundaries, transmembrane regions (TM), and the MPER regions of Gn and Gc are indicated

release of proteases and proteolytic degradation. Finally, S2 cells produce glycoproteins with simple, high-mannose sugars [22], which recapitulates the *N*-glycans observed in authentic hantaviruses [23] and which plays an important structural role in the spike [13].

A few remarks are needed about the constructs. Gn^B is formed by a β-sandwich followed by an amphipathic α-helical hairpin (Fig. 2a). The latter is a membrane proximal region (Gn-MPER) that interacts with the viral membrane through a hydrophobic surface [13]. Both regions interact to make a compact Gn^B tetramer. In solution, the Gn^B tetramer is sticky and aggregates easily, complicating its purification and subsequent handling. An alternative is to remove the Gn-MPER region and produce a construct containing only the β-sandwich domain. This construct produces much better but is monomeric in solution. A similar consideration can be made for the Gc ectodomain. At its C-terminal end, between the transmembrane region and a conserved CX_4C motif, there is an amphipathic helix. This region is known as Gc-MPER (Fig. 2b) and interacts with the viral membrane [13]. Although it is possible to produce the Gc ectodomain containing this region, when it is removed, a larger amount of protein is obtained. The heterodimer Gn/Gc is difficult to purify because the complex cannot be reconstituted *in vitro* from its components and the wild-type form is too labile to be purified. The approach that we have followed to purify the complex is to connect both subunits using a flexible linker joining the C-terminus of Gn^H (we have indicated the boundary between Gn^H and Gn^B in various hantaviruses in Fig. 2a) with the N-terminus of Gc. The sequence of the linker (GGSGLVPRG SGGGSGGGS**WSHPQFEK**GGGTGGGTLVPRGSGTGG) contains two thrombin cleavage sites (underlined) at either end and a strep-tag sequence (in bold) in the middle, separated by flexible GGGS or GGGT repeats. We have introduced these motifs to facilitate the crystallization of the complex but for some applications, it may be necessary to remove them. Based on the structure of the complex, we estimate that the minimum linker length needed to join Gn^H and Gc in the prefusion configuration is about 25 amino acids, assuming that the linker is flexible and does not interfere with the folding of the complex. Using the linker showed above and the protocol detailed in this chapter, we have succeeded in purifying the Gn^H/Gc heterodimers from PUUV, ANDV, HNTV, SNV, Maporal (MPRV), and Choclo viruses (CHOV).

2 Materials

2.1 Plasmids

1. pMT/BiP vector series (*see* **Note 1**).
2. pCoPURO plasmid (Addgene, #17533 [24]) for selection of stable cell lines (*see* **Note 2**).

2.2 Cells and Medium

1. Drosophila S2 cells.
2. Growing medium: SFM4Insect medium supplemented with 1% penicillin/streptomycin (*see* **Note 3**).
3. Selection medium: Growing medium supplemented with 8 μg/mL of puromycin.
4. Freezing medium: 90% fetal bovine serum, 10% DMSO.
5. Effectene transfection kit.
6. Trypan Blue Solution 0.4%.
7. 5 μM $CdCl_2$ or 500 μM $CuSO_4$.

2.3 Other Materials

1. Cell culture flasks 25 cm^2, 75 cm^2, and 150 cm^2.
2. Cell scrapers.
3. Freezing container.
4. Cryo tubes.
5. Inverted light microscope.
6. Automated cell counter.
7. 1 L or 4 L glass spinner flasks or 5 L Erlenmeyer flasks with ordinary magnetic stirrers.
8. Incubator at 28 °C without CO_2.
9. Sterile disposable bottle top filter, membrane PES 0.2 μm filter.
10. Tangential ultrafiltration devices.
11. FPLC system.
12. Class II laminar flow hood.

2.4 Protein Purification Buffers

1. Washing buffer: 10 mM Tris-HCl, pH 8.0, 150 mM NaCl, 1 mM EDTA.
2. Elution buffer: 10 mM Tris-HCl pH 8.0,150 mM NaCl, 1 mM EDTA, 2.5 mM desthiobiotin.
3. Gel filtration buffer: 10 mM Tris-HCl pH 8.0, 100 mM NaCl.

3 Methods

3.1 Thawing S2 Cells

1. Thaw cells quickly in a 37 °C water bath and transfer them to a centrifuge tube containing 5 mL of growing medium. Centrifuge them for 5 min at 200 g.
2. Resuspend the cell pellet in 15 mL of growing medium and transfer the cell suspension to a 75 cm^2 flask. Let the cells recover for 2–3 days.

3.2 Passaging of S2 Cells

1. Use a pipette to wash down the surface of the flask and dislodge any adherent cells. Sometimes it can be useful to gently tap the side of the bottle with the palm of the hand to dislodge the cells before washing them with the pipette.
2. Determine the cell density and viability using an automated cell counter (*see* **Note 4**). To determine viability, mix equal volumes of cells and trypan blue solution. As the membranes of dead cells are leaky, they will take up the dye and be colored blue. The device will be able to distinguish between live and dead cells and obtain a viability percentage. If an automatic cell counter is not available, it is possible to estimate the cell density and viability using a hemacytometer and the trypan blue dye method. Passaging of S2 cells should be performed when the culture density reaches about 10^7 cells/mL.
3. Dilute the cells in growing medium to a final concentration of 2×10^6 cells/mL and seed them into a new flask (*see* **Note 5**).

3.3 Transfection and Selection of Stable Cell Lines

1. Seed 15×10^6 cells in 15 mL of growing medium into a 75 cm^2 flask (final density equivalent to 10^6 cells/mL) and incubate them overnight at 28 °C (*see* **Notes 6** and **7**).
2. Mix 2.0 μg of the expression plasmid with 0.1 μg of the selection plasmid (*see* **Notes 8** and **9**) into a sterile 1.5 mL eppendorf. Dilute the DNA mixture (minimal DNA concentration of 0.1 μg/μL) with Effectene Buffer EC, to a total volume of 150 μL. Add 16 μL Effectene Enhancer and mix by vortexing for 1 s.
3. Incubate the mixture at room temperature for 5 min and then spin it down for a few seconds to remove drops from the top of the tube.
4. Add 20 μL Effectene Transfection Reagent to the DNA–Enhancer mixture. Mix by pipetting up and down five times or by vortexing for 10 s.
5. Incubate the samples for 15 min at room temperature (15–25 °C) to allow the formation of the transfection complex.

6. Add 1 mL of growing medium to the tube containing the transfection complexes. Mix by pipetting up and down and add the transfection complexes drop-wise onto the cells. Do this gently to avoid dislodging the adhered cells.
7. Incubate cells at 28 °C for 24 h.
8. Carefully place the flask upright, trying not to dislodge the adhered cells. Transfer the supernatant into a fresh tube and add 10 mL of selection medium to the flask. Spin down the supernatant at 200 g for 3 min. Resuspend the cell pellet in 5 mL of selection medium and put the cells back into the same flask.
9. Incubate cells at 28 °C for at least 5 days. Monitor them using a microscope. At first, the cells should appear healthy and dense. Over time, non-transfected cells die resulting in a reduction of cell density (*see* **Note 10**).
10. Resuspend the cells tapping the flask and pipetting up and down. Transfer the 15 mL of the cell suspension to a 150 cm^2 flask and add 15 mL of selection medium.
11. Incubate cells at 28 °C for at least 3 days. Monitor them under the microscope. If the transfection has worked correctly, the number of cells should double every 3 days and the stable cell line would be established (*see* **Note 11**).
12. Transfer 10 ml of cell suspension into a 150 cm^2 flask containing 20 ml of selection medium. Repeat 2 times to prepare 2 more 150 cm^2 flasks. Incubate them at 28 °C for at least 3 days (*see* **Note 12**).

3.4 Expansion of Cell Culture

1. Determine the cell density and viability using an automated cell counter. When the cell density in the 150 cm^2 flasks reaches around 1–2 × 10^7 cells/mL, split the cells into nine 150 cm^2 flasks transferring 10 mL cell suspension and 20 mL fresh selection medium to each flask (270 mL in total). One flask will be used to prepare a stock of frozen cells as described in Subheading 3.6. The remaining flasks are dedicated to the preparation of the large culture in either spinner or Erlenmayer flasks (*see* **Note 13**).
2. After 3–5 days, look at the cells under the microscope to confirm that they are growing healthy and measure cell density and viability. Resuspend the cells by tapping the flask and by pipetting up and down to wash the flask surface (*see* **Note 14**).
3. To maintain the cell line, transfer 5 mL of the cell suspension to a 150 cm^2 flask and add 25 mL of selection medium.

4. Transfer the remaining 235 mL of cell suspension into a spinner flask or an Erlenmeyer and add about 250 mL of growing medium to a total volume of about 500 mL (*see* **Notes 13** and **15**). Grow the cells at 28 °C. If an Erlenmeyer flask is used, use an agitation rate of 130 rpm.
5. After 3–5 days, count the cells. When the cell density reaches 3–5 × 10^7 cells/mL, add 500 mL of growing medium to a total volume of about 1 L (*see* **Note 16**). Induce the culture by adding 5 μM $CdCl_2$ or 500 μM $CuSO_4$ (*see* **Note 17**).
6. After 3–10 days, collect the cell solution and centrifuge at 6000 g for 30 min at 15 °C to separate the cells and supernatant (**Note 18**). Discard the cell pellet.

3.5 Protein Purification

1. Concentrate the supernatant fraction by using a tangential ultrafiltration device. Starting from a volume of 1 to 4 L, concentrate up to 50–100 mL. Adjust the pH with 1/10 of the volume of 1 M Tris-HCl (pH 8) and 120 μl of biotin-blocking solution for each 50 mL of concentrated supernatant.
2. Centrifuge the concentrated supernatant at 40,000 g for 30 min at 4 °C and filter it using a 0.22 μm top filter (*see* **Note 19**). Load the filtered supernatant into a 50 mL loop and plug it into an FPLC system.
3. Pass the supernatant through a StrepTrap chromatography column pre-equilibrated in washing buffer.
4. Wash the column with 10 bed volumes of washing buffer until the OD_{280} reaches baseline and then elute the protein complex with 10 bed volumes elution buffer.
5. Concentrate the fractions containing the eluted protein to 0.5 mL using 20 mL concentrators (cut-off 10 kDa) (*see* **Note 20**).
6. Pre-equilibrate the size exclusion chromatography column (Superdex 200 10/300) with 60 mL of gel filtration buffer.
7. Load the sample using a 500 μL loop and run the gel filtration column at a suitable flow rate (0.4 mL/min). Monitor the absorbance signal at 280 nM and collect fractions. Hantavirus glycoproteins usually elute in two peaks, a minor dimeric fraction and a major monomeric fraction eluting respectively at about 11 and 12.5 mL (Fig. 3).
8. Analyze the fractions using SDS-Polyacrylamide gel electrophoresis (SDS-PAGE) (Fig. 3) (*see* **Note 21**).
9. Pool the fractions of interest and concentrate using 6 mL concentrators (cut-off 10 kDa). Check concentration by measuring the optical density at 280 nM using a Nanodrop device (*see* **Note 22**). Prepare aliquots and flash freeze them in liquid nitrogen.

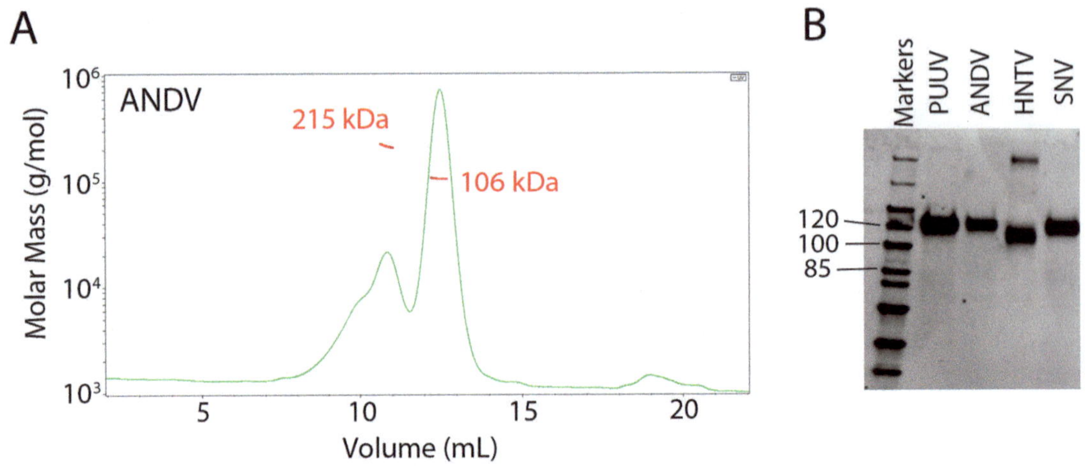

Fig. 3 Protein purification. (**a**) SEC elution volume profiles of Gn^H/Gc from ANDV at pH 8.0. The left axis indicates the molecular mass (kDa) determined by MALS, with the values for each species indicated on the corresponding peak. (**b**) SDS-PAGE of the monomeric peaks of Gn^H/Gc from PUUV, ANDV, HNTV, and SNV, as indicated

3.6 Freezing S2 Cells

1. Freezing of S2 cells should be performed when the culture density reaches about 10^7 cells/mL.
2. Dislodge any adherent cells by tapping the flask and pipetting up and down to wash down the surface of the flask.
3. Spin down the cells for 5 min at 200 g and resuspend the cell pellet in cold freezing medium. Cell density should be higher than 1.5×10^7 cell/mL in the freezing medium (*see* **Note 23**).
4. Aliquot resuspended cells in 1.5 mL aliquots in cryo vials and freeze them at −80 °C for 24–48 h in a "Mr. Frosty".
5. Transfer cryovials to liquid nitrogen for long-term storage.

4 Notes

1. To optimize protein production and help in the purification, we cloned synthetic codon-optimized genes for expression in Drosophila cells into a modified pMT/BiP plasmid, which translates the protein in frame with a Thrombin cleavage site (underlined) and a double strep-tag (in bold) at the C-terminal end of the sequence (GSGLVPRGSGGSGGSAG **WSHPQFEK**GGGSGGGSGGGS**WSHPQFEK**).
2. The pCoPuro plasmid confers resistance to puromycin. There is also the possibility of using pCoBlasto or pCoHygro, which confer resistance to blasticidin S hydrochloride and hygromycin B, respectively.

3. Another culture medium that can also be used is Insect Xpress.
4. The TC20 automated cell counter counts cells within a range of 10^4–10^7 cells/mL. S2 cells can grow at densities above this range, so it may be necessary to dilute them. For your calculation note that the counter takes into consideration just the 1: 2 dilution in Trypan Blue solution.
5. As a rule of thumb, cells are diluted 1 to 5 or 6. Higher dilution rates may decrease cell viability.
6. A quicker alternative that works in our hands is to seed 21×10^6 cells 1 h before transfection.
7. Under optimal working conditions, incubate the cells at 28 °C. S2 cell incubations can also be performed at room temperature, omitting the need for an incubator. However, the incubation time between passages must be corrected accordingly.
8. Transfections should be made with DNA of the highest purity. Contaminants may kill the cells and interfere with the formation of the transfection complexes, thus decreasing transfection efficiency. However, the purer the DNA, the more expensive the process. We generally purify the resistance plasmid using a Midi-prep kit and the different expression plasmids using a miniprep kit. Transfections performed with these plasmids have good efficiencies. Purification of the expression plasmid by the use of a Midi Kit can increase transfection efficiency in difficult cases.
9. The number of inserted gene copies in the genome, and therefore the level of protein expression, can be manipulated by varying the ratio of expression plasmid to resistance plasmid [25].
10. The next steps simultaneously allow the selection of the stably transfected population and cell amplification. The presence of a large number of cell clusters in suspension indicates that the cells are in a good state and are ready to be amplified. The first two passages after transfection are the most critical ones. If the cells are too diluted, they will not survive.
11. In most cases, the stable S2 cell transfectants maintain their expression level for a very large number of passages. However, sometimes, the quantity or quality of secreted protein worsens with the number of passages. We recommend freezing aliquots of the stable S2 transfectants immediately after the stable lines are generated. To do so, please follow the instructions in Subheading 3.6 (Freezing S2 cells).
12. S2 cells begin to grow by adhering to the bottom of the flask and then, as density increases, they detach from the bottom and form clusters in suspension. Depending on the cell line, this process can be faster or slower. It may be possible to have

many cells in suspension and still have many cells attached to the bottom of the plate. In this case, it is advisable to recover these cells as well using a cell scraper.

13. Eight 150 cm^2 flasks give the right number of cells to start the culture in a 4 L spinner. Although for a 1 L spinner six 150 cm^2 flasks are sufficient, the use of eight flasks allows to reach the time of induction more quickly.
14. For S2 stable transfectants, 100% viability is not a prerequisite. A viability index above 90% is considered acceptable.
15. The presence of the selection antibiotic (*see* **Note 2**) is required for each passaging step to keep selection, but it can be omitted during amplification for protein production to save costs. However, if amplification is done immediately after the establishment of the cell line, we strongly advise using selection medium during the first production.
16. If a larger final volume is needed, wait 3–5 days and dilute the cells 1:2 o 1:3. Repeat as many times as necessary until the desired volume is reached. The final amount of $CdCl_2$ or $CuSO_4$ used for the induction has to be corrected accordingly.
17. The expression of the proteins is under the control of the metallothionein (MT) promoter, which is inducible by adding either $CdCl_2$ or $CuSO_4$. Although the former is a more effective inducer, the latter is to be preferred for its lower toxicity.
18. The optimal induction time depends on each protein and the concentration and type of inductor ($CdCl_2$ or $CuSO_4$) and must be determined for each stable S2 cell transfectant. This can be done on a small scale by collecting 1 mL of supernatant every day for 10 days after induction. After centrifugation at 20,000 g for 30 min at 4 °C in a benchtop centrifuge, run a Western blot using streptactin-HRP for detection. We have found that the optimal induction conditions for hantavirus glycoproteins are 5 μM $CdCl_2$ for 5–6 days.
19. The filtration step is not compulsory but allows the preservation of the affinity purification columns for longer.
20. For concentration, we centrifuge at 4000 g for 15 min at 4 °C. For high volumes, multiple centrifugation cycles may be necessary. It is best to avoid prolonged centrifugation steps that might induce protein aggregation. In any case, it is advisable to resuspend the protein with a pipette between two successive centrifugations.
21. SDS-PAGE analysis is essential to check the purity and the quality of the eluted proteins. Usually, we run it in reducing conditions, i.e., adding 100 mM of fresh DTT in the loading buffer, but in some cases can be essential to run it in non-reducing conditions to detect the possible presence of disulfide-linked oligomers.

22. To estimate the protein concentration more precisely, we advise to determine the theoretical molar extinction coefficient from the protein sequence.
23. From one 150 cm^2 flask it's possible to freeze between 3 and 4 cryo-vials.

References

1. Watson DC, Sargianou M, Papa A, Chra P, Starakis I, Panos G (2014) Epidemiology of Hantavirus infections in humans: a comprehensive, global overview. Crit Rev Microbiol 40(3):261–272. https://doi.org/10.3109/1040841X.2013.783555
2. Guardado-Calvo P, Rey FA (2017) The envelope proteins of the bunyavirales. Adv Virus Res 98:83–118. https://doi.org/10.1016/bs.aivir.2017.02.002
3. Acuna R, Cifuentes-Munoz N, Marquez CL, Bulling M, Klingstrom J, Mancini R, Lozach PY, Tischler ND (2014) Hantavirus Gn and Gc glycoproteins self-assemble into virus-like particles. J Virol 88(4):2344–2348. https://doi.org/10.1128/JVI.03118-13
4. Mittler E, Dieterle ME, Kleinfelter LM, Slough MM, Chandran K, Jangra RK (2019) Hantavirus entry: perspectives and recent advances. Adv Virus Res 104:185–224. https://doi.org/10.1016/bs.aivir.2019.07.002
5. Jangra RK, Herbert AS, Li R, Jae LT, Kleinfelter LM, Slough MM, Barker SL, Guardado-Calvo P, Roman-Sosa G, Dieterle ME, Kuehne AI, Muena NA, Wirchnianski AS, Nyakatura EK, Fels JM, Ng M, Mittler E, Pan J, Bharrhan S, Wec AZ, Lai JR, Sidhu SS, Tischler ND, Rey FA, Moffat J, Brummelkamp TR, Wang Z, Dye JM, Chandran K (2018) Protocadherin-1 is essential for cell entry by new world hantaviruses. Nature 563(7732):559–563. https://doi.org/10.1038/s41586-018-0702-1
6. Guardado-Calvo P, Bignon EA, Stettner E, Jeffers SA, Perez-Vargas J, Pehau-Arnaudet G, Tortorici MA, Jestin JL, England P, Tischler ND, Rey FA (2016) Mechanistic insight into bunyavirus-induced membrane fusion from structure-function analyses of the hantavirus envelope glycoprotein Gc. PLoS Pathog 12(10):e1005813. https://doi.org/10.1371/journal.ppat.1005813
7. Mittler E, Wec AZ, Tynell J, Guardado-Calvo P, Wigren-Bystrom J, Polanco LC, O'Brien CM, Slough MM, Abelson DM, Serris A, Sakharkar M, Pehau-Arnaudet G, Bakken RR, Geoghegan JC, Jangra RK, Keller M, Zeitlin L, Vapalahti O, Ulrich RG, Bornholdt ZA, Ahlm C, Rey FA, Dye JM, Bradfute SB, Strandin T, Herbert AS, Forsell MNE, Walker LM, Chandran K (2022) Human antibody recognizing a quaternary epitope in the Puumala virus glycoprotein provides broad protection against orthohantaviruses. Sci Transl Med 14(636):eabl5399. https://doi.org/10.1126/scitranslmed.abl5399
8. Engdahl TB, Kuzmina NA, Ronk AJ, Mire CE, Hyde MA, Kose N, Josleyn MD, Sutton RE, Mehta A, Wolters RM, Lloyd NM, Valdivieso FR, Ksiazek TG, Hooper JW, Bukreyev A, Crowe JE Jr (2021) Broad and potently neutralizing monoclonal antibodies isolated from human survivors of new world hantavirus infection. Cell Rep 36(3):109453. https://doi.org/10.1016/j.celrep.2021.109453
9. Engdahl TB, Crowe JE Jr (2020) Humoral immunity to hantavirus infection. mSphere 5(4). https://doi.org/10.1128/mSphere.00482-20
10. Willensky S, Bar-Rogovsky H, Bignon EA, Tischler ND, Modis Y, Dessau M (2016) Crystal structure of glycoprotein C from a hantavirus in the post-fusion conformation. PLoS Pathog 12(10):e1005948. https://doi.org/10.1371/journal.ppat.1005948
11. Guardado-Calvo P, Rey FA (2021) The viral class II membrane fusion machinery: divergent evolution from an ancestral heterodimer. Viruses 13(12). https://doi.org/10.3390/v13122368
12. Guardado-Calvo P, Rey FA (2021) The surface glycoproteins of hantaviruses. Curr Opin Virol 50:87–94. https://doi.org/10.1016/j.coviro.2021.07.009
13. Serris A, Stass R, Bignon EA, Muena NA, Manuguerra JC, Jangra RK, Li S, Chandran K, Tischler ND, Huiskonen JT, Rey FA, Guardado-Calvo P (2020) The hantavirus surface glycoprotein lattice and its fusion control mechanism. Cell 183(2):442–456 e416. https://doi.org/10.1016/j.cell.2020.08.023
14. Bignon EA, Albornoz A, Guardado-Calvo P, Rey FA, Tischler ND (2019) Molecular organization and dynamics of the fusion protein Gc at the hantavirus surface. elife 8. https://doi.org/10.7554/eLife.46028

15. Rissanen I, Stass R, Krumm SA, Seow J, Hulswit RJ, Paesen GC, Hepojoki J, Vapalahti O, Lundkvist A, Reynard O, Volchkov V, Doores KJ, Huiskonen JT, Bowden TA (2020) Molecular rationale for antibody-mediated targeting of the hantavirus fusion glycoprotein. elife 9. https://doi.org/10.7554/eLife.58242
16. Li S, Rissanen I, Zeltina A, Hepojoki J, Raghwani J, Harlos K, Pybus OG, Huiskonen JT, Bowden TA (2016) A molecular-level account of the antigenic hantaviral surface. Cell Rep 16(1):278. https://doi.org/10.1016/j.celrep.2016.06.039
17. Rissanen I, Stass R, Zeltina A, Li S, Hepojoki J, Harlos K, Gilbert RJC, Huiskonen JT, Bowden TA (2017) Structural transitions of the conserved and metastable hantaviral glycoprotein envelope. J Virol 91(21). https://doi.org/10.1128/JVI.00378-17
18. Rissanen I, Krumm SA, Stass R, Whitaker A, Voss JE, Bruce EA, Rothenberger S, Kunz S, Burton DR, Huiskonen JT, Botten JW, Bowden TA, Doores KJ (2021) Structural basis for a neutralizing antibody response elicited by a recombinant hantaan virus Gn immunogen. mBio 12(4):e0253120. https://doi.org/10.1128/mBio.02531-20
19. Duehr J, McMahon M, Williamson B, Amanat F, Durbin A, Hawman DW, Noack D, Uhl S, Tan GS, Feldmann H, Krammer F (2020) Neutralizing monoclonal antibodies against the Gn and the Gc of the Andes virus glycoprotein spike complex protect from virus challenge in a preclinical hamster model. mBio 11(2). https://doi.org/10.1128/mBio.00028-20
20. Garrido JL, Prescott J, Calvo M, Bravo F, Alvarez R, Salas A, Riquelme R, Rioseco ML, Williamson BN, Haddock E, Feldmann H, Barria MI (2018) Two recombinant human monoclonal antibodies that protect against lethal Andes hantavirus infection in vivo. Sci Transl Med 10(468). https://doi.org/10.1126/scitranslmed.aat6420
21. Li S, Rissanen I, Zeltina A, Hepojoki J, Raghwani J, Harlos K, Pybus OG, Huiskonen JT, Bowden TA (2016) A molecular-level account of the antigenic hantaviral surface. Cell Rep 15(5):959–967. https://doi.org/10.1016/j.celrep.2016.03.082
22. Kim YK, Shin HS, Tomiya N, Lee YC, Betenbaugh MJ, Cha HJ (2005) Production and N-glycan analysis of secreted human erythropoietin glycoprotein in stably transfected Drosophila S2 cells. Biotechnol Bioeng 92(4):452–461. https://doi.org/10.1002/bit.20605
23. Shi X, Elliott RM (2004) Analysis of N-linked glycosylation of hantaan virus glycoproteins and the role of oligosaccharide side chains in protein folding and intracellular trafficking. J Virol 78(10):5414–5422. https://doi.org/10.1128/jvi.78.10.5414-5422.2004
24. Iwaki T, Figuera M, Ploplis VA, Castellino FJ (2003) Rapid selection of Drosophila S2 cells with the puromycin resistance gene. BioTechniques 35(3):482–484, 486. https://doi.org/10.2144/03353bm08
25. Johansen H, van der Straten A, Sweet R, Otto E, Maroni G, Rosenberg M (1989) Regulated expression at high copy number allows production of a growth-inhibitory oncogene product in Drosophila Schneider cells. Genes Dev 3(6):882–889. https://doi.org/10.1101/gad.3.6.882

Chapter 2

Production and Purification of Filovirus Glycoproteins

Madeleine Noonan-Shueh, M. Javad Aman, and Shweta Kailasan

Abstract

Ebola (EBOV) and Marburg (MARV) viruses cause hemorrhagic fever disease in humans and non-human primates (NHPs) with case-fatality rates as high as 90%. The 2013–2016 Ebola virus disease (EVD) outbreak led to over 28,000 cases and 11,000 deaths and took an enormous toll on the economy of West African nations, in the absence of any vaccine or therapeutic options. Like EVD, there have been at least 6 outbreaks of MVD with ~88% case-fatality and the most recent cases emerging in Equatorial Guinea in February 2023. These outbreaks have spurred an unprecedented global effort to develop vaccines and therapeutics for EVD and MVD and led to an approved vaccine (ERVEBO™) and two monoclonal antibody (mAb) therapeutics for EBOV. In contrast to EVD, therapeutic options against Marburg and another Ebola-relative Sudan virus (SUDV) are lacking. The filovirus glycoprotein (GP), which mediates host cell entry and fusion, is the primary target of neutralizing antibodies. In addition to its pre- and post-fusion trimeric states, the protein is highly glycosylated making production of pure and homogeneous trimers on a large scale, a requirement for subunit vaccine development, a challenge. In efforts to address this roadblock, we have developed a unique combination of structure-based design, selection of expression system, and purification methods to produce uniform and stable EBOV and MARV GP trimers at scales appropriate for vaccine production.

Key words Filovirus, Marburg virus, Ebola glycoprotein, GP, Filovirus vaccine, MVD, EVD

1 Introduction

Ebola virus belongs to the family *Filoviridae* and is a negative-stranded, enveloped virus that causes a severe hemorrhagic fever (Ebola virus disease or EVD) in humans and non-human primates [1, 2]. Since its initial discovery in the 1970s in the content of Africa, six species of EBOV have been isolated, namely, Zaire (EBOV), Sudan, Côte d'Ivoire, Reston (REBOV), Bundibugyo (BDBV), and Bombali [2–4]. Zaire, Sudan, and Bundibugyo EBOVs have been the causative agents of large outbreaks in Africa with a human case-fatality rate of 25–90% [3]. Similar to EVD, Marburg virus (MARV) has also caused several outbreaks since

Steven B. Bradfute (ed.), *Recombinant Glycoproteins: Methods and Protocols*, Methods in Molecular Biology, vol. 2762,
https://doi.org/10.1007/978-1-0716-3666-4_2,

1976 with a case-fatality of 88% and includes several isolates such as Angola, Musoke, Ci67, Popp as well as Ravn virus (RAVV) [4].

All filoviruses express glycoproteins (GP) on the virion surface which are mainly responsible for attachment, fusion, and host cell entry. Filoviruses attach to target cells via lectins or phosphatidylserine receptors and are engulfed by micropinocytosis [5]. The filovirus GP monomer when cleaved post-translationally by furin in producer cells yields two subunits, GP1 and GP2, that remain covalently linked by a natural disulfide bond except for MARV GPs which lack the disulfide bond [6]. Three GP1-GP2 dimers form a trimer on the viral surface. GP1 contains the receptor-binding site (RBS) and mucin-like domains (MLD) while GP2 contains the machinery required to fuse virus and host membranes in infection. Numerous neutralizing antibodies against EBOV, SUDV, BDBV, and MARV targeting these functional domains have been shown to have therapeutic effect resulting in the GP as the main target for vaccine development [7–14].

Filovirus GPs have been expressed in mammalian and insect cell lines with varying degrees of success [1]. GPs are highly glycosylated, and the glycosylation profile is variable depending on the expression system [1]. One of the major challenges of the field has been expressing the GP in a stable trimeric form. Multiple crystal structures are now available for GP trimers of EBOV [15], SUDV [16], and MARV [6] and reveal key insights into residues that are important to stabilize the trimeric structure.

Here, we describe the expression of filovirus GPs such as EBOV and MARV with high yield and purity in Drosophila Schneider 2 (S2) cells. Briefly, S2 cells are transiently transfected with a plasmid expressing EBOV or MARV GP lacking the MLD and transmembrane (TM) domains [17]. With a gene for antibiotic resistance downstream of GP gene in the plasmid, we can select for transfected cells using the antibiotic (e.g., Zeocin) over several passages to produce stably expressing S2 cells. The presence of a canonical signal sequence at the N-terminus results in the secretion of the GP in the supernatant of the cells allowing easy harvest by centrifugation. GP trimers are then purified using a highly selective resin called Strep-Tactin® resin which has high affinity for the eight amino acid affinity tag (Trp-Ser-His-Pro-Gln-Phe-Glu-Lys) included at the C-terminal end of the GP sequence called Strep-tag®. Briefly, the process comprises of four steps which include transient transfection, stable selection using antibiotic, scale-up of pooled stably expressing cells, and purification of protein using an affinity resin. In the first step, plasmids can be used to set up a small-scale transient transfection to screen for expression using western blot. Once expression is confirmed, a stable cell line can be generated that constitutively expresses the GP (Step 2). This polyclonal pool of stably-GP expressing cells may be used to generate large-scale cultures (Step 3) that can be purified by FPLC (Step 4).

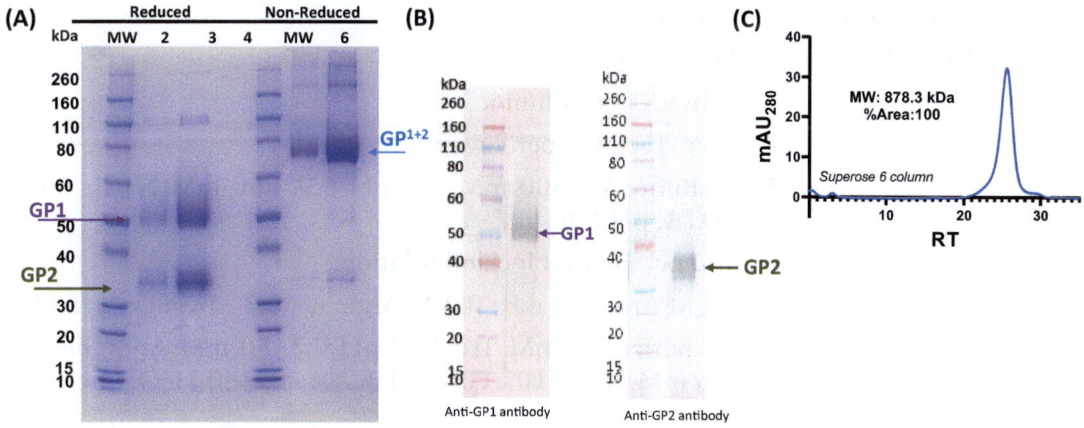

Fig. 1 (**a**) SDS-PAGE demonstrating 1 μg and 5 μg (lanes 2,3) under denaturing and reducing condition and 1 μg and 5 μg (lanes 6,7) under denaturing and non-reducing conditions of GP. MW denotes Novex Sharp prestained protein marker. (**b**) Western blot detection of GP at 200 ng with anti-GP1 antibody (right) and anti-GP2 antibody (left). (**c**) HPLC shows homogeneous peak at expected MW. Theoretical MW of Filovirus GPs falls between 300 and 450 Kda without glycosylation

Downstream characterization demonstrates that the resulting product is pure, homogenous by HPLC, and recognized by highly neutralizing filovirus antibodies available in the literature (Fig. 1a–c).

2 Materials

2.1 S2 Cell Culture

1. Erlenmeyer cell culture flasks.
2. EX-Cell 420 Serum-Free Medium for insect cells.
3. Trypan blue.
4. Incubator shaker.
5. Water bath.

2.2 Transfection and Harvest

1. Plasmid encoding protein of interest with Strep-Tag.
2. Erlenmeyer cell culture flasks (125 mL-2 L).
3. T25 and T75 cell culture flasks.
4. 0.15 M NaCl.
5. Polyethylenimine (PEI) at a stock concentration of 2 mg/mL.
6. Celfectin II Reagent.
7. Heat Inactivated FBS.
8. Zeocin.
9. Ultracentrifuge.
10. Bottle-top vacuum filter.
11. Drosophila S2 cell system (Expres2ion Biotechnologies or similar).

2.3 Purification with Strep-Tactin Resin

1. TFF cartridge.
2. AKTA system.
3. Cytiva XK16 column.
4. Strep-Tactin Superflow resin.
5. 5× Buffer W: 500 mM Tris pH 8.0, 750 mM NaCl, 5 mM EDTA, pH 8.0.
6. BioLock Biotin blocking solution.
7. 0.4 M Arg/Glu mix: 0.2 M Arginine, 0.2 M Glutamic acid.
8. GP buffer: 25 mM Tris HCl pH 7.3, 50 mM Arg/Glu mix, 150 mM NaCl, 10% Glycerol in distilled (di) H_2O.

2.4 SDS-Page

1. Bolt 4–12% Bis Tris Plus Gels.
2. Gel tank.
3. Power supply.
4. Heat block.
5. 20× Bolt™ MOPS SDS running buffer or similar.
6. 4× sample buffer, reducing.
7. 4× sample buffer, non-reducing.
8. Novex Sharp Pre-stained protein marker or similar.
9. 1× DPBS.

3 Methods

3.1 S2 Cell Culture

1. **Day 0**: Prepare a 125 mL shake flask with 20 mL of EX-Cell 420 medium. Thaw a vial of S2 cells in 25 °C water bath. Using a sterile 2 mL pipette, transfer cells to shake flask. Add 1 mL of medium to vial, pipetting up and down, avoiding bubbles and transfer to shake flask. Let the cells sit at room temperature for 5 min then transfer the flask to an incubator shaker set at 25 °C at 115 rpm.
2. **Day 2–3**: Count the cells. Record cell count and viability. If cell count is greater than 12E6 cells/mL, add 7 mL of EX-Cell 420 medium.
3. **Day 4**: Count the cells. Split the cells at 8E6 cells/mL in EX-Cell 420 medium.
4. Passage cells every 4 days at 8E6 cells/mL, scaling up the culture volume every two passages if necessary.

3.2 Transient Transfection (Step 1)

1. **Day 0**: Pre-split S2 cells to 8E6 cells/mL in 500 mL Ex-Cell medium.
2. **Day 1**: Dilute 1.2 mg of plasmid DNA in 25 mL of 0.15 M NaCl. Pipet up and down to thoroughly mix. Add 3.7 mL of PEI and gently mix. Incubate the mixture at room temperature for 8 min (*see* **Note 1**).
3. Add mixture to cells while gently swirling the flask.
4. Incubate in shaking incubator set at 25 °C at 115 rpm.
5. After 5 h, add 500 mL of Ex-Cell 420 medium.
6. **Day 8**: Seven days post-transfection, collect a sample of supernatant. Proceed to harvest and purification (*see* **Note 2**).

3.3 Stable Transfection and Selection (Step 2)

1. **Day 0**: Split S2 cells to 8E6 cells/mL in 5 mL of media in a T25 flask.
2. Add 25 μg of DNA with 187.5 μL serum-free media (3× volume of transfection reagent). Add 62.5 μL Celfectin II transfection reagent (12.5 μL reagent/mL of culture). Add the diluted DNA to the diluted Celfectin II and incubate for 5 min at RT (*see* **Note 3**).
3. Add mixture dropwise to cells. Let the culture stand for 5 min in the hood. Swirl the flask gently and place in a 37 °C incubator. After 5 h, add 500 μL of FBS.
4. **Day 1:** Add stock Zeocin for a final concentration of 1000 μg/mL.
5. **Day 4–24**: Monitor the cell viability by counting every 3–4 days (*see* **Note 4**). For splitting, centrifuge the cell suspension at 1200 rpm for 3 min. Remove and replace half of the supernatant with fresh 10% FBS + ExCell supplemented with 1000 μg/mL Zeocin. Save a sample from the supernatant and label it with the passage number (*see* **Note 5**).
6. Once the viability starts increasing and cell diameter recovers, usually around passage 5/6, transfer 6 mL of cells to a T75 and add 4 mL of 10% FBS + ExCell 420 *without* selection agent (*see* Fig. 2a, b).
7. **Day 25**: The next day, add an additional 5 mL of media.
8. **Day 26** or later: Transfer 15 mL of the resuspended culture into a 125 mL shake flask and add 10 mL of ExCell 420 media.
9. Continue counting and monitoring the viability every 3–4 days. Split each passage to 8E6 cells/mL, centrifuging if needed to promote good viability (*see* **Note 6**).
10. Continue to passage cells until they have recovered (>95% viability) and proceed to scale up the culture to the desired volume (*see* **Note** 7).

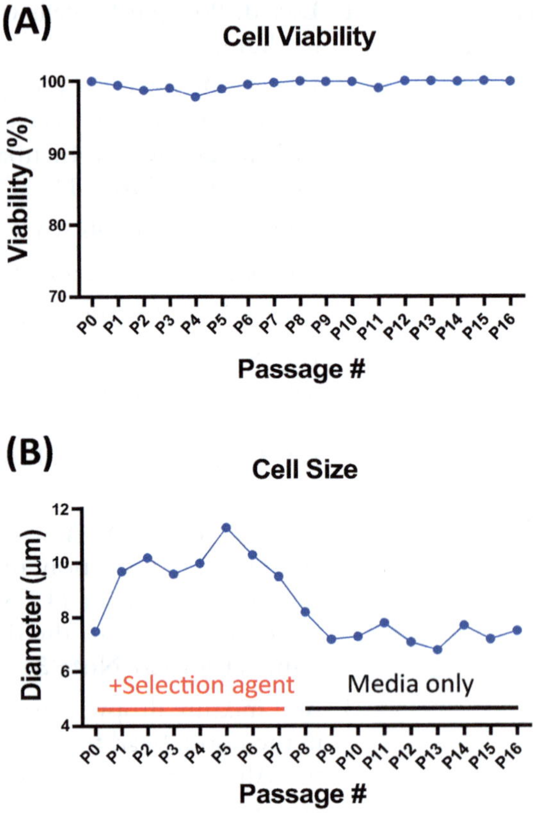

Fig. 2 (**a**) Cell viability remains high throughout the stable line generation. Viability drops slightly when selection agent is introduced around P4 and when the culture is transferred to a shake flask (Top). (**b**) Cell size increases as selection agent is introduced and returns to the initial size once transferred to the shake flask (Bottom)

3.4 Harvest (Step 3)

1. Harvest supernatant by centrifuging at 5000 rpm for 20 min at 4 °C.
2. Taking care not to disturb the pellet, decant supernatant in an appropriately sized container.
3. Filter supernatant with a 0.22 μm low protein binding bottle-top vacuum filter.
4. Supernatant can be frozen at −80 °C until purification.

3.5 Purification with Strep-Tactin Resin (Step 4)

1. Thaw and concentrate the supernatant ten-fold using a 30 kDa cut-off cartridge by Tangential fast flow (TFF). Raise the pH to 8.0 by adding 5× Buffer W. Add 600 μL of IBA BioLock biotin-blocking solution.
2. Filter the resulting solution with a 0.22um filter and load onto a Cytiva XK16 column prepacked with 10 mL of Strep-Tactin

Superflow resin (*see* **Note 8**). The flow rate for all steps should not exceed 4 mL/minute.

3. Wash the column with 150 mL of 1× Buffer W.
4. Elute the sample with 30CV of 0–100% gradient of Buffer W plus 2.5 mM D-(+)-Biotin.
5. Collect peak fractions (5 mL each) and analyze via SDS-PAGE.
6. Combine and dialyze pure fractions into GP Buffer.
7. Sterile-filter final sample before preforming downstream characterization assays.

3.6 SDS-Page

1. Prepare 20 μL of sample to load onto the gel. Add up to 15 uL of each fraction and dilute in 1× DPBS if applicable. Add 5 μL of 4× Sample Buffer, reducing.
2. Heat samples at 100 °C for 5 min using heat block.
3. Prepare 1× running buffer by adding 25 mL of stock 20× Bolt MOPS SDS to 475 mL of diH_2O. Pour running buffer into the gel tank up to fill line mark and check for leaks. Remove Bolt 4–12% Bis Tris Plus Gel from package, place into gel tank, and remove comb gently.
4. Briefly vortex and spin the samples prior to loading into the wells of the gel. Load 5 μL of Novex Sharp Protein Ladder in the first lane. Load 15 μL of each sample into the lanes.
5. Run the gel at 165 V for 45 min.
6. Following electrophoresis, remove gel(s) from the gel box and pry open plates. Rinse the gel(s) briefly in diH_2O.
7. Stain and image.

4 Notes

1. This transfection step can be scaled up or down using the same ratio between cells, DNA, and transfection reagent. After incubating for 8 ± 2 min, the solution becomes cloudy.
2. Prior to proceeding to purification, screen for expression of the protein by western blot using purified protein controls (IBT Bioservices, IBT-0501-025, IBT-0502-015, IBT-0513-015). Load the culture supernatant neat and at several dilutions and blot with anti-GP1 and anti-GP2 antibodies (IBT Bioservices, IBT-0203-025).
3. This transfection can be scaled up or down using the same ratio of cells, DNA, and transfection reagent.
4. The diameter of the cells will increase in the range of 9–12 μm as they are transfected and selected with Zeocin.

5. Aliquots of supernatant for each passage can be saved and used to track protein expression by western blot.
6. To maintain cell viability >90%, keep cells at the same volume for at least two consecutive passages before scaling to a larger volume. Cultures may be centrifuged at 1200 rpm for 3 min and resuspended to get rid of cell debris and improve viability.
7. Once the cells recover, stocks should be frozen down and stored in liquid nitrogen for future use. The culture should be >90% viability. **Day 1**: Count cells and calculate the amount of cell suspension needed to freeze 250E6 cells per vial in 1 mL of freezing media (10% DMSO +90% ExCell 420 medium). Centrifuge calculated volume of cells at 1200 rpm for 3 min. Decant all the supernatant and keep aside. On ice, make 10% DMSO + ExCell media freeze media, using half volume of the supernatant and half volume from fresh media. Resuspend cells in freeze media and aliquot 1 mL/vial. Freeze overnight at −80 °C. **Day 2**: Transfer vials to liquid nitrogen tank within 2–3 days to maintain viability.
8. We elected to use the Strep-Tactin system for ease of purification. Other purification tags may be used in lieu of the Strep-Tag. We have also successfully purified filovirus glycoproteins with Histidine tag utilizing a NiNTA™ column.

References

1. Beer B, Kurth R, Bukreyev A et al (1999) Characteristics of filoviridae: marburg and ebola viruses. Naturwissenschaften 86(1):8–17
2. Hensley LE, Alves DA, Geisbert JB et al (2011) Pathogenesis of Marburg hemorrhagic fever in cynomolgus macaques. J Infect Dis 204(Suppl 3):S1021–S1031
3. Towner JS, Sealy TK, Khristova ML et al (2008) Newly discovered ebola virus associated with hemorrhagic fever outbreak in Uganda. PLoS Pathog 4(11):e1000212
4. Brauburger K, Hume AJ, Mühlberger E et al (2012) Forty-five years of Marburg virus research. Viruses 4(10):1878–1927
5. Saeed MF, Kolokoltsov AA, Albrecht T et al (2010) Cellular entry of ebola virus involves uptake by a macropinocytosis-like mechanism and subsequent trafficking through early and late endosomes. PLoS Pathog 6(9):e1001110
6. Hashiguchi T, Fusco ML, Bornholdt ZA et al (2015) Structural basis for Marburg virus neutralization by a cross-reactive human antibody. Cell 160(5):904–912
7. Howell KA, Qiu X, Brannan JM et al (2016) Antibody treatment of Ebola and Sudan virus infection via a uniquely exposed epitope within the glycoprotein receptor-binding site. Cell Rep 15(7):1514–1526
8. King LB, Fusco ML, Flyak AI et al (2018) The marburgvirus-neutralizing human monoclonal antibody MR191 targets a conserved site to block virus receptor binding. Cell Host Microbe 23(1):101–109
9. Wang Y, Howell KA, Brannan J et al (2021) Prominent neutralizing antibody response targeting the ebolavirus glycoprotein subunit interface elicited by immunization. J Virol 95(8):e01907–e01920
10. Fusco ML, Hashiguchi T, Cassan R et al (2015) Protective mAbs and cross-reactive mAbs raised by immunization with engineered Marburg virus GPs. PLoS Pathog 11(6): e1005016
11. Holtsberg FW, Shulenin S, Vu H (2016) Pan-ebolavirus and pan-filovirus mouse monoclonal antibodies: protection against Ebola and Sudan viruses. J Virol 90(1):266–278
12. Keck ZY, Enterlein SG, Howell KA et al (2016) Macaque monoclonal antibodies targeting novel conserved epitopes within filovirus glycoprotein. J Virol 90(1):279–291

13. West BR, Wec AZ, Moyer CL et al (2019) Structural basis of broad ebolavirus neutralization by a human survivor antibody. Nat Struct Mol Biol 26(3):204–212
14. Bornholdt ZA, Ndungo E, Fusco ML et al (2016) Host-primed Ebola virus GP exposes a hydrophobic NPC1 receptor-binding pocket, revealing a target for broadly neutralizing antibodies. MBio 7(1):e02154–e02115
15. Lee JE, Fusco ML, Hessell AJ et al (2008) Structure of the Ebola virus glycoprotein bound to an antibody from a human survivor. Nature 454(7201):177–182
16. Bale S, Dias JM, Fusco ML et al (2012) Structural basis for differential neutralization of ebolaviruses. Viruses 4(4):447–470
17. Jeffers SA, Sanders DA, Sanchez A (2002) Covalent modifications of the ebola virus glycoprotein. J Virol 76(24):12463–12472

Chapter 3

Modification of N-Linked Glycan Sites in Viral Glycoproteins

Nicholas J. Lennemann, Lochlain Corliss, and Wendy Maury

Abstract

Post-translational modification of proteins by the addition of sugar chains, or glycans, is a functionally important hallmark of proteins trafficked through the secretory system. These proteins are termed glycoproteins. Glycans are known to be important for initiating signaling through binding of cell surface receptors, facilitating protein folding, and maintaining protein stability. For pathogens, glycans can also mask vulnerable protein regions from neutralizing antibodies. Thus, there is a need to develop methods to decipher the role of specific glycans attached to proteins in order to understand their biological role. Here, we describe established methods for identifying glycosylated residues and understanding their role in protein synthesis and function using viral glycoproteins as a model.

Key words Glycoprotein, Glycosylation, Glycan, N-linked glycan

1 Introduction

Glycosylation is a common post-translational modification with greater than 50% of eukaryotic proteins being decorated with these moieties [1]. For N-linked glycosylation events, proteins to be glycosylated are translocated via a signal peptide into the endoplasmic reticulum (ER) lumen where glycosyltransferases catalyze the attachment of N-acetyl glucosamine (GlcNAc) and mannose moieties to an asparagine residue of the conserved amino acid sequon, N-X-S/T with X representing any amino acid other than proline [2]. Within the lumen of the ER, these high-mannose precursor glycans serve to promote protein folding and stability [2]. As glycoproteins traffic through the secretory pathway into the Golgi, high-mannose glycans are modified by a number of host glycan modification enzymes to produce hybrid and complex glycans that contain an array of additional sugars attached to the glcNAc-mannose core (Fig. 1) [3]. Additionally, within the Golgi, glycoproteins can be decorated with a second type of post-translational glycosylation event on Ser and Thr residues, termed O-linked glycans; however, these moieties have not been shown to

Steven B. Bradfute (ed.), *Recombinant Glycoproteins: Methods and Protocols*, Methods in Molecular Biology, vol. 2762, https://doi.org/10.1007/978-1-0716-3666-4_3,

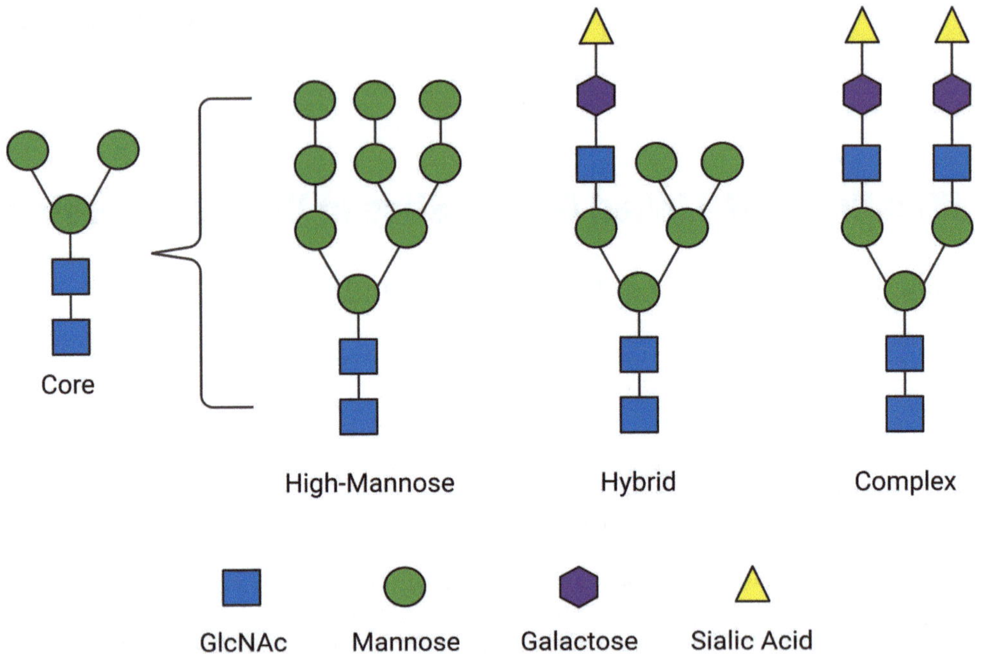

Fig. 1 Cartoons of N-linked glycans. The N-linked glycan core precursor is attached to asparagine residues where it is further modified by glycan-modifying enzymes into high-mannose, hybrid, and complex species. Illustration made with BioRender

occur at predictable motifs [4]. These sugars typically occupy less physical space than N-linked glycans and are present in high numbers in individual proteins to promote rigidity and protein function [4].

Apart from promoting protein folding and stability, glycan modifications serve a variety of other functions [5–9]. Glycans provide protection from proteolytic cleavage and serve as critical signaling components by the binding of cell surface lectin receptors [7, 10]. The importance of this post-translational modification is highlighted by its importance in viral pathogenesis. Enveloped viruses encode glycoproteins that contain a glycosylated ectodomain that is exposed to the extravirion environment and mediate cell binding and entry. Thus, these proteins are major targets of the host immune system. However, glycans are well-known to shield these essential proteins from antibody-dependent neutralization [5, 8, 9, 11–13]. Additionally, N-linked glycans on viral glycoproteins have been shown to be critical factors for modulating cellular tropism through the binding of cell surface lectin receptors, such as DC-SIGN [14, 15]. However, many glycoproteins contain more than one glycosylation site that can facilitate different biological functions [5–7].

N-linked glycans have been shown to be attached to a conserved sequon; however, not all sites are glycosylated, and the composition of high-mannose, hybrid, and complex glycans present on a glycoprotein is dependent upon both the number of glycans present on the protein and the cellular environment [1, 16]. Given the significant role of glycans on protein biology and function, it is important to understand the biological relevance of these modifications. Here, we present methods to predict and manipulate N-link glycosylation sites (NGS) to explore the biological function of these critical post-translational modifications, using the heavily glycosylated Ebola virus (EBOV) glycoprotein as a model.

2 Materials

2.1 Reagents

1. Glycoprotein expression plasmid and empty plasmid.
2. Custom oligonucleotides.
3. 10 mM dNTPs.
4. Pfu Turbo DNA polymerase (2.5 U/μL).
5. 10× Cloned Pfu DNA polymerase reaction buffer.
6. *Dpn*I restriction enzyme.
7. Q5 High-Fidelity DNA polymerase (2 U/μL).
8. 5× Q5 Reaction buffer.
9. Appropriate restriction enzymes.
10. Quick Calf Intestinal Phosphatase (CIP).
11. 0.8% Agarose gel in TAE.
12. Gel/PCR DNA Fragment Extraction Kit.
13. 2× NEB HiFi Assembly Master Mix.
14. Competent *E. coli* DH5α (*see* **Note 1**).
15. Antibiotics for plasmid selection.
16. Taq 2× Master Mix.
17. GeneJet Miniprep Kit.
18. 150 mM NaCl.
19. Polyethylenimine (PEI) transfection reagent (1 mg/mL, *see* **Note 2**).
20. Transduction reagent (*see* **Note 3**).
21. Protease inhibitor tablets.
22. Endoglycosidase H (Endo H).
23. PNGase F.
24. 10× GlycoBuffer 2.
25. 10× GlycoBuffer 3.
26. 10× Glycoprotein denaturing buffer.

27. 10% NP-40.
28. Pre-cast 4–20% TGX Gels (*see* **Note 4**).
29. Protein ladder.
30. 0.45 μm Nitrocellulose.
31. Immunoblot filter paper.
32. Primary/secondary antibodies to detect protein of interest via immunoblot.
33. Accutase.

2.2 Buffers

1. 1X Tris Acetate-EDTA (TAE): 40 mM tris acetate, 1 mM EDTA; pH 8.3; store at room temperature.
2. 6× DNA Loading Dye: 19.8 mM Tris HCl, 6 mM EDTA, pH 8.0, 0.48% SDS (w/v), 30% glycerol (v/v), 0.05% bromophenol blue (w/v).
3. 1 L Luria broth (LB) agar: 10 g tryptone, 10 g sodium chloride, 5 g yeast extract, 15 g agar; pH 7.0; store at 4 °C.
4. 1 L Luria broth (LB): 10 g tryptone, 10 g sodium chloride, 5 g yeast extract; pH 7.0; store at 4 °C.
5. Cell lysis buffer: 150 mM sodium chloride, 50 mM Tris-HCL, 1% NP-40, 0.5% sodium deoxycholate, protease inhibitor tablet (1 per 50 mL); pH 8; store at 4 °C.
6. 1X Phosphate buffered saline (PBS): 10 mM sodium phosphate, 1.9 mM potassium phosphate, 137 mM sodium chloride, 2.7 mM potassium chloride; pH 7.4; store at room temperature.
7. 1X SDS-PAGE running buffer: 250 mM Tris base, 1920 mM glycine, 1% SDS; pH 8.3; store at room temperature.
8. 1X Transfer buffer: 25 mM Tris base, 192 mM glycine, 20% methanol; pH 8.3; store at room temperature.
9. PBS-T: 10 mM sodium phosphate, 1.9 mM potassium phosphate, 137 mM sodium chloride, 2.7 mM potassium chloride, 0.1% tween; pH 7.4; store at room temperature.
10. 6× SDS-PAGE Loading Dye: 375 mM Tris-HCl, 9% SDS (w/v), 50% glycerol (v/v), 0.05% bromophenol blue (w/v). Add 10% 2-mercaptoethanol fresh before diluting into sample.
11. Blocking Buffer: 10% dried non-fat milk diluted in PBS.
12. Primary antibody dilution buffer: 5% bovine serum albumin diluted in PBS-T.
13. Secondary antibody dilution buffer: 5% dried non-fat milk diluted in PBS-T.

2.3 Cell Lines and Media

1. Human embryonic kidney (HEK) 293 T cells.
2. Vero cells (African green monkey kidney cells).
3. Cell growth media: Dulbecco's modified Eagle's medium, 10% fetal bovine serum, 1 × penicillin/streptomycin.

2.4 Equipment

1. Thermal cycler.
2. Horizontal gel electrophoresis apparatus.
3. Cell culture equipment: Class II biosafety cabinet, humidified incubators (CO_2 and temperature controlled), and sterile glass/plastic consumables.
4. Polyacrylamide gel electrophoresis apparatus.
5. Protein transfer apparatus.
6. Immunoblot detection equipment.
7. Flow cytometer.

3 Methods

3.1 In Silico Protein Analysis

While NGS are easily identified in the primary sequence of a protein, not all proteins enter the secretory pathway, and not all of these sites are eventually modified with a glycan in secreted proteins. There are numerous NGS prediction servers that will identify the presence of a signal peptide that mediates the translocation of the protein into the secretory pathway where glycosylation occurs [17]. While these prediction servers are a valuable starting point, they are not completely accurate. Thus, follow-up studies are required to determine which NGS are indeed glycosylated. One of the initial follow-up studies is to identify surface-exposed NGS by exploring the three-dimensional protein structure. This can be achieved using previously published data or using the highly accurate AlphaFold protein structure prediction server [18]. The combination of these tools provides a strong foundation to build validation studies.

1. Identify viral glycoprotein sequence.
2. Run sequence through NetNGlyc 1.0 to identify predicted glycan sites (Fig. 2) [17].
3. Identify surface exposed sequons from existing structural data or run viral glycoprotein sequence through AlphaFold to obtain a predicted structure (MIT-Collab, *see* **Note 5**) [18].

3.2 Design of Mutagenesis Primers

1. Change identified codon to encode an alanine (*see* **Note 6**, normally mutate position S/T).
2. Design forward primer by selecting 15 nucleotides to the 5′ and 15 nucleotides to the 3′ end of the mutation.

```
Name:  Bombali_ebolavirus_GP          Length:  672
MILQVPEKRHQRTVLFIWLVILFQRAVSVPLGVIHNSTLQVSDIDKFVCHDKLTSTNQLRSIGLNLEGNGIATDVPSATK     80
RWGFRAGVPPKVVGYEAGEWAENCYNLEIKKPDGSECLPMAPEGIRGFPRCRYVHKVSGTGSCESGFAFHKEGAFFLYDR    160
LASTIIYRGTTFAEGVVAFIILPKAEKNFLQPPLTQGPTNTTNDPSSMYHSTTLEYETTRFGTNRSAFWFKVDNLTFVQL    240
ESRFTPQFLVELNETIYIEGKRSNTTGRLIWQVNSRVDTDVGEWAFWENKKNLKKSFPREELSLTAVPRAADSEHDAHPP    320
EYTPGPDSNPTINDNTELVTEDPAHLVQLQRQGRKEILPTTIPQAIEREPPAAQHDNPRNSPTPPSPIESDITDSTQAED    400
LTHTDDPSTINSATEEPLPEVGETTQVPRDPEGPRRTQPTPPTGQPEQPSDNTMTPGQIHSESAAPMGERSIDGPGLLTN    480
TLAGVARLITGAGRAKRESPEIRGAKCNPNLHYWTTHEESAAAGLAWIPYFGPAAEGIYTEGLMQNQNELICGLRQLANE    560
TTQALQLFLRSTTELRTFSILNRKAIDFLLQRWGGTCRILGPDCCIEPHDWTKNITDRIDQIIHDFVDKPLPDQSNNDNW    640
WTGWRQWIPAGIGVVGVIAAFIALICICKIIC
...................................N............................................     80
................................................................................    160
.......................................N.......................N.........N......    240
............N..........N........................................................    320
........N.......................................................................    400
................................................................................    480
..............................................................................N.    560
.....................................................N..........................    640
................................                                                    720

(Threshold=0.5)
----------------------------------------------------------------------
SeqName         Position  Potential   Jury     N-Glyc
                                     agreement  result
----------------------------------------------------------------------
Bombali_ebolavirus_GP   36 NSTL   0.7580     (9/9)   +++
Bombali_ebolavirus_GP  200 NTTN   0.6908     (8/9)   +
Bombali_ebolavirus_GP  224 NRSA   0.5748     (9/9)   ++
Bombali_ebolavirus_GP  234 NLTF   0.6924     (9/9)   ++
Bombali_ebolavirus_GP  253 NETI   0.6799     (9/9)   ++
Bombali_ebolavirus_GP  264 NTTG   0.6780     (8/9)   +
Bombali_ebolavirus_GP  329 NPTI   0.5646     (8/9)   +     WARNING: PRO-X1.
Bombali_ebolavirus_GP  559 NETT   0.5346     (5/9)   +
Bombali_ebolavirus_GP  614 NITD   0.5062     (5/9)   +
----------------------------------------------------------------------
```

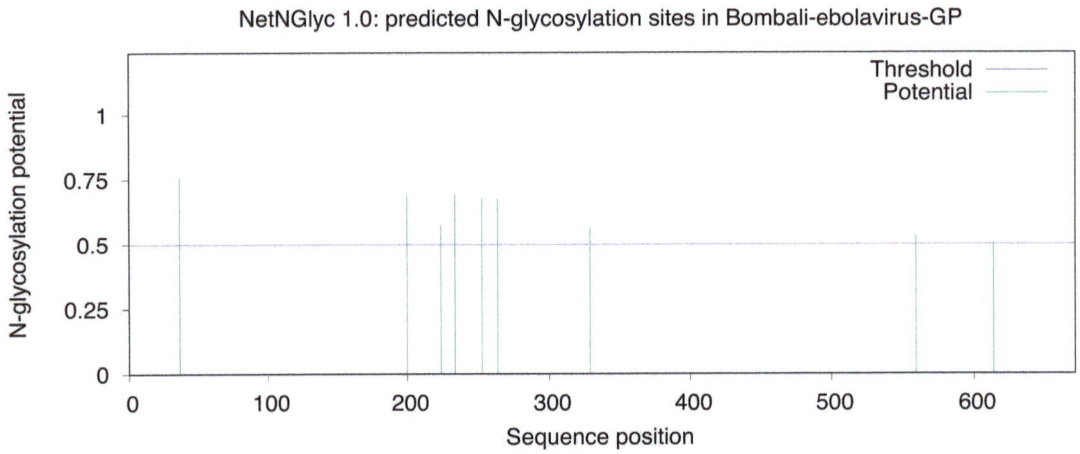

Fig. 2 Example NetNGlyc-1.0 output. The Bombali ebolavirus glycoprotein amino acid sequence was analyzed by NetNGlyc. Red/blue text highlights N-X-ST sequons with asparagine residues in red. Middle, table of identified sites with the likelihood of being glycosylated based on neural network analysis data shown below

3. Design reverse primer by generating the reverse complement of the forward primer.
4. Determine primer melting temperatures (T_m, *see* **Note 7**).

3.3 Site-Directed Mutagenesis

Multiple methods and kits exist for the generation of point mutations in plasmids. Here, we present two methods that have been successful in our laboratories.

3.3.1 Method 1: Quick Change Protocol

This is a well-known method used in our laboratory that utilizes complimentary primers encoding one to three nucleotide changes that are used in a long PCR reaction to amplify the entire plasmid [5, 6, 19]. The parental plasmid template is then removed via restriction enzyme digestion with *Dpn*I. This enzyme only digests methylated DNA, which is derived from bacteria (template), but not the de novo DNA sequence from the PCR. This reaction leaves only the PCR product intact. This product is then transformed into *E. coli* and recovered plasmid is screened via sequencing. While this method has been extensively used in the literature, it can be time-consuming and may require screening/sequencing of several clonal plasmids.

1. Prepare mutagenesis PCR: 25 ng glycoprotein expression plasmid, 1× Cloned Pfu DNA polymerase reaction buffer, 200 μM dNTPs, 0.4 μM forward primer, 0.4 μM reverse primer, 1 μL Pfu Turo DNA polymerase, and H_2O up to 50 μL.
2. Perform PCR (Table 1).
3. Digest methylated template by adding 2 μL *Dpn*I for 4 h at 37 °C.
4. Optional: Run 10 μL of reaction on a 0.8% agarose in TAE to look for PCR product (*see* **Note 8**).
5. Transform 5 μL of reaction into competent DH5α *E. coli*.
6. Plate transformation on desired antibiotic LB Agar plates (*see* **Note 9**).
7. Grow colonies overnight at 37 °C.

Table 1
Quick change site-directed mutagenesis PCR

Step	Temperature (°C)	Time
1 – Initial Melt	95	30 s
2 – Melt	95	30 s
3 – Anneal	T_m – 5 °C	30 s
4 – Elongation	68	1 min/kb
Repeat steps 2–4 × 19		
5 – Elongation	68	5 min
6 – Hold	12	∞

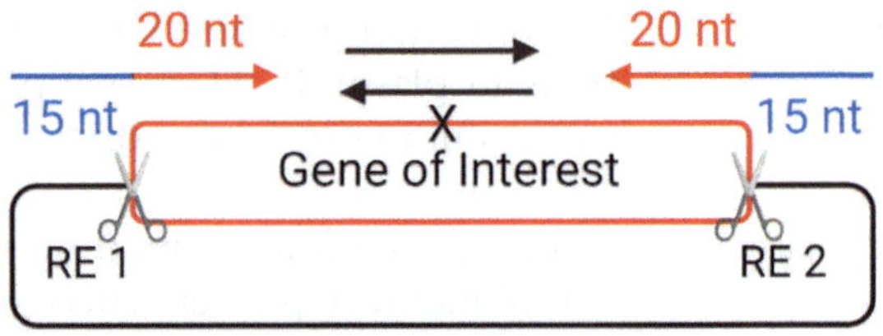

Fig. 3 Strategy for designing primers to perform multi-fragment assembly for mutagenesis. The X stands for the mutation that will be introduced into the sequon. Illustration made BioRender

8. Inoculate 5 mL of LB + antibiotic with individual colonies to grow overnight at 37 °C.
9. Perform miniprep.
10. Submit plasmid for sequencing.

3.3.2 Method 2: Multi-fragment Assembly Mutagenesis

This more recent method used in our laboratory takes advantage of technologies that allow for the efficient assembly of multiple DNA fragments (Fig. 3) [20]. Initially, unique restriction enzyme (RE) sites flanking the gene of interest are identified. Next, 5′ and 3′ primers are designed to have homology to the vector and respective termini of the gene of interest. These primers are used in separate reactions with the complimentary primers designed in Method 1 to generate two fragments that have homology at the termini to allow for seamless assembly with a digested vector. While this method requires more reagents, it saves time by exploiting faster PCR reaction times and highly efficient colony screening methods.

1. Identify unique restriction enzyme sites that flank the region of the gene of interest in the viral glycoprotein expression plasmid (*see* **Note 10**).
2. Design forward primer that contains 15 nucleotides of homology with the 5′ end of the vector and ~21 nucleotides of the 5′ end of the target gene.
3. Design reverse primer that contains 15 nucleotides of homology with the 3′ end of the vector and ~21 nucleotides of the 3′ end of the target gene.
4. Prepare vector: Digest empty expression plasmid with appropriate restriction enzymes for 2 h at 37 °C.
5. Optional: Quick CIP vector via the addition of 1 μL of Quick CIP (*see* **Note 11**).
6. Prepare PCR reaction for 5′ mutant fragment (5′A): 10 ng glycoprotein expression plasmid template, 200 uM dNTPs, 1.25 μL forward primer, 1.25 μL reverse primer with mutation, 5 μL 5× Q5 polymerase reaction buffer, bring reaction to

Table 2
Quick change site-directed mutagenesis PCR

Step	Temperature (°C)	Time
1 – Initial Melt	98	30 s
2 – Melt	98	10 s
3 – Anneal	T_m + 3 °C	30 s
4 – Elongation	72	15–30 s/kb
Repeat steps 2–4 × 34		
5 – Elongation	72	5 min
6 – Hold	12	∞

24.8 μL with H_2O, and add 0.2 μL Q5 High Fidelity DNA polymerase.

7. Prepare PCR reaction for 3′ mutant fragment (3′A): 10 ng glycoprotein expression plasmid template, 200 uM dNTPs, 1.25 μL forward primer, 1.25 μL reverse primer with mutation, 5 μL 5× Q5 polymerase reaction buffer, bring reaction to 24.8 μL with H_2O, and add 0.2 μL Q5 High Fidelity DNA polymerase.
8. Perform PCR reactions (Table 2).
9. Run samples on 0.8% agarose gel (TAE).
10. Cut out products of appropriate size.
11. Perform gel extraction with IBI kit, elute in 30 μL.
12. Prepare HiFi assembly mixture on ice in thin-walled PCR tube: 1 μL vector, 0.5 μL 5′A, 0.5 μL 3′A, 2 μL 2× NEB HiFi Assembly Master Mix (*see* **Note 12**).
13. Incubate reaction at 55 °C for 30 min (*see* **Note 13**).
14. Place reaction on ice or freeze at −20 °C.
11. Transform 2 μL of reaction into competent *E. coli*.
12. Plate transformation on appropriate LB agar + antibiotic plates (*see* **Note 7**).
13. Grow colonies overnight at 37 °C.
14. Resuspend colonies into 30 μL of H_2O.
15. Screen for colonies containing the assembled plasmid via colony PCR: 6.5 μL H_2O, 5 μL resuspended colony, 0.5 μL forward primer from **step 6**, 0.5 μL reverse primer from **step 7**, 12.5 μL Taq 2× Master Mix.
16. Perform colony PCR (Table 3).
17. Run PCR on 0.8% agarose gel to identify positive colonies.

Table 3
Quick change site-directed mutagenesis PCR

Step	Temperature (°C)	Time
1 – Initial Melt	95	5 min
2 – Melt	95	30 s
3 – Anneal	$T_m - 5$ °C	30 s
4 – Elongation	68	1 min/kb
Repeat steps 2–4 × 29		
5 – Elongation	68	5 min
6 – Hold	12	∞

18. Inoculate 5 mL of LB + antibiotic with 20 uL of resuspended colony confirmed via PCR to grow overnight at 37 °C.
19. Perform miniprep.
20. Submit plasmid for sequencing.

3.4 Assessment of NGS Present on Parental and Mutant Constructs and Enzymatic Removal of Glycans

Determination of the broad types of N-linked glycans on individual proteins can be assessed using specific glycosidases. EndoH and PNGase F are commonly used to distinguish between the presence of high-mannose glycans and hybrid/complex glycans [21]. Here, we describe methods for enzymatic removal of glycans from denatured protein samples. These methods can be used in combination with mutagenesis to understand how manipulation of individual sites impacts the glycosylation status of other NGS.

1. Prepare HEK 293 T cell suspension at 250,000 cells/0.45 mL in growth media in a 24-well tissue culture plate (*see* **Note 14**).
2. Prepare plasmid dilutions (empty, wild-type parental control, and mutants): 500 ng in 25 μL 150 mM NaCl.
3. Prepare PEI dilution: 1 μL PEI (1 mg/mL stock) in 25 μL 150 mM NaCl.
4. Add PEI dilution to plasmid dilution, mix by pipetting 3–5×.
5. Incubate at room temperature for 15 min.
6. Apply transfection mixture dropwise.
7. Incubate in humidified chamber at 37 °C with 5% CO_2 for 48–72 h.
8. Chill 24-well plate on ice for 10 min.
9. Replace supernatant with 150 μL of cold cell lysis buffer.
10. Incubate on ice with occasional rocking for 10 min.

11. Clarify cell lysates by centrifugation at 12,000× *g* for 10 min at 4 °C.
12. Samples can be store at −20 °C until needed.
13. Prepare Endo H digestion to remove high-mannose glycans only: up to 9 μL cell lysate, 1 μL 10× glycoprotein denaturing buffer; heat to 100 °C for 10 min; add 2 μL 10× GlycoBuffer 3, H_2O, and 1–5 μL Endo H for a total reaction volume of 20 μL; incubate at 37 °C for 1 h.
14. Prepare PNGase F digestion to remove all N-linked glycans: up to 9 μL cell lysate, 1 μL 10× glycoprotein denaturing buffer; heat to 100 °C for 10 min; add 2 μL 10× GlycoBuffer 2, 2 μL 10% NP-40, 6 μL H_2O, and 1 μL PNGase F (or H_2O as a control) for a total reaction volume of 20 μL; incubate at 37 °C for 1 h.
15. Prepare samples for immunoblot by adding 4 μL 6× SDS-PAGE Loading Dye.
16. Boil samples for 7–10 min.
17. Load protein ladder and 24 μL of samples into appropriate wells.
18. Run protein gel until dye front reaches the bottom.
19. Transfer protein from the gel to 0.45 μm nitrocellulose.
20. Block membrane in blocking buffer for 30 min.
21. Rinse membrane ×3 in PBS-T.
22. Incubate membrane with primary antibody in primary antibody dilution buffer for 1 h.
23. Rinse membrane ×5 in PBS-T.
24. Incubate membrane with secondary antibody diluted in secondary antibody dilution buffer for 30–60 min.
25. Wash membrane for 5 min ×3.
26. Develop membrane using equipment appropriate for the secondary antibody conjugation.
27. Interpret immunoblot to determine protein glycosylation status (see example interpretation in Fig. 4).

3.5 Production of VSV Pseudovirions Bearing Wild-type or Mutant EBOV GP and the Effect of Glycans on Viral Glycoprotein-mediated Entry

1. Reverse transfect HEK 293 T as in Subheading 3.4 **steps 1–6**.
2. At 16–24 h post transfection, inoculate cells with VSVΔG-GFP bearing the VSV glycoprotein.
3. At 2–4 h post inoculation, gently wash cells ×2 with 2 mL of PBS.
4. Replace wash buffer with 2 mL growth media for 24–48 h.
5. Harvest supernatants with 3 mL sterile syringe and filter through 0.45 μm syringe filter.

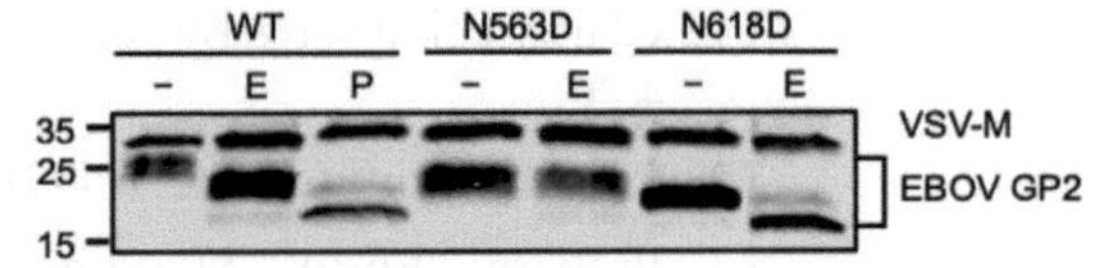

Fig. 4 Determination of site-specific glycans by enzymatic removal of N-linked glycans. Immunoblot of VSV matrix (VSV-M) and EBOV GP subunit 2 (GP2, containing 2 predicted NGS at residues N653 and N618) from supernatants containing VSV pseudovirions. Lanes 1–3: wild-type glycoprotein virions. Untreated samples show a broad band for GP2, which is indicative of a glycoprotein. Endo H (E)-treated sample migrates more efficiently through the gel indicating the presence of at least one high mannose glycan. PNGase F (P)-treated sample migrates more efficiently than both WT and Endo H samples, indicating the additional presence of a complex glycan. Lanes 4 and 5: GP2 bearing a N563D mutation to abolish glycosylation at this site. Glycosylation of N563D is not influenced by Endo H treatment, indicating the remaining glycan at residue 618 is a complex glycan. Lanes 6 and 7: GP2 bearing a N618D mutation to abolish glycosylation at this site. Endo H treatment of N618D results in protein migration equivalent to PNGase F treatment of WT, indicating the remaining glycan at residue 618 is a high-mannose glycan, which is consistent with the interpretation of results from Lanes 1–5

6. Aliquot and store pseudovirions at −80 °C.
7. Seed 50,000 Vero cells per well in a 48-well format (**Note 15**).
8. After 18–24 h, add serial dilutions of pseudovirions to Vero cells in DMEM supplemented with 1.5% FBS and 1× P/S (**Note 16**).
9. Incubate pseudovirions with cells in a cell culture incubator for 18–24 h.
10. Prepare cells for flow cytometry: remove media, detach cells with 100 μL Accutase, and transfer to tubes appropriate for a flow cytometer.
11. Using a flow cytometer, determine the percentage of GFP-positive cells per μL of pseudovirus-containing supernatant to assess infectivity (*see* **Note 17**).

4 Notes

1. There is a wide variety of competent cells used for cloning. Use the appropriate strain for your expression plasmid. Additionally, there are many protocols for generating laboratory stocks of competent *E. coli*. Our laboratory has had great success using the *Mix and Go E. coli* Transformation Kit and Buffer Set (Zymo Research) for generating laboratory stocks. When using an ampicillin-resistant plasmid, these cells do not require

a heat shock. Plasmid is added to thawed competent cells on ice and can be immediately spread on a pre-warmed LB-Amp plate.

2. Use the transfection reagent that works best for the cell line used in individual assays. PEI is an inexpensive reagent that works well for transfecting a number of common cell lines: HEK 293 T, baby hamster kidney cells (BHK-21), U2OS, and Cos-7.
3. This protocol is directed toward mutagenesis and characterization of viral glycoproteins and takes advantage of the VSVΔG-GFP pseudotyping system [22–24]. However, other pseudotyping systems are available, including feline immunodeficiency virus [25]. Lastly, the mutagenesis protocols outlined here can be implemented for any glycoprotein. Thus, the characterization assays would have to be developed accordingly.
4. This protocol utilizes pre-cast gels for SDS-PAGE and the PAGE apparatus from NEB. However, it is possible to use gels poured in the laboratory and other electrophoresis equipment.
5. AlphaFold has been shown to predict protein structures with high efficiency [18]. In the absence of a solved protein structure, this tool can provide a starting point to identify surface-exposed sequons that are accessible for the glycosylation machinery of the cell. However, this is only a predicted model and requires validation via mutagenesis of predicted sites.
6. Mutation of the serine or threonine in the sequon to alanine is a good starting point, given the modest change in amino acid characteristics. However, it is also possible to mutate the asparagine residue to glutamine or aspartate.
7. There are several primer T_m calculators available online. We typically use NEB T_m Calculator, which provides the anneal temperatures for primer sets based on the polymerase being used.
8. Often times, the PCR product is visible as a faint band on an agarose gel, with smearing due to digested template. However, the lack of a band does not always suggest an unsuccessful PCR.
9. Plating volume will depend on the efficiency of the competent cells used.
10. Ideally, the restriction sites used to insert the gene of interest into the empty vector would be used for designing the multi-fragment assembly primers. This increases the confidence in colony PCR results. However, it is possible to use other unique sites within the gene of interest on either side of the sequence to be mutated.

11. Quick CIP will remove 5′ and 3′ phosphates from the digested vector to prevent re-ligation, which will reduce the number of negative colonies.

12. The protocol provided on the NEB website differs from that provided here and suggests a 20 μL reaction volume. However, we are consistently successful in scaling this reaction down to 4 μL.

13. A 30 min incubation is typically long enough for a 3-fragment assembly. However, for large constructs (>8 kbp), we extend the reaction time to 60 min.

14. Glycosylation patterns of glycoproteins can differ between cell types, due to varying expression of glycan-modifying enzymes and should be considered when designing individual experiments [16]. Additionally, manipulation of the number of asparagines modified by glycans can skew the species of glycans present on the protein, which needs to be considered when interpreting results [1, 26].

15. Vero cells are used here because they support EBOV GP-mediated entry independent of glycosylation of GP [27]. However, other target cells can be used to assess the importance of GP glycosylation on entry. Alternatively, a poorly permissive cell line, such as HEK 293 T cells for EBOV GP studies, can be used to explore the role of individual lectin receptors on NGS mutant GP-mediated entry through exogenous expression [5].

16. This step can be modified to investigate the sensitivity of NGS mutants to inhibitors and neutralizing antibodies [5–7].

17. Titer can be determined by: (% GFP-positive cells X # cells per well)/mL of pseudovirus. Accurate titer can be derived from the linear portion of the serial dilution curve [19, 23].

References

1. Apweiler R, Hermjakob H, Sharon N (1999) On the frequency of protein glycosylation, as deduced from analysis of the SWISS-PROT database. Biochim Biophys Acta Gen Subj 1473:4–8
2. Stanley P, Moremen KW, Lewis NE et al (2022) Glycosphingolipids. In: Varki A, Cummings RD, Esko JD et al (eds) Essentials of glycobiology, 4th edn. Cold Spring Harbor, New York
3. Rini JM, Moremen KW, Davis BG et al (2022) Glycosyltransferases and glycan-processing enzymes. In: Varki A, Cummings RD, Esko JD et al (eds) Essentials of glycobiology, 4th edn. Cold Spring Harbor, New York
4. Varki A (1993) Biological roles of oligosaccharides: all of the theories are correct. Glycobiology 3:97–130
5. Lennemann NJ, Rhein BA, Ndungo E et al (2014) Comprehensive functional analysis of N-linked glycans on ebola virus GP1. MBio 5: e00862
6. Lennemann NJ, Walkner M, Berkebile AR et al (2015) The role of conserved N-linked glycans on ebola virus glycoprotein 2. J Infect Dis 212: S204–S209
7. Wong AC, Sandesara RG, Mulherkar N et al (2010) A forward genetic strategy reveals destabilizing mutations in the Ebolavirus

glycoprotein that alter its protease dependence during cell entry. J Virol 84:163–175

8. Wang CC, Chen JR, Tseng YC et al (2009) Glycans on influenza hemagglutinin affect receptor binding and immune response. Proc Natl Acad Sci U S A 106:18137–18142
9. Cook JD, Lee JE (2013) The secret life of viral entry glycoproteins: moonlighting in immune evasion. PLoS Pathog 9:e1003258
10. Hanover JA (2001) Glycan-dependent signaling: O-linked N-acetylglucosamine. FASEB J 15:1865–1876
11. Walls AC, Tortorici MA, Frenz B et al (2016) Glycan shield and epitope masking of a coronavirus spike protein observed by cryo-electron microscopy. Nat Struct Mol Biol 23:899–905
12. Von Bülow S, Sikora M, Blanc FEC et al (2023) Antibody accessibility determines location of spike surface mutations in SARS-CoV-2 variants. PLoS Comput Biol 19:e1010822
13. Fenouillet E, Gluckman JC, Jones IM (1994) Functions of HIV envelope glycans. Trends Biochem Sci 19:65–70
14. Tortorella D, Gewurz BE, Furman MH et al (2000) Viral subversion of the immune system. Annu Rev Immunol 18:861–926
15. Monteiro JT, Lepenies B (2017) Myeloid C-type lectin receptors in viral recognition and antiviral immunity. Viruses 9:59
16. Croset A, Delafosse L, Gaudry JP et al (2012) Differences in the glycosylation of recombinant proteins expressed in HEK and CHO cells. J Biotechnol 161:336–348
17. Gupta R, Brunak S (2002) Prediction of glycosylation across the human proteome and the correlation to protein function. Pac Symp Biocomput, pp 310–322
18. Jumper J, Evans R, Pritzel A et al (2021) Highly accurate protein structure prediction with AlphaFold. Nature 596:583–589
19. Lennemann NJ, Dillard J, Ruggio N et al (2021) A naturally occurring polymorphism in the base of Sudan virus glycoprotein decreases glycoprotein stability in a species-dependent manner. J Virol 95:e0107321
20. Corliss L, Holliday M, Lennemann NJ (2022) Dual-fluorescent reporter for live-cell imaging of the ER during DENV infection. Front Cell Infect Microbiol 12:1042735
21. Maley F, Trimble RB, Tarentino AL et al (1989) Characterization of glycoproteins and their associated oligosaccharides through the use of endoglycosidases. Anal Biochem 180: 195–204
22. Takada A, Robison C, Goto H et al (1997) A system for functional analysis of Ebola virus glycoprotein. Proc Natl Acad Sci U S A 94: 14764–14769
23. Brouillette RB, Maury W (2017) Production of filovirus glycoprotein-pseudotyped vesicular stomatitis virus for study of filovirus entry mechanisms. Methods Mol Biol 1628:53–63
24. Schnell MJ, Buonocore L, Kretzschmar E et al (1996) Foreign glycoproteins expressed from recombinant vesicular stomatitis viruses are incorporated efficiently into virus particles. Proc Natl Acad Sci U S A 93:11359–11365
25. Sinn PL, Coffin JE, Ayithan N et al (2017) Lentiviral vectors pseudotyped with filoviral glycoproteins. Methods Mol Biol 1628:65–78
26. Williams R, Ma X, Schott RK et al (2014) Encoding asymmetry of the N-glycosylation motif facilitates glycoprotein evolution. PLoS One 9:e86088
27. Kondratowicz AS, Lennemann NJ, Sinn PL et al (2011) T-cell immunoglobulin and mucin domain 1 (TIM-1) is a receptor for Zaire ebolavirus and Lake Victoria marburgvirus. Proc Natl Acad Sci U S A 108:8426–8431

Chapter 4

Production of Influenza Virus Glycoproteins Using Insect Cells

Madhumathi Loganathan, Benjamin Francis, and Florian Krammer

Abstract

The baculovirus/insect cell expression system is a very useful tool for reagent and antigen generation in vaccinology, virology, and immunology. It allows for the production of recombinant glycoproteins, which are used as antigens in vaccination studies and as reagents in immunological assays. Here, we describe the process of recombinant glycoprotein production using the baculovirus/insect cell expression system.

Key words Influenza virus, Hemagglutinin, Neuraminidase, Protein production, Insect cells, Baculovirus, Viral glycoproteins

1 Introduction

Recombinant influenza virus glycoproteins refer to artificially produced versions of the glycoproteins found on the surface of the influenza virus. These glycoproteins, called hemagglutinin (HA) and neuraminidase (NA), are responsible for the virus's ability to enter and exit host cells and are the main targets of the adaptive immune response [1]. Recombinant influenza virus proteins are typically produced by expressing the genes that encode for HA or NA in cells of a production system, such as bacteria [2, 3], yeast [4], insect cells [5], or mammalian cells followed by protein purification.

One of the advantages of using recombinant glycoproteins is that they can be designed to reflect the natural conformation of the viral proteins, making them effective immunogens that can elicit a strong immune response. In addition, recombinant glycoproteins can be engineered to eliminate unwanted features, e.g., metastability of viral fusion proteins [6, 7], while retaining their ability to stimulate an immune response. Recombinant influenza virus glycoproteins are an important tool for studying the biology of the

Steven B. Bradfute (ed.), *Recombinant Glycoproteins: Methods and Protocols*, Methods in Molecular Biology, vol. 2762,
https://doi.org/10.1007/978-1-0716-3666-4_4,

virus, immune responses to it and for developing new approaches to prevent and treat influenza virus infections.

The baculovirus expression system, which uses recombinant baculoviruses that infect insect cells and force them to produce the desired protein, is a widely used method for producing large quantities of recombinant proteins [8]. It has several advantages over other expression systems, such as bacterial and mammalian cells. One of the main advantages over bacterial systems is that insect cells are capable of post-translational modifications, such as glycosylation, that are similar to those found in mammalian cells. The system can therefore be used to express large and complex proteins or even multiple units of protein complexes at the same time. However, production costs are often lower than for mammalian expression systems and yields are often higher. This makes the baculovirus expression system a popular choice for producing recombinant proteins including for use in biopharmaceuticals.

To use the baculovirus expression system, a gene encoding the protein of interest is first cloned into a baculovirus transfer vector, which is then used to generate a recombinant baculovirus genome. This genome is then used to rescue infectious, recombinant baculovirus in insect cells. As a next step, the rescued virus is used to infect more insect cells, from which the recombinant protein is harvested. The protein is then either purified from the cell supernatant if the protein is secreted and soluble (e.g., soluble versions of HA or NA), or from cell lysates if the protein is located within the cell or in the cell membrane (e.g., influenza virus nucleoprotein or matrix protein).

The baculovirus expression system has been widely used to express functional recombinant influenza virus hemagglutinin (HA) and neuraminidase (NA) proteins [9–11]. In order to obtain soluble trimers, the HA is usually expressed as ectodomain with the transmembrane domain substituted by a trimerization domain (e.g., a T4 foldon or a leucine zipper) [5, 12, 13] and a purification tag (hexahistidine tag, strep tag II, etc.) is usually added at the C-terminus. Expression of non-trimerized HA can lead to misfolding and loss of epitopes, especially in the stalk domain of HA [5]. NA is typically expressed as NA head only construct (the head domain includes most of the protein and the enzymatic side) that is held together by an N-terminal tetramerization domain (e.g., from the human vasodilator stimulating phosphoprotein [14], from the bacterial tetrabrachion [15] or from the measles virus phosphoprotein [16]) while the authentic cytoplasmic tail, transmembrane domain, and the helical stalk domain are removed. A typical design for expression of soluble NA includes a signal peptide followed by a purification tag, followed by the tetramerization domain and then the head domain of the NA. If the tetramerization domain is left out and NA is expressed as monomer, it often loses enzymatic activity and induces non-protective immunity when used as antigen [17]. Of note, NA cleaves sialic acid off from N-linked glycans [18]. This can lead to toxicity and low expression levels in mammalian cells but since insect cells used for the baculovirus expression lack sialic acid, the enzyme is inert in the system and no toxicity is observed. In addition to recombinant, soluble versions of HA and NA, membrane-bound wild type versions can be expressed as well. These proteins can either be expressed separately and extracted from the membrane and purified (this is how the influenza vaccine Flublok is manufactured) [19, 20] or expressed together with, e.g., the influenza virus matrix protein (M1) which leads to the formation of virus-like particles (VLPs) [21]. Of note, when expressed in insect cells, 'regular' HA is not proteolytically cleaved between HA1 and HA2 and may be more stable and less prone to flip from the prefusion to the postfusion conformation. However, HAs from highly pathogenic H5 and H7 strains that include a polybasic cleavage site which is recognized by furin-like proteases will be processed into HA1 and HA2 in insect cells. To avoid that, it is recommended to remove the polybasic cleavage site in expression constructs. Another important point is that cellular supernatants from the baculovirus system contain baculoviruses. When protein is purified, the virus is removed efficiently. However, when VLPs are purified, baculovirus is typically co-purified [22]. This is not a safety concern since these viruses cannot replicate in mammalian cells. However, they do enter mammalian cells and trigger Toll-like receptor 9 (TLR9) leading to stimulation of innate immunity [23]. This can have a positive adjuvant effect for vaccine formulations [22] but can also be problematic for intranasal vaccines since it can produce short-lived unspecific protection which can mask real vaccine-induced adaptive immune-based protection [24, 25].

The baculovirus system is relatively flexible and can be used with different *Spodoptera frugiperda* and *Trichoplusia ni* insect cell lines, some of which are better suited than others for protein secretion [26]. But the yields and protein quality can be influenced also by media choices and passage history. It is therefore suggested to try expression with different cell lines and media combinations to find out what works best for a specific laboratory. In addition, different systems for generating recombinant baculoviruses are available and a number of promotors can be used as well. Furthermore, many different methods can be used for downstream purification including stringent purification using a combination of different steps as well as low key solutions that include simple gravity flow purification. What we describe here is what has been working well in our laboratory for the last 10 years but may not be the optimal choice for others.

In summary, the baculovirus system is a versatile and useful system for expression or recombinant HA and NA. It can be used in many different settings to produce reagents for immunology assays as well as antigens for vaccine development. What is described below is a protocol for the expression of secreted HA trimers and NA tetramers.

2 Materials

2.1 General Supplies (Used Throughout Protocol)

1. 1000 μL, 200 μL, 20 μL, and 10 μL micropipette tips.
2. P1000, P200, P20, and P2 micropipettes.
3. 1X phosphate buffer saline (PBS). Store at room temperature (RT).
4. Molecular biology grade sterile purified water. Store at RT.
5. Molecular biology grade ethanol, 200 proof. Store at RT.
6. NanoDrop to measure the concentration of DNA.
7. Eppendorf Thermomixer to apply heat of up to 96 °C to Eppendorf tubes.
8. Aluminum foil.

2.2 Polymerase Chain Reaction and Colony PCR

1. Q5® High Fidelity DNA Polymerase kit or similar. Store at −20 °C.
2. cDNA template from RT-PCR of an influenza virus genome or synthesized gene or plasmid with sequence encoding HA or NA of interest, desalted. Dilute with sterile molecular biology grade water and store at −20 °C.
3. Forward and reverse primers specific to the HA or NA insert, desalted. Dilute with molecular biology grade sterile purified water and store at −20 °C.
4. *Taq* DNA Polymerase 2× Master Mix or similar. Store at −20 °C.
5. Glycerol (for glycerol bacteria stock). Before use, dilute to 40% in molecular biology grade sterile purified water.

2.3 PCR Purification Kit, Gel Extraction Kit, and Spin Miniprep Kits

The kits used are from QIAgen®. Kits from different vendors can be used, however the methods section outlines the steps for QIAgen® PCR Purification, Gel Extraction, and Spin Miniprep kits specifically. The gel extraction and PCR purification kits use the same columns for filtration. All reagents are included in the respective kits.

1. Buffer PE (Wash buffer) for PCR purification, gel extraction, and miniprep. Add 100% ethanol before use and store at RT (*see* **Note 1**).
2. QIAquick spin columns for PCR purification and gel extraction.

3. Buffer QG (solubilization buffer) for gel extraction. Store at RT.
4. 100% isopropanol for gel extraction aliquoted into 50 mL conical tubes.
5. RNase A for miniprep. Store at RT before use.
6. Buffer P1 (Resuspension buffer) for miniprep. Store at RT before use. Add 200 μL RNase A to the buffer before use. Once RNase A is added, make sure to store the buffer at 4 °C (*see* **Note 2**).
7. Buffer P2 (Lysis buffer) for miniprep. Store at RT.
8. Buffer N3 (Neutralization buffer) for miniprep. Store at RT.
9. QIAprep 2.0 Spin Miniprep Columns for miniprep.

2.4 Restriction Digestion, Ligation, and Transformation

1. T4 DNA ligase. Store at −20 °C.
2. T4 ligase buffer. Store at −20 °C (*see* **Note 3**).
3. *Bam*HI-HF restriction enzyme. Store at −20 °C.
4. *Not*I-HF restriction enzyme. Store at −20 °C.
5. *Hind*III-HF restriction enzyme. Store at −20 °C.
6. *Xba*I restriction enzyme. Store at −20 °C.
7. rCutsmart Buffer. Store at −20 °C.
8. Modified pFastBacDual Transfer Plasmids (modified to include a signal peptide, a 6xhis tag, a VASP tetramerization domain and a thrombin cleavage site for NA cloning and a thrombin cleavage site, T4 trimerization domain and 6xhis tag for HA cloning). For HA, we abbreviate the plasmid pHA. For NA, we abbreviate the plasmid pNA. Store at −20 °C. (To obtain modified pFastBacDual Transfer Plasmids, request from the corresponding author).
9. Calf Intestinal Alkaline Phosphatase (CIP). Store at −20 °C.
10. 10X Tris-Acetate -EDTA (TAE) Buffer. Store at RT. Before use, dilute to 1× by combining 100 mL of buffer and 900 mL of distilled water.
11. Agarose powder. Store at RT.
12. Gel casting combs (large and small) for casting wells into DNA gels.
13. LB + Agar Medium capsules for pouring plates for bacterial culture. Store at RT.
14. XL10-Gold Competent cells. Store at −80 °C.
15. Super Optimal Broth with Catabolite Repression (SOC) Medium. Store at RT.
16. DH10Bac Competent cells. Store at −80 °C.

17. Kanamycin, powder. Store at −20 °C.
18. Tetracycline, powder. Store at −20 °C.
19. Gentamycin, liquid. Store at RT in the dark.
20. 5-Bromo-4-chloro-3-indolyl β-D-galactopyranoside (X-Gal) powder. Store at −20 °C.
21. Isopropyl β-D-1-thiogalactopyranoside (IPTG) powder. Store at −20 °C.

2.5 Midiprep Kit

The kit outlined is from Invitrogen™. Different kits can be used, but keep in mind that the steps outlined in the methods section are from the Invitrogen™ Midiprep Kit. The kit included all of the following:

1. HiPure Midiprep Filter Columns.
2. Equilibration buffer. Store at RT.
3. RNase A. Store at RT.
4. Resuspension Buffer. Store at RT. Before use, add 1.5 mL of RNase A to the buffer (*see* **Note 4**).
5. Lysis buffer. Store at RT. If cloudy, incubate at 37 °C to dissolve the precipitate that formed in the buffer.
6. Precipitation buffer. Store at RT.
7. Wash buffer. Store at RT.
8. Elution buffer. Store at RT.

2.6 Transfection

1. *Spodoptera frugiperda* Sf9 cells.
2. Fetal bovine serum (FBS). Store at −20 °C (*see* **Note 5**).
3. 100X Penicillin-Streptomycin at 10,000 U/mL of penicillin and 10,000 μg of streptomycin (Pen-Strep). Store at −20 °C for long term (greater than 1 week), 4 °C for short term (1 week).
4. *Trichoplusia ni* medium formulation Hink (TNM-FH) insect cell culture liquid media supplemented with 0.6 g/L of L-glutamine. Store at 4 °C.
5. 3% FBS and 10% FBS TNM-FH media. Make 1000 mL of each at a time, both supplemented with 30 mL or 100 mL of FBS respectively, 10 mL of Pen-Strep, and 10 mL of Pluronic F-68 (also called Poloxamer 88). Store at 4 °C.
6. Sf9 cells. Culture at 27 °C.
7. Cellfectin II transfection reagent. Store at 4 °C.

2.7 Sodium Dodecyl Sulfate Polyacrylamide Gel Electrophoresis (SDS PAGE)

1. Sample loading buffer: To an Eppendorf tube, combine 950 μL of 2× Laemmli sample buffer, and 50 μL of β-mercaptoethanol (BME). Mix well (*see* **Note 6**).
2. 10X Tris/Glycine SDS (TGS) Buffer. Store at RT. Before use, dilute to 1X by combining 100 mL of buffer and 900 mL of distilled water (*see* **Note 7**).
3. 10 well precast Tris/Glycine gel. Store at 4 °C.
4. SDS PAGE cassette and power unit.

2.8 Western Blotting

1. Transfer stack: contains a nitrocellulose transfer membrane, filter paper, absorbent pads, copper anode and cathode, and a plastic tray. Store at RT.
2. iBlot2® Blotting Device (Thermofisher Cat. No. IB21001).
3. Non-fat dry milk powder. Can be stored at RT.
4. Tween 20. Store at RT.
5. Primary and secondary antibodies. Anti-hexahistidine mAb and anti-mouse IgG γ chain respectively.
6. Alkaline Phosphatase (AP) Conjugate Substrate Kit: Contains 25× AP Color Development Buffer, AP Color Reagent A (5-bromo-4-chloro-3′-indolyphosphate p-toluidine salt), and AP Color Reagent B (nitro-blue tetrazolium chloride). Store AP color reagents A and B at −20 °C. Store the AP Color Development Buffer at RT.
7. Plate roller.

2.9 Protein Purification

1. *Trichoplusia ni* (High Five™) cells.
2. FBS. Store at −20 °C (*see* **Note 5**).
3. Express Five™ Serum Free Media. Store at 4 °C.
4. L-glutamine 200 mM. Store at −20 °C for long term (greater than 1 week), and at 4 °C for short term (1 week).
5. Autoclaved 2800 mL flat bottom Fernbach flasks.
6. Disposable 50 mL conical tubes or reusable 500 mL centrifugation bottles.
7. Nickel-nitriloacetic acid (Ni-NTA) agarose beads. Store at 4 °C.
8. Polypropylene filter columns.
9. Wash buffer: To 4 L of distilled water, add 31.74 g of sodium phosphate monobasic monohydrate ($NaH_2PO_4 \cdot H_2O$) (57.5 mM), 70.16 g of sodium chloride (NaCl) (300 mM), and 5.44 g of imidazole (19.9 mM). The pH of this buffer should be set to 8. Dissolve completely and filter using vacuum filtration through 0.22 μm Stericup filters (Sigma Cat. No. S2GPU10RE). Store at RT.

10. Elution Buffer: To 4 L of distilled water, add 31.74 g of sodium phosphate monobasic monohydrate (57.5 mM), 70.16 g of sodium chloride (300 mM), and 64 g of imidazole (235 mM). The pH of this buffer should be set to 8. Dissolve completely and filter using vacuum filtration through 0.22 μm Stericup filters from Millipore Sigma.
11. Amicon™ Ultra-15 Centrifugal Filter Units (Sigma. Cat. No. UFC9030), 30 kDa molecular weight cutoff (MWCO) (*see* **Note 8**).
12. Bradford reagent for protein concentration. Store at 4 °C in the dark.
13. Semi microvolume disposable cuvettes for spectrophotometer.

3 Methods

To generate the recombinant glycoprotein of interest, DNA cloning is performed to produce the baculovirus transfer vector and baculovirus genome (bacmid) necessary for recombinant baculovirus rescue.

3.1 Template/Primer Ordering and PCR

A few considerations regarding the template and primer design must be made before the cloning procedure can begin. First, it is important to remove the transmembrane and endodomains of the HA and NA since secretion is not possible while these domains are present. Second, the inclusion of the signal peptide shown in Figs. 2, and 3 is necessary as it allows for proper secretion of the protein and an efficient purification process downstream. For HA, the original signal peptide should be included in the constructs, for NA the signal peptide from baculovirus gp64 is included in the pNA plasmid. Third, considering the above, cut-off points must be established on the HA and NA sequences and primers must be designed accordingly, considering the restriction enzyme cut sites (Fig. 1), to ensure in frame cloning.

1. Engineer or request modified pFastBacDual Transfer Plasmids (pHA or pNA).
2. Order the template (synthesized gene) of the desired HA or NA or run an RT PCR of the HA or NA gene segment from isolated from an influenza virus preparation to use as a template.
3. Design and order primers that will properly amplify these templates in a PCR reaction (*see* **Note 9**).
   (a) Designed primers must include the restriction sites for the respective HA or NA plasmids before the gene of interest. For pHA, *Bam*HI-HF and *Not*I-HF are used. For pNA,

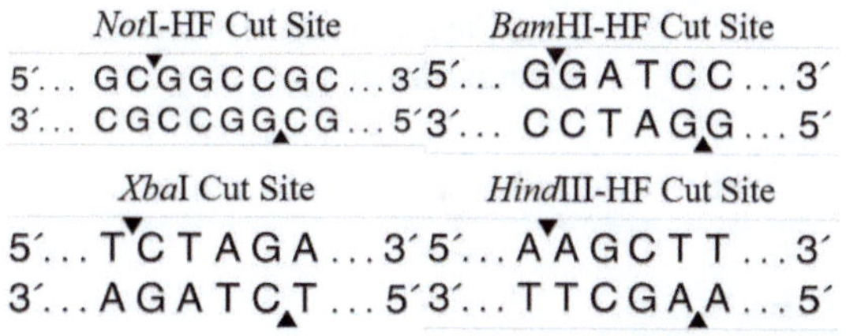

Fig. 1 Restriction enzyme cut sites for enzymes used for NA and HA cloning into the modified pFastBacDual transfer plasmids

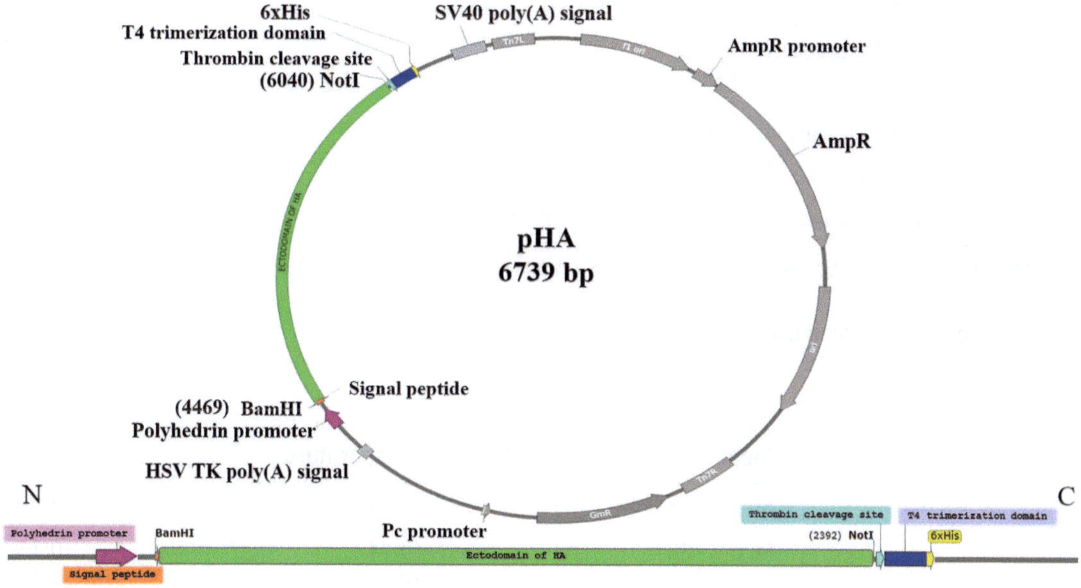

Fig. 2 Vector map of pHA

*Hind*III-HF and *Xba*I are used. Of course, if these cleavage sites already exist in the insert, you need to use alternatives with compatible ends or use recombination-based cloning.

4. Predilute primers with sterile molecular biology grade water from the stock tube to 100 μM, then dilute primers to 10 μM in a separate Eppendorf tube. Dilute template to 100 μM with sterile molecular biology grade water and start the PCR reaction using 40 μL of molecular biology grade water, 4 μL of both diluted 10 μM forward and reverse primers, 1.5 μL of the HA or NA template, and 50 μL of Q5® High-Fidelity 2X Master Mix. Set up the PCR reaction settings according to the specific DNA polymerase used. For example, for HA and NAs, we find that Q5® High-Fidelity 2X Master Mix is working well with the settings below:

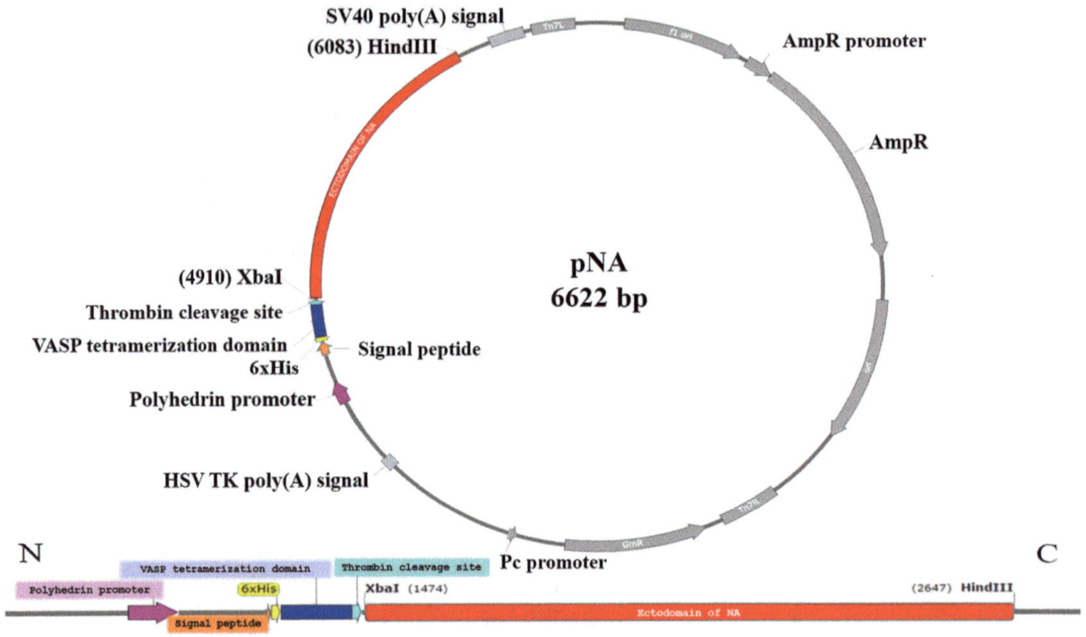

Fig. 3 Vector map showing pNA

Step	Temperature	Time
Step 1 (1×)	98 °C	2:00
Step 2 (35×)	98 °C	0:15
	60 °C	0:15
	72 °C	1:30
Step 3 (1×)	72 °C	2:00
	4 °C	Hold

3.2 PCR Purification (Using QIAquick® PCR Purification Kit)

1. Once the reaction is complete, add the whole reaction to an Eppendorf tube containing 500 μL of buffer PB (binding buffer) and mix well. Then, add the binding buffer and PCR reaction mixture to a QIAquick spin column and centrifuge in a tabletop microcentrifuge for 1 min at 14,000 rcf.
2. After centrifugation, remove the supernatant and add 700 μL of buffer PE (wash buffer) to the column. Spin again for 1 min at 14,100 rcf. Remove the supernatant and centrifuge at the same speed for 2 min to remove as much wash buffer as possible from the column.
3. Discard the supernatant and remove the column. Place the column into a sterile Eppendorf tube and add 40 μL of sterile molecular biology-grade water to begin the elution. After

2 min, centrifuge the column for 1 min at 14,100 rcf on the tabletop microcentrifuge. Use a NanoDrop or similar to determine the concentration of the PCR-purified product.

3.3 Restriction Digestion of Vector and Gene of Interest (Qiagen)

1. To the Eppendorf tube containing the PCR purified product, add 1 μL of both restriction enzymes (2 μL total), 3 μL of molecular biology-grade water and 5 uL of rCutsmart buffer. For HA, use *Bam*HI-HF and *Not*I-HF. For NA, use *Xba*I and *Hind*III. Incubate this Eppendorf tube at 37 °C overnight.
2. For digesting pHA or pNA vectors, digest 6 μg of the vector, use the same amount enzymes as **step 1**, 2 μl of buffer and fill to 20 μL using sterile molecular biology grade water. Incubate overnight at 37 °C.
3. The next day, pour a 1% agarose DNA gel by dissolving agarose in 1X TAE, inserting large gel casting combs into the gel, and adding 10–15 μL of DNA gel stain. While the gel is setting, add 1 μL of Quick CIP to the ongoing vector restriction digestion (but not the insert digestion) incubating at 37 °C to stop the reaction. Once Quick CIP is added, incubate the reaction at 37 °C for 30 min.
4. After 30 min have passed and the gel has hardened, remove the combs carefully. Add 10 μL of loading dye to the digested insert and/or vector and pipet into the wells. Add 20 uL of 1 kb + DNA Ladder to a fresh well and close the cassette. Run the DNA gel at 130 V for 25–30 min.
5. View the DNA gel under UV light, and using a handheld razor blade or a scalpel, cut the correct sized band (~1.7 kb for HA, ~1.4 kb for NA, and ~ 5 kb for the vector), of both the insert and the vector and place in separate Eppendorf tubes (*see* **Note 10**). These gel slices will be used for gel extraction.

3.4 Gel Extraction (Using Qiagen Gel Extraction Kit)

1. To the gel slice, add 500 μL of Buffer QG. Dissolve the gel by incubating it at 56 °C and periodically check the dissolving process.
2. Once fully dissolved, add 500 μL of isopropanol to the tube and mix well by gently pipetting up and down. Once mixed, pipet the mixture into a QIAquick spin column and spin down for 1 min at 14,100 rcf in a tabletop microcentrifuge. Discard the supernatant and repeat this process until all the dissolved gel with added isopropanol has been centrifuged through the column.
3. Add 700 μL of Buffer PE (wash buffer) and centrifuge at 14,100 rcf for 1 min. Discard the flowthrough and spin the column empty for another 2 min at 14,100 rcf to fully remove the wash buffer.

4. Elute the DNA by adding 40 μL of sterile molecular biology grade water, waiting 2 min, and transferring the spin column to a new Eppendorf tube before centrifugation. Centrifuge the spin column in the new Eppendorf tube for 1 min at 14,100 rcf rpm. Label the Eppendorf tubes accordingly (I for insert, and pHA or pNA for the respective vector) and leave in 4 °C for storage.

3.5 Ligation, Transformation into XL10-Gold Competent Cells (Agilent), and Plating

1. To begin ligation, fill an ice bucket with ice and place your vector, insert, T4 ligase buffer, and T4 DNA ligase on ice. Add to a fresh Eppendorf tube the specified insert and vector, in a molar ratio of 1:3, 1 μL of T4 ligase buffer, and 1 μL of T4 DNA ligase and bring to a final volume of 10 μL with molecular biology-grade water is needed. Add the T4 DNA ligase last. Place this Eppendorf tube on ice for 10 min, before removing it from the ice and incubating it at room temperature for 3 h.
2. After 3 h, retrieve a vial of XL10-Gold competent cells, and place them on ice to thaw for 3–5 min. Once they have thawed, add 3 μL of the ligation reaction to the XL10-Gold competent cells and incubate for 10 min on ice.
3. After incubating for 10 min on ice, heat shock the XL10-Gold cells plus ligase reaction for 30 s at 42 °C. Immediately after, place the heat-shocked competent cells on ice for 2 min.
4. After 2 min, add 500 μL of SOC medium to the competent cells and incubate and shake at 37 °C for 1 h.
5. During incubation, either pour LB Agar+Ampicilin plates or retrieve 1 plate from storage if they already have been prepared.
6. After 1 h, retrieve competent cells from the 37 °C incubator and centrifuge in a tabletop microcentrifuge at 2500 rcf for 8 min.
7. Once the centrifuge is finished, turn on the Bunsen burner. Under flame, remove the supernatant from the tube, leaving 40–50 μL of supernatant to resuspend the pellet. Resuspend the pellet completely, and pipet the mixture onto an LB Agar +Ampicilin plate. Sterilize a metal spreader and use it to make sure the competent cells are evenly distributed across the plate. Place the plate into a 37 °C incubator and leave overnight (Fig. 4).

3.6 Colony PCR, and Miniprep Set Up

1. Retrieve the plates from the incubator and examine for colonies. If colonies are present, count and mark eight of them at random. These will be used to run a colony PCR.
2. Fill an ice bucket with ice and thaw Taq 2X Master Mix, sterile molecular biology grade water, 30 μL of forward pHA/pNA screening primers, and 30 μL of reverse pHA/pNA screening primers. The sequences for these primers are as follows:

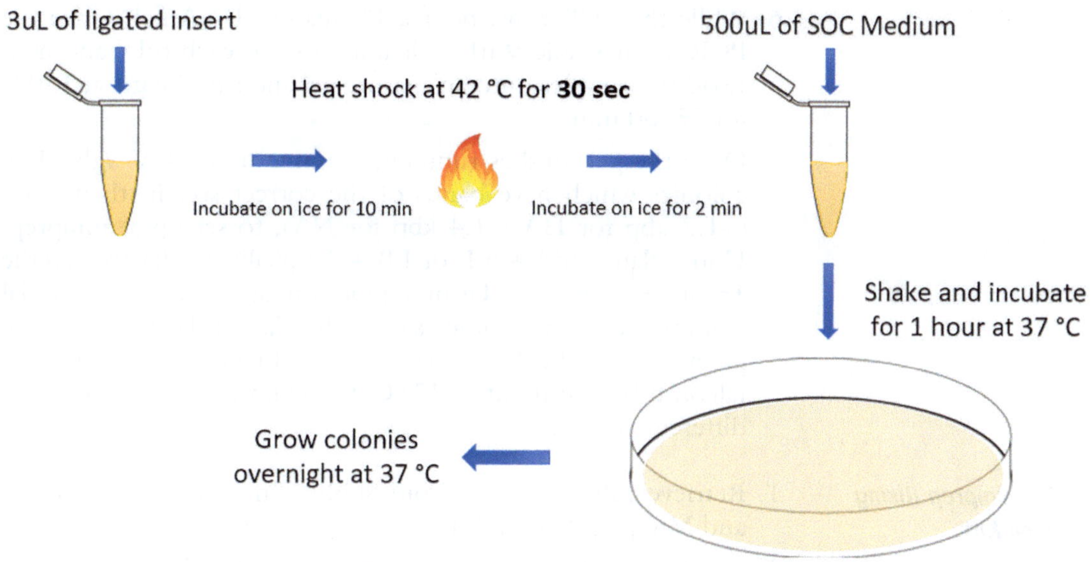

Fig. 4 Flow chart outlining the steps to transformation. It is extremely important that heat shock is only kept at 30 s. Any more or less will negatively affect transformation efficiency

Forward Screen: CGT AAC AGT TTT GTA ATA AAA AAA CCT ATA AAT ATT

Reverse Screen: CTA CAA ATG TGG TAT GGC TGA TTA TGA T

3. Once all components are thawed, add to a sterile Eppendorf tube 500 μL of Taq 2X Master Mix, 440 μL of sterile molecular biology grade water, 30 μL of forward pHA/pNA screen primers, and 30 μL of reverse pHA/pNA screen primers.
4. Retrieve eight PCR tubes and label them 1 through 8. Add 30 μL of the now prepared master mix to each tube. Finally, using a pipet tip, scrape half of a numbered colony and place into the corresponding numbered PCR tube. Do this for all colonies (1 through 8).
5. Place the PCR tubes into the thermocycler and run the colony PCR cycle:

Step	Temperature	Time
Step 1 (1×)	95 °C	5:00
Step 2 (35×)	95 °C 55 °C 68 °C	0:15 0:15 2:00
Step 3 (1×)	68 °C 4 °C	5:00 Hold

6. While the PCR runs, pour a 1% agarose DNA gel. Once the PCR finishes, add 9 uL of loading dye to each tube and mix. Load the samples onto an agarose gel and run the gel at 130 V for 25–30 min.
7. Once the gel finishes running, examine under UV light. Use colonies which have bands of the correct size in their lanes (~1.7 kbp for HA, ~1.4 kbp for NA), to set up a miniprep. Under flame add 4 mL of LB + Ampicilin media to a sterile 15 mL Falcon tube. Using a pipet tip, scrape colony material from the respective colonies from the plate and drop the whole pipet tip into the 15 mL falcon tube. Incubate and shake the falcon tube overnight at 37 °C. Grow up three positives from different colonies.

3.7 Miniprep (Using Qiagen Kit)

1. Retrieve cultures set up from Subheading 3.6 "Colony PCR, and Miniprep Set Up" from the incubator.
2. Before starting the miniprep, create a glycerol cryostock. Do this by adding 500 μL of a sterile 40% glycerol solution and 500 μL of the bacteria culture to a cryo tube. Store at −80 °C.
3. Centrifuge down the remaining part of the bacterial culture in an Eppendorf tube.
4. Using 250 μL of resuspension buffer, resuspend the bacteria pellet (*see* **Note 11**). Next, add 250 μL of lysis buffer and invert to mix. Incubate this mix for 5 min. Next, add 350 μL of neutralization buffer.
5. Spin the Eppendorf tube at 11,300 rcf for 10 min in a tabletop microcentrifuge to separate supernatant and bacteria cell debris. After the spin is finished, transfer the supernatant into a QIAprep spin miniprep column.
6. Spin this column down at 14,100 rcf for 1 min. Discard the supernatant and add 700 μL of wash buffer. Centrifuge again at 14,100 rcf for 1 min. Discard the supernatant and centrifuge the column again for 2 min at 14,100 rcf to remove any extra wash buffer.
7. Retrieve a fresh sterile Eppendorf tube and transfer the column to the Eppendorf tube. Add 40 μL of sterile molecular biology grade water and incubate for 2 min at room temperature. After 2 min, centrifuge at 14,100 rcf for 1 min and use a NanoDrop to determine the concentration.
8. Once the concentration has been confirmed, sequence plasmid using the forward and reverse pHA/pNA screening primers. This miniprep will be used in the next step after sequencing results arrive showing the correct insert sequence with no mutations. Do only move plasmids with completely confirmed insert sequences forward to the next step.

3.8 Transformation into DH10Bac (Invitrogen) Competent Cells

In DH10Bac competent cells, the insert from the modified pFast-Bac Dual vector is recombined into a bacmid via Tn7 transposons.

1. Retrieve a vial of DH10Bac competent cells and thaw them for 3–5 min on ice. After thawing, add 10–50 ng of the miniprep from the previous step and place the competent cells back on the ice for another 10 min.
2. After incubating on ice for 10 min, heat shock at 42 °C for exactly 30 s, then place them back on ice for another 2 min.
3. After 2 min, add 500 μL of SOC medium to the DH10Bac competent cells and incubate and shake at 37 °C for 4 h.
4. During incubation, dilute tetracycline to 10 μg/mL, kanamycin to 50 μg/mL, IPTG to 0.1 mM, gentamycin to 5 μg/mL, and X-gal to 20 μg/mL in 500 mL of LB + Agar.
5. Pour LB + Agar plates with kanamycin, tetracycline, gentamycin, IPTG, and X-Gal (KTGIX) or retrieve premade plates from storage. The reagents are light-sensitive and so they should be poured with lights out in the hood. The plates should be stored wrapped in aluminum foil to avoid light.
6. After the 4 h incubation is finished, centrifuge the DH10Bac competent cells at 2500 rcf for 8 min in a tabletop microcentrifuge.
7. When the centrifuge is finished, turn on the Bunsen burner and remove the supernatant. Leave 40–50 μL of supernatant in the Eppendorf tube in order to resuspend the pellet.
8. Resuspend the pellet under the flame, and pipet onto the prepared LB + Agar plates with kanamycin, tetracycline, gentamycin, IPTG, and X-Gal (KTGIX).
9. Using a metal spreader, spread the bacteria evenly across the plate. Once spread evenly, incubate the plate at 37 °C for 2 days. The plates should be stored wrapped in aluminum foil to avoid light.
10. Once the plate has incubated for 2 days, examine the plate for white colonies. If there are white colonies, turn on the flame and restreak that white colony on another LB + Agar plate with KTGIX and incubate for 2 days at 37 °C.
11. After 2 days, prepare LB broth supplemented with kanamycin (50 μg/mL), tetracycline (10 μg/mL), and gentamycin (5 μg/mL) (KTG).
12. Pick a single white colony and drop it into the media using a 20 μL micropipette tip. Incubate this and shake it for 1 day. On the next day, transfer this to 50 mL of the same LB broth with KTG and incubate and shake for another day. After the next day, use this bacterial culture to perform a midiprep.

3.9 Midiprep (Using Invitrogen Kit)

1. To begin, check to see if the bacterial culture from the previous step has grown. If yes, using the Invitrogen™ Midiprep Kit, equilibrate the column by adding 15 mL of equilibration buffer to the column and allowing it to flow through using gravity.
2. Centrifuge the bacterial culture which contains the bacmid at 4000× *g* for 10 min (*see* **Note 12**). Once centrifuged, discard the supernatant and resuspend the pellet in 10 mL of resuspension buffer (*see* **Note 11**).
3. Next, pipet in 10 mL of lysis buffer. Mix the conical tube by inverting it several times and then let it sit for 5 min to let the lysis complete.
4. Finally, add 10 mL of precipitation buffer to stop the lysis reaction. Mix this well by inverting.
5. Pour the contents from **step 4** into the column that was equilibrated in **step 1**, and let it flow through. Once it has completely flowed through, throw away the filtration column and pipet in 20 mL of wash buffer. When the wash buffer has completely flowed through, retrieve a new 15 mL conical tube and position the column over the top. Add 5 mL of elution buffer and collect the flowthrough in the 15 mL conical tube.
6. To the eluate, add 3.5 mL of isopropanol to precipitate the bacmid and spin at 4000× *g* for 1 h. After 1 h has passed, check to see if there is a small white pellet in the bottom of the 15 mL centrifuge tube (*see* **Note 13**).
7. Use a Pasteur pipet and a vacuum line and aspirate out the supernatant, being careful not to disturb the pellet, and add back in 3 mL of 70% ethanol. Centrifuge for another 30 min at 4000× *g* to wash the pellet. Then, remove the supernatant again via aspiration.
8. Air dry the bacmid pellet for about 10 min, then resuspend the pellet by adding 100 μL of sterile molecular biology-grade water. Use a NanoDrop to determine the concentration of the bacmid. We expect the concentration to be approximately 300–400 ng/μL. If the concentration is in this range, sequence midiprep using the forward and reverse pHA/pNA sequencing primers.
9. Once sequencing results arrive and the correct sequence is confirmed, move on to the next step (*see* **Notes 14** and **15**). It is also very important to note that bacmids are relatively fragile. Breaks in the dsDNA induced by freezing or shear stress may lead to difficulties rescuing baculovirus and freeze-thawing and shear stress should therefore be avoided. Also, some commercial prep kits are less suitable than others for bacmid preparations. Testing different brands may help to find an optimal one.

3.10 Sf9 Cell Transfection

All of 3.10 Sf9 Cell Transfection must be done in a laminar flow hood to avoid contamination of the culture and the baculovirus (BV) stock.

1. Add 40 mL of 10% TNM-FH media to a T175 cm^2 flask 100% confluent with Sf9, clap the cells off and then transfer 1 mL of the Sf9 cell suspension to a 6-well plate. Let the Sf9 cells in the well of the 6-well plate sit for 20 min to adhere to the plate. Cells should have a density of approximately 2×10^5 cells/ cm^2 [11].
2. While the Sf9 cells are settling in the 6-well plate, supplement 500 mL of fresh TNM-FH media with 5 mL of Pen-Strep. From this bottle of TNM-FH media with 5 mL of Pen-Strep, pipet out 150 μL and add to an Eppendorf tube. Add 2 μg of your bacmid prepared from the midiprep to this Eppendorf tube.
3. To another different sterile Eppendorf tube, add 150 mL of the same media and 6 uL of Cellfectin II.
4. Combine the Eppendorf tubes from **steps 6** and 7 and incubate for 30 min.
5. After 20 min have passed and after Sf9 cells are adhered to the 6-well plate, aspirate out the media by tilting the plate carefully and using a Pasteur pipet and vacuum aspirator.
6. Add 2 mL of TNM-FH to the well. Add the media from the Eppendorf tube in **step 4** to the well dropwise with a 1000 uL micropipette.
7. Incubate plate with new media, midi DNA + Cellfectin II, and Sf9 cells overnight. In the morning of the next day, aspirate out the media by tilting the plate carefully.
8. Add 2 uL of 3% FBS TNM-FH media and incubate the plate for 5–7 days. After the 5–7-day incubation, harvest the supernatant (this is going to be the P0 stock) and perform a Western blot on the cell pellet (*see* **Note 16**).

3.11 Western Blotting

A Western blot is comprised of three steps: Separation, Transfer, and Detection. They are outlined below.

3.11.1 Separation

1. Using a micropipette, detach as many cells as possible from the 6-well plate from the transfection step by pipetting up and down, and pipet into a 15 mL conical tube.
2. Centrifuge this tube at 4 °C for 10 min at 3720 rcf.
3. Separate the supernatant into another 15 mL conical tube, label with protein name and P0 (Passage #0) and keep at 4 °C. This will be used to amplify more BV stock.

4. Resuspend a small amount of the cell pellet that was kept in the original 15 mL conical tube in 20 μL of PBS. Transfer it to an Eppendorf tube along with 20 μL of prepared SDS PAGE sample loading buffer.
5. To prepare the gel box, dilute 10× TGS to 1× with distilled water. The total volume should be 1000 mL of 1× TGS. Retrieve the precast Tris/Glycine Gel and remove the comb (*see* **Note 17**). Remove the sticker at the bottom of the precast gel. Then, lock in the gel into the electrode assembly and block the other side with a buffer dam. Place the electrode assembly into the gel tank and fill to the "2 gels" mark on the side of the tank. Even though 2 gels are not being ran, the buffer still must be filled to the "2 gel" mark for the electricity to flow properly. Make sure to fill the inside of the electrode assembly to the top with 1× TGS.
6. Once the cell pellet has finished incubating, load it into the Tris/Glycine Gel and run at 140 V for 35–40 min.
7. After the run is finished, remove the TGX precast gel from the gel box and pry it open using metal tweezers, being careful not to break the TGX gel inside. Keep this aside for the next steps of the Western blot.

3.11.2 Transfer

This step will outline the procedure used for an iBlot®2 dry blotting system. However, if you do not have access to the equipment, follow the steps for a wet transfer.

1. After setting aside the Tris/Glycine Gel, open one packet of the iBlot 2® NC Mini Stacks (Invitrogen).
2. To set up the stack properly, keep the bottom stack in the plastic tray and remove the top stack and set aside. Place the plastic tray with the bottom stack on the blotting surface of the iBlot 2® Transfer device. Pick up protein gel and place it on the transfer membrane of the bottom stack.
3. Soak filter paper in dH_2O and place it on top of the gel. Use a roller to remove any air bubbles. Place the top stack over the filter paper and remove air bubbles using a plate roller.
4. Finally, place an absorbent pad over the top stack making sure electrical contact points are in the proper position and close the lid (Fig. 5).
5. Input the correct settings on the iBlot 2® Transfer device:

Step 1	Step 2	Step 3
20 V 01:00	23 V 04:00	25 V 02:00

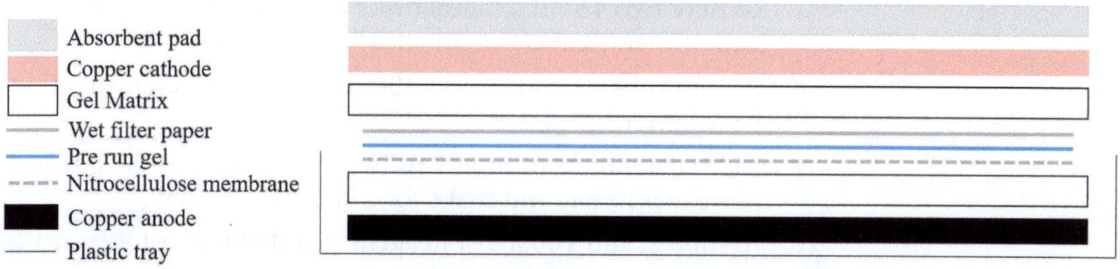

Fig. 5 Schematic depicts the layout of the transfer stack. The Tris/Glycine Gel is placed directly on the nitrocellulose membrane. A wet filter paper is then laid flat above the precast gel, then the rest of the top stack is sandwiched on top. The absorbent pad is placed last to absorb extra liquid produced during blotting

6. After 7 min when the run is complete, remove the whole tray. While taking apart the stack, only touch the corners as touching the middle may contaminate or damage the nitrocellulose membrane, which is what will be used in the next steps.

3.11.3 Detection

1. Prepare blocking solution by adding 1.5 g non-fat milk powder to 50 mL of PBS. In a small container, cover the nitrocellulose membrane in blocking solution and shake the membrane for 30 min at 80 rpm on an orbital shaker.
2. Remove the blocking solution by pouring it out of the side of the container, only touching the corners of the nitrocellulose membrane with tweezers.
3. Prepare the primary antibody solution, which uses a hexahistidine mAb, by adding 0.5 g of non-fat milk powder to 50 mL of PBS and diluting the primary antibody 1:3000 in the PBS and milk solution. Pour this solution into the small container with the nitrocellulose membrane and incubate for 1 h shaking at 80 rpm.
4. After 1 h, discard the primary antibody solution by pouring it out of the side of the container. Wash the membrane three times with 1× PBS with 0.1% Tween 20, discarding the 1× PBS 0.1% Tween 20 solution after each wash.
5. Dilute the secondary antibody, which is an anti-mouse IGg γ-chain, 1:3000 in 50 μL of PBS, and incubate the nitrocellulose membrane for 1 h with the secondary antibody mixture on the shaker at 80 rpm.
6. After 1 h, discard the secondary antibody solution by pouring it out the side of the container. Wash the membrane again three times with 1× PBS with 0.1% Tween 20, discarding the 1× PBS 0.1% Tween 20 solution after each wash. Set them aside and prepare the alkaline phosphatase (AP) conjugate substrate kit (Bio-Rad).

7. Retrieve two 15 mL conical tubes and add 9.5 mL of dH_2O in each of them. Label one tube "1" and one tube "2".
8. In both tubes 1 and 2, add 400 μL of 25× AP color development buffer.
9. Add the contents of tube 1 to the container with the nitrocellulose membrane and shake for 1 min at 120 rpm.
10. To tube 2, add 100 μL of Reagent A and 100 μL of Reagent B.
11. After 1 min discard the contents of tube 1 from the container and then add the contents of tube 2 on top of the membrane slowly (*see* **Note 18**). Shake at 160 rpm until bands are visible on the nitrocellulose membrane. The band size should be around 66 kDa for an HA construct, and around 60 kDa for an NA construct.

3.12 Sf9 Infection and P0 Amplification

This step will only be completed once there is protein expression confirmed by the Western blot in Subheading 3.11.

1. Bring the 15 mL conical tube kept at 4 °C labeled with the protein name and "P0" into the laminar flow hood, along with a 90% confluent T175 cm^2 flask of Sf9 cells and 1000 mL of TNM-FH media supplemented with 3% FBS, 10 mL of Pen-Strep, and 10 mL of Pluronic F-68 (3% TNM-FH media).
2. Using a Pasteur pipet, aspirate out all the media in the flask, being careful not to disturb the Sf9 cells that are adhered to the inside of the flask.
3. Into the flask, add 30 mL of 3% TNM-FH media and 300 μL of the P0 BV stock to infect the Sf9 cells. Incubate the flask at 27 °C for 7 days.
4. Every day, inspect the T175 cm^2 flask for signs of infection. Signs of infection include enlarged nuclei, a granular appearance, and larger than normal cells as well as floating, detached cells. If the cells are mostly dead earlier than 7 days, the flask can be harvested then. Otherwise harvest on day 7.
5. Shake the flask to resuspend the infected Sf9 cells, pipet into a 50 mL conical tube and centrifuge at 3720 rcf for 10 min at 4 °C. Harvest the supernatant in a separate 50 mL conical tube and label with correct BV stock name and passage number. Discard the pellet.
6. Repeat this process twice to generate a high-titer P3 stock with a large volume. Only P3 stocks should be used for expression. Also, recombinant baculovirus should not be grown or used beyond P3 since "wild type-like" viruses without or with mutated inserts may outcompete the correct recombinant virus.

All the following procedures are specified for one 2800 mL Fernbach flask for expression. Scale up as necessary.

3.13 Cell Harvesting and Baculovirus Infection

1. Collect High Five™ cells from six T175 cm^2 flasks by clapping flasks lightly and unite two T175 cm^2 flasks' volumes into 50 mL conical tubes each. Save one empty T175 cm^2 flask for later to collect used.
2. Centrifuge conical tubes at 3720 rcf for 10 min at 4 °C to separate cells and supernatant.
3. Bring centrifuged 50 mL conical tubes into the laminar flow hood. Discard supernatant in a used T175 cm^2 flask.
4. Pipet 8 mL of desired baculovirus (BV) stock into one 50 mL tube with the cell pellet. Resuspend cell pellet in 8 mL of BV stock using a 10 mL serological pipet, then transfer resuspended pellet into a second 50 mL tube with another cell pellet. Repeat this step until 3 tubes worth of cell pellet are fully resuspended in 8 mL of BV stock. Leave the resuspended mixture at room temperature for 15 min.
5. While the resuspended mixture is incubating, supplement one 1000 mL bottle of Express Five™ SFM with 10 mL Pen-Strep and 10 mL of L-glutamine solution in the laminar flow hood. After supplementation, the concentration of L-glutamine will be 2 mM and of Pen-Strep will be 100 U/mL (Penicillin) and 100 μg/mL.
6. After the 15-min incubation period is over, retrieve an autoclaved 2800 mL sterile Fernbach flask having the opening covered in foil and bring it into the laminar flow hood. Sterily pour 500 mL of Express Five™ SFM supplemented with Pen-Strep and L-glutamine into the Fernbach flask. Then pour resuspended BV and cell pellet mixture into the flask. Cover with the same aluminum foil that was on the autoclaved Fernbach flask (and has been placed upside down in the hood to keep it sterile) and shake at 67 rpm on an orbital shaker and 27 °C for 72 h.

3.14 Nickel Bead Incubation

1. Remove Fernbach flask from the 27 °C shaker and pour the mixture from the Fernbach flask into an autoclaved 500 mL centrifuge bottle. Set Fernbach flask aside for later. If there is no access to a centrifuge that fits 500 mL bottles, pour mixture into 50 mL conical tubes.
2. Spin down the 500 mL centrifuge bottle at 4 °C and 3720 rcf for 10 min. Alternatively, spin down 50 mL conical tubes at 4 °C and 3720 rcf for 10 min.
3. To wash Ni^{2+} NTA beads, shake bottle to resuspend Ni^{2+} NTA bead sediment and pipet out 8 mL of beads into a 50 mL conical tube. Fill the rest of the tube up to 50 mL with PBS. Then spin down the tube at 3720 rcf for 10 min.

4. Once Ni^{2+} NTA beads have been pelleted, discard the PBS and keep the Ni^{2+} NTA bead pellet in the 50 mL conical tube.
5. To wash the Fernbach flask that was set aside at **step 1**, rinse 3 times with around 300 mL of distilled water. After washing, pour supernatant with protein into the washed Fernbach flask from a 500 mL centrifuge bottle, leaving around 100 mL in the bottle. Alternatively, leave two 50 mL tubes with supernatant in them.
6. Resuspend the Ni^{2+} NTA bead pellet by filling the 50 mL conical tube with the supernatant from **step 5**. Pour the resuspended Ni^{2+} NTA beads into the Fernbach flask. Do this step twice to ensure that all the beads are resuspended.
7. Once all the supernatant from the protein/cell mixture and the nickel beads have been properly combined, place the Fernbach flask back on the shaker and shake at RT or 4 °C (preferred) for 3 h.

3.15 Filtration, Washing, Elution, and Concentration

1. After protein and Ni^{2+} NTA bead mixture have shaken at RT for 3 h, set up ring stands with polypropylene filter columns. Make sure to place a collection bucket under the columns to catch the flowthrough. Fill columns to the top with the protein and Ni^{2+} NTA bead mixture using a serological pipet. To begin filtration, remove the small caps from the bottom of the columns and apply light pressure to the top of the column to allow mixture to flow through. Refill column from the top until you have filtered all the protein/ Ni^{2+} NTA bead mixture through the column.
2. After filtration, wash the Fernbach flask with 25 mL of wash buffer to remove any remaining Ni^{2+} NTA beads from the bottom of the flask. Pipet this wash buffer into the columns and let it flow through completely. Wash three more times with 25 mL of wash buffer each.
3. After washing, use the small caps to cap the bottom of the columns. Then, pipet in 4 mL of elution buffer. Let the elution buffer sit for 5 min.
4. While the elution buffer is in the columns, discard the flow through and fill the bucket with ice. Place an uncapped 50 mL conical tube in the ice. After 5 min, remove caps from the columns and allow the elution buffer to flow through.
5. While elution buffer is flowing through, fill the 30 kDa molecular weight cut-off (MWCO) centrifugal filters to the top with PBS, and centrifuge at 3720 rcf for 10 min to equilibrate.
6. Once elution buffer has flowed through, cap the bottom of the columns again, and repeat the elution process one more time. At the end of elution, there should be approximately 16 mL of eluate in the 50 mL conical tube on ice.

7. Retrieve the 30 kDa MWCO centrifugal filters, discard the PBS flow through, and fill them with the ~16 mL of elution buffer from the 50 mL conical tube. To concentrate the proteins, centrifuge filled 30 kDa MWCO centrifugal filters for 30 min at 3720 rcf.
8. When the first round of centrifugation has finished (and the majority of volume went through the membrane, leaving only a few hundred μL above the filter), discard the flowthrough, and re-fill the filter to the top again with PBS. Centrifuge two more times at 3720× *g* for 30 min. For each step, if liquid does not flow through properly, spin down tube again for 15 min. Continue doing so until the liquid has flown through. Refill only once the majority of volume went through the membrane.

3.16 Protein Harvesting, Quantification, and Aliquoting

1. After the third round of centrifugation, remove the 30 kDa MWCO centrifugal filters and, using a 200 μL micropipette, pipet out all the liquid from the top of the filter and put into either an Eppendorf tube or a 15 mL conical tube on ice, depending on the amount of filtration tubes used and amount of liquid from each filter (if the same protein was purified via several columns, it can be pooled).
2. Fill an Eppendorf tube with 1000 μL of Bradford reagent. This will be used as the blank. Fill another Eppendorf tube with 1000 μL of Bradford reagent and then 5 μL of protein from the 15 mL conical tube. Mix Bradford reagent and protein thoroughly.
3. Transfer both Eppendorf tubes to a semi-micro volume cuvette and bring to a spectrophotometer. Blank the spectrophotometer first using the cuvette with Bradford reagent alone, then measure the 595 nm absorbance of the Bradford reagent and protein sample. In parallel, also measure different concentrations of a protein standard (e.g., BSA) and draw a standard curve (e.g., using Microsoft excel). Determine the linear range of the curve and calculate the best fit using linear regression. Then use the absorbance measured for your sample to calculate the concentration. If the absorbance did not fall into the linear range, predilute the sample in PBS until the absorbance you measure is in the linear range allowing for calculation of the concentration.
4. After determining the concentration of the protein, aliquot 53 μL into labeled Eppendorf tubes and place each tube on ice. After aliquoting, place Eppendorf tubes into a cryobox and store at −80 °C for storage. Make sure one tube can be used for running a protein gel to determine the quality of the protein (*see* **Note 19**).

3.17 SDS PAGE for Quality Check

1. To check the quality of protein, either thaw an Eppendorf tube of the protein from the −80 °C freezer on ice or retrieve a thawed tube from 4 °C storage and place on ice. Set the Eppendorf Thermomixer to 96 °C.
2. To standardize protein concentration, use the concentration previously determined and calculate how much volume is needed to pipet out 2–4 μg of protein.
3. Prepare reducing reagent and gel box the same way as listed in "3.8 Western Blotting".
4. Add 20 μL of prepared SDS-PAGE sample buffer to the protein and PBS mixture prepared previously, and place in the 96 °C heat block for 10 min.
5. After 10 min, remove the Eppendorf tubes from the heat block and load the protein into an empty well. Do not forget to also load a protein ladder into an empty well. Run the protein gel for 35–40 min at 140 V.
6. Once the run is finished, remove the gel from the tank and crack it open. Place the Tris/Glycine Gel in a small plastic container and fill with distilled water. Microwave the container with the Tris/Glycine Gel and distilled water for 10 to 15 s to remove any extra 1× TGS buffer. Pour out the distilled water and repeat this step two more times. When finished, empty the distilled water out and fill the container with SimplyBlue™ Safe Stain (or any other SDS PAGE gel stain). Shake the gel with this stain for 1 h (*see* **Note 20**), then remove the gel stain and fill the container with distilled water to de-stain. Discard and refill the container with distilled water twice before letting the container shake with distilled water overnight.
7. When viewing the gel and determining the quality, look for clear and correctly sized bands (66 kDa for HA, 60 kDa for NA), and no other extra bands in the well. Incorrectly sized bands, multiple bands, or smears indicate degradation of the protein sample (Fig. 6).

4 Notes

1. Make sure to dilute with 24 or 40 mL of 100% ethanol depending on the size of the wash buffer bottle.
2. After a while, or if the buffer is mistakenly left out at RT, the RNase activity begins to decline. To continue using the buffer without having to order a new kit, we find that another 200 uL of RNase A can be added to the buffer instead.
3. Upon receiving the ligase buffer, aliquot the ligase into 10 μL aliquots and the buffer into 20 μL aliquots before freezing at −20 °C for long-term storage. DNA ligase and ligase buffer are

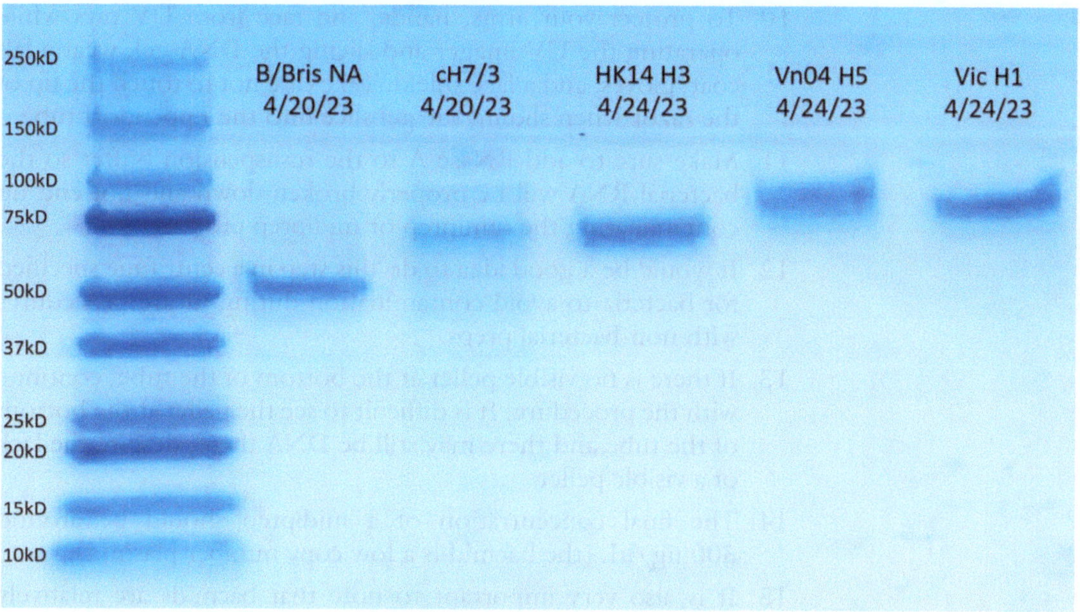

Fig. 6 SDS PAGE gel obtained after purifying protein. Each well is labeled with a shorthand name for each protein and the date the protein was purified. Proteins are a neuraminidase from B/Brisbane/60/2008 (B/Bris NA), a chimeric hemagglutinin with a head domain from A/Shanghai/1/2013 (H7N9) and a stalk domain from A/Hong Kong/4801/2014 (H3N2) denoted cH7/3, and wild type hemagglutinins from A/Hong Kong/4801/2014 (H3N2, HK14 H3), A/Vietnam/1203/04 (H5N1, Vn04 H5), and A/Victoria/2570/2019 (H1N1, Vic H1)

temperature sensitive so putting them through multiple freeze/thaw cycles degrades the quality of the reagents. Aliquoting before use avoids this problem.

4. Though not required, it is a good idea to store this buffer at 4 °C once 1.5 mL RNase A is added to avoid degradation of the enzyme over the long term.
5. Thaw FBS and aliquot serum into 50 mL conical tubes, before freezing again. This avoids multiple freeze/thaw cycles and possible contamination from opening and closing the same bottle of FBS every time it needs to be used.
6. BME has a very pungent odor and is toxic. Mix with the Laemmli Sample Buffer in a fume hood to avoid smelling the odor directly from the BME container.
7. We find that this buffer can be used for multiple protein gels and does not have to be thrown out after one use.
8. We find that these filtration columns are the best to use after testing multiple different brands of 30 kDa centrifugal filters.
9. Although not required, codon optimization for insect cells may be useful when ordering the primers and gene blocks.

10. To protect your arms, hands, and face from UV rays while operating the UV imager and slicing the DNA gel, wear a lab coat, gloves, and a face shield. Take care not to touch the tip of the razor when sliding the gel slice into the Eppendorf tube.
11. Make sure to add RNase A to the resuspension buffer so the bacterial RNA will be properly broken down and not end up contaminating the miniprep or midiprep product.
12. It would be a good idea to do this step in a centrifuge specified for bacteria to avoid contamination during future procedures with non-bacterial preps.
13. If there is no visible pellet at the bottom of the tube, continue with the procedure. It is difficult to see the pellet at the bottom of the tube and there may still be DNA there, despite the lack of a visible pellet.
14. The final concentration of a midiprep should be around 300 ng/μL (the bacmid is a low copy number plasmid).
15. It is also very important to note that bacmids are relatively fragile. Breaks in the dsDNA induced by freezing or shear stress may lead to difficulties rescuing baculovirus and freeze--thawing and shear stress should therefore be avoided. Also, some commercial prep kits are less suitable than others for bacmid preparations. Testing different brands may help to find an optimal one.
16. It is important to change tips when aspirating media from the different wells and to ensure that pipettes do not touch them when adding media. At this step, bacmids could be carried over leading to mixed populations/cross-contamination of recombinant baculoviruses when more than one construct is handled at once.
17. When taking the combs out of the gel, be careful not to shift the wells. Pull the comb straight out, and not off to the side, to avoid this.
18. We find that pouring this solution over the top of the nitrocellulose membrane slowly yields the best results. Pour for about 5 s.
19. The Eppendorf tube that will be used for running a protein gel can be kept at 4 °C if you plan to run the protein gel the day of or on the following day. Otherwise, store this protein at −80 °C until use.
20. To stain faster, you can microwave the gel for 10–15 s.

Acknowledgments

The Krammer laboratory receives support for work in the immunology, virology, therapeutics, and vaccine space from the NIAID Centers of Excellence for Influenza Research and Response (CEIRR, 75N93021C00014) and Collaborative Influenza Vaccine Innovation Centers (CIVICs, 75N93019C00051) contracts as well as NIAID grants and contracts R01 AI146101, U19 AI168631, R01 AI154470, U19 AI162130, R01 AI137146, U01 AI144616, HHSN272201800048C and U19 AI118610. Additional support comes from FluLab and the Bill and Melinda Gates Foundation. The laboratory is also supported in whole or in part with Federal funds from the National Cancer Institute, National Institutes of Health, under U54 CA260560 and under Contract number 75N91021F00001 via 21X092F1 Mod 01. The content of this publication does not necessarily reflect the views or policies of the Department of Health and Human Services, nor does mention of trade names, commercial products, or organizations imply endorsement by the U.S. Government.

References

1. Krammer F, Smith GJD, Fouchier RAM et al (2018) Influenza. Nat Rev Dis Primers 4:3. https://doi.org/10.1038/s41572-018-0002-y
2. Khurana S, Verma S, Verma N et al (2011) Bacterial HA1 vaccine against pandemic H5N1 influenza virus: evidence of oligomerization, hemagglutination, and cross-protective immunity in ferrets. J Virol 85:1246–1256. JVI.02107-10 [pii]. https://doi.org/10.1128/JVI.02107-10
3. Aguilar-Yáñez JM, Portillo-Lara R, Mendoza-Ochoa GI et al (2010) An influenza a/H1N1/2009 hemagglutinin vaccine produced in Escherichia coli. PLoS One 5:e11694. https://doi.org/10.1371/journal.pone.0011694
4. Saelens X, Vanlandschoot P, Martinet W et al (1999) Protection of mice against a lethal influenza virus challenge after immunization with yeast-derived secreted influenza virus hemagglutinin. Eur J Biochem 260:166–175
5. Krammer F, Margine I, Tan GS et al (2012) A carboxy-terminal trimerization domain stabilizes conformational epitopes on the stalk domain of soluble recombinant hemagglutinin substrates. PLoS One 7:e43603. PONE-D-12-16229 [pii]. https://doi.org/10.1371/journal.pone.0043603
6. McLellan JS, Chen M, Joyce MG et al (2013) Structure-based design of a fusion glycoprotein vaccine for respiratory syncytial virus. Science 342:592–598. https://doi.org/10.1126/science.1243283
7. Hsieh CL, Goldsmith JA, Schaub JM et al (2020) Structure-based design of prefusion-stabilized SARS-CoV-2 spikes. Science 369:1501. https://doi.org/10.1126/science.abd0826
8. Mishra V (2020) A comprehensive guide to the commercial baculovirus expression vector systems for recombinant protein production. Protein Pept Lett 27:529–537. https://doi.org/10.2174/0929866526666191112152646
9. Krammer F, Grabherr R (2010) Alternative influenza vaccines made by insect cells. Trends Mol Med 16:313–320. S1471-4914(10)00071-7 [pii]. https://doi.org/10.1016/j.molmed.2010.05.002
10. Cox MM, Hollister JR (2009) FluBlok, a next generation influenza vaccine manufactured in insect cells. Biologicals 37:182–189. https://doi.org/10.1016/j.biologicals.2009.02.014
11. Margine I, Palese P, Krammer F (2013) Expression of functional recombinant hemagglutinin and neuraminidase proteins from the novel H7N9 influenza virus using the baculovirus expression system. J Vis Exp. https://doi.org/10.3791/51112
12. Stevens J, Corper AL, Basler CF et al (2004) Structure of the uncleaved human H1 hemagglutinin from the extinct 1918 influenza virus. Science 303:1866–1870. 1093373 [pii]. https://doi.org/10.1126/science.1093373

13. Weldon WC, Wang BZ, Martin MP et al (2010) Enhanced immunogenicity of stabilized trimeric soluble influenza hemagglutinin. PLoS One 5:e12466. https://doi.org/10.1371/journal.pone.0012466
14. Xu X, Zhu X, Dwek RA et al (2008) Structural characterization of the 1918 influenza virus H1N1 neuraminidase. J Virol 82:10493–10501. JVI.00959-08 [pii]. https://doi.org/10.1128/JVI.00959-08
15. Schmidt PM, Attwood RM, Mohr PG et al (2011) A generic system for the expression and purification of soluble and stable influenza neuraminidase. PLoS One 6:e16284. https://doi.org/10.1371/journal.pone.0016284
16. Strohmeier S, Amanat F, Zhu X et al (2021) A novel recombinant influenza virus neuraminidase vaccine candidate stabilized by a measles virus phosphoprotein tetramerization domain provides robust protection from virus challenge in the mouse model. mBio 12: e0224121. https://doi.org/10.1128/mBio.02241-21
17. McMahon M, Strohmeier S, Rajendran M et al (2020) Correctly folded – but not necessarily functional - influenza virus neuraminidase is required to induce protective antibody responses in mice. Vaccine 38:7129–7137. https://doi.org/10.1016/j.vaccine.2020.08.067
18. Krammer F, Fouchier RAM, Eichelberger MC et al (2018) NAction! How can neuraminidase-based immunity contribute to better influenza virus vaccines? MBio 9:1. https://doi.org/10.1128/mBio.02332-17
19. Dalakouras T, Smith B, Platis D et al (2006) Development of recombinant protein-based influenza vaccine. Expression and affinity purification of H1N1 influenza virus neuraminidase. J Chromatogr A 1136:48–56. S0021-9673(06)01808-5 [pii]. https://doi.org/10.1016/j.chroma.2006.09.067
20. Buckland B, Boulanger R, Fino M et al (2014) Technology transfer and scale-up of the Flublok recombinant hemagglutinin (HA) influenza vaccine manufacturing process. Vaccine 32:5496–5502. https://doi.org/10.1016/j.vaccine.2014.07.074
21. Krammer F, Nakowitsch S, Messner P et al (2010) Swine-origin pandemic H1N1 influenza virus-like particles produced in insect cells induce hemagglutination inhibiting antibodies in BALB/c mice. Biotechnol J 5:17–23
22. Margine I, Martinez-Gil L, Chou YY, Krammer F (2012) Residual baculovirus in insect cell-derived influenza virus-like particle preparations enhances immunogenicity. PLoS One 7:e51559. https://doi.org/10.1371/journal.pone.0051559
23. Abe T, Hemmi H, Miyamoto H et al (2005) Involvement of the Toll-like receptor 9 signaling pathway in the induction of innate immunity by baculovirus. J Virol 79:2847–2858
24. Abe T, Takahashi H, Hamazaki H et al (2003) Baculovirus induces an innate immune response and confers protection from lethal influenza virus infection in mice. J Immunol 171:1133–1139
25. Abe T, Kaname Y, Wen X et al (2009) Baculovirus induces type I interferon production through toll-like receptor-dependent and -independent pathways in a cell-type-specific manner. J Virol 83:7629–7640. JVI.00679-09 [pii]. https://doi.org/10.1128/JVI.00679-09
26. Krammer F, Schinko T, Palmberger D et al (2010) Trichoplusia ni cells (High Five) are highly efficient for the production of influenza A virus-like particles: a comparison of two insect cell lines as production platforms for influenza vaccines. Mol Biotechnol 45:226–234. https://doi.org/10.1007/s12033-010-9268-3

Chapter 5

SARS-CoV-2 S-Protein–Ace2 Binding Analysis Using Surface Plasmon Resonance

Jason Baardsnes and Béatrice Paul-Roc

Abstract

Surface plasmon resonance (SPR) allows for the label-free determination of the binding affinity and rate constants of bimolecular interactions. Here, we describe the method used for the analysis of the Ace2–SARS-CoV2 S-protein interaction using indirect capture of the S-protein onto the SPR surface, and flowing monomeric Ace2. This method will allow for the determination of the rate constants for affinity, with additional analysis that is achievable using S-protein capture levels in conjunction with the sensorgram response for relative activity benchmarking.

Key words Label-free, Surface plasmon resonance (SPR), Ace2, SARS-CoV-2 S-protein, Affinity (KD), Binding kinetics, Binding, Association, Dissociation

1 Introduction

The COVID-19 pandemic that began in December 2019 has seen unprecedented levels of interest in the production and analysis of SARS-CoV-2 S-protein for research and vaccine production [1]. One of the most desirable critical quality attributes (CQA) of S-protein is the affinity (K_D) of S-protein for human angiotensin-converting enzyme 2 (Ace2), the receptor that has been recruited by SARS CoV-2 for viral entry into host cells [2, 3]. This value can be used as a parameter to measure the relative infectivity of different S protein variants [4–6] and used as a quality attribute to track similarity between recombinant productions.

Here, we describe how the affinity (K_D) of S-protein is determined using a surface plasmon resonance (SPR) label-free technology. SPR is a robust and widely used technology that measures the

The original version of the chapter has been revised. A correction to this chapter can be found at https://doi.org/10.1007/978-1-0716-3666-4_21

Steven B. Bradfute (ed.), *Recombinant Glycoproteins: Methods and Protocols*, Methods in Molecular Biology, vol. 2762, https://doi.org/10.1007/978-1-0716-3666-4_5,

rate constants and affinity of bimolecular binding interactions between proteins and other molecules in real-time and label-free [7–11]. For the S-protein–Ace2 interaction, the SPR setup uses an indirect capture approach, where the S-protein of interest is captured on to the SPR surface via an antibody against the C-terminus, allowing the receptor binding domains of the S-protein to be correctly oriented for interaction with the flowing Ace2 monomer. The National Research Council Canada (NRC) has generated recombinant S-protein fused to the trimerization domain from human resistin to generate S-protein trimer that mimics its native conformation on the surface of the viral particle [12]. This domain has been recruited as a tag to capture the S-protein onto the SPR surface using immobilized anti-resistin antibodies. Once captured onto the surface, monomeric Ace2 is flown over the captured S-protein trimer at varied concentrations to generate binding sensorgrams for the determination of rate constants and derive the affinity (K_D) for the interaction. Flowing monomeric Ace2 in this manner is crucial for generating 1:1 binding stoichiometry with the receptor binding domain subunit of the S-protein, and will allow accurate and reproducible determination of the K_D with the 1:1 binding model in the analysis software. In the analysis software, successive rounds of incremental changes to the rate constants are used until mathematical convergence is found for the computational model to the experimental data based on a minimum of the squared residuals [7, 13–15]. The two rate constants determined in the 1:1 binding model are the association constant and the dissociation constant. The association rate (k_a (1/Ms)) is concentration dependent and is determined from the portion of the sensorgrams during each Ace2 injection where the response is increasing. The dissociation rate (k_d (1/s)) is concentration independent and is determined post-Ace2 injection and measures the decay of the S-protein–Ace2 complex. The affinity constant, K_D, is derived from the ratio, k_d/k_a, and is in molar units (M).

We selected S-protein for capture on to the SPR surface, as if trimeric S-protein was flown over immobilized Ace2, there will be avidity due to multivalency of the flowing protein component. Additionally, many commercial sources of Ace2 will be only available as a Fc-dimer and need to be used with caution. Flowing these multimeric constructs instead of monomeric Ace2 generally results in poor sensorgram fits due to dimer not allowing a true 1:1 binding interaction. Avidity from S-protein trimer flowing or Ace-Fc dimer will enhance binding and lead to low K_D values due to markedly reduced off-rates compared to monomer that are dependent on the surface density of the binding partner [16–18], and because of this, it will be difficult to discriminate between S-protein variants with small K_D changes and obtain reproducible results between experiments. Therefore, using these combinations should be avoided.

In this method we will describe the steps using the Biacore T200 to generate reliable K_D values as best determined with a 1:1 interaction, but it can be applicable to any label-free analysis instrument with dual-channel referencing capabilities and sensor chips that are amenable to amine coupling.

2 Materials

Generally, we use the commercially available buffers and solutions because the volumes required are very small, and the commercial lots are convenient to be tracked for quality assurance and the developmental operating procedures we use. Commercial solutions also do not require filtering before use.

1. Biacore T200 SPR instrument (Cytiva Inc.) (*see* **Note 1**).
2. Sensor chip Biacore CM-5, Cytiva BR10034 (*see* **Note 2**).
3. Running buffer: PBS with 3.4 mM EDTA and 0.05% Tween 20. Add 136 mL 500 mM EDTA pH 8 to 1000 mL 20× PBS with 1% Tween (sterile), dilute to 20 L in a carboy with distilled and deionized water (ddH20) (*see* **Note 3**).
4. 70% glycerol BiaNormalizing solution.
5. 100 uM N-hydroxysuccinimide (NHS): Dissolve 115 mg NHS in 10 mL ddH20, store 100 uL aliquots in 0.7 mL plastic Biacore tubes at −80 °C. (*see* **Note 4**).
6. 400 uM (1-ethyl-3-(3-dimethylamino-propyl)carbodiimide hydrochloride) (EDC): Dissolve 750 mg in 10 mL ddH2O, store 100 uL aliquots in 0.7 mL plastic Biacore tubes at −80 °C (*see* **Note 4**).
7. 1 M Ethanolamine (*see* **Note 4**).
8. 10 mM NaOAc pH 4.5 buffer.
9. 50 mM NaOH, Cytiva BR100358.
10. 10 mM Glycine pH 1.5 buffer, Cytiva BR100354.
11. Recombinant Anti-resistin sdAb-Fc fusion [19] stock >0.1 mg/mL in DPBS or other compatible buffer (*see* **Note 5**).
12. Recombinant trimerized S-protein [12] > 0.1 mg/mL in DPBS or other compatible buffer.
13. Recombinant monomeric Ace2 [12] stock >0. 5 mg/mL in DPBS or other compatible buffer (*see* **Note 6).**

3 Methods

In our lab we use the Cytiva Biacore T200 SPR instrument. This is a workhorse SPR in that it can do almost any type of SPR experiment in a low to medium-throughput fashion, and is one of the most common SPR research instruments. With the T200, the antibody

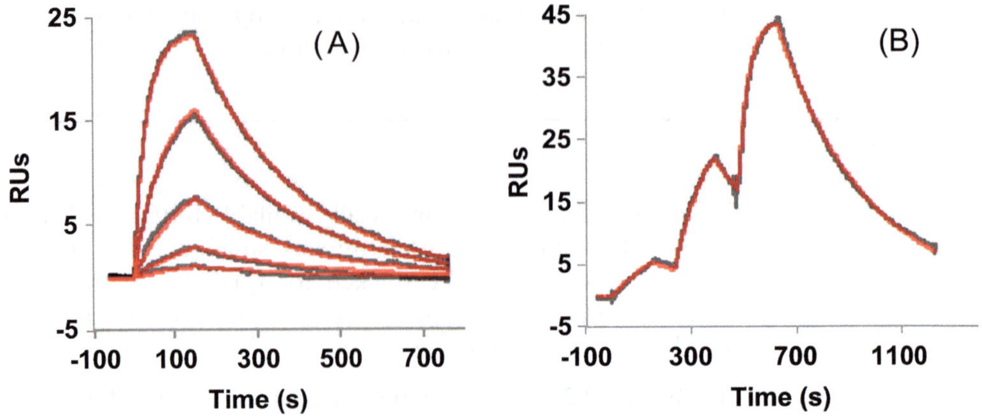

Fig. 1 Traditional multi-cycle kinetics (MCK) uses a single concentration of analyte in each injection cycle (**a**), compared to single-cycle kinetics (SCK) (**b**) that uses multiple increasing concentrations of analyte with only a single dissociation in each injection cycle. SCK saves considerable time compared to MCK with less consumption of the captured protein on the SPR surface

capture surface is made using the Immobilization Wizard in the Biacore Control software. Once this is done, the experimental method for generating the Ace2-S-protein binding sensorgrams is created within the Method File. For the Ace2 binding interaction with the captured S-protein, it can be either injected in a multi-cycle kinetics mode or using single-cycle kinetics [20] as detailed below (Fig. 1). Specific details of the Biacore software will be left to the Biacore manual. The parameters for the various steps in the method to generate quality Ace2–S-protein binding data can be used with other label-free instruments with analogous software.

3.1 Creation of the SPR Capture Antibody Surface

1. Use PBST running buffer at 25 °C with the auto sampler compartment set to 10 °C. Remove a CM-5 sensor chip from storage at 4 °C, and allow to come to room temperature. Click on the sensor chip icon in the T200 Biacore Control software and follow the instructions to dock the sensor chip in the instrument. Give the sensor chip a unique name along with the chip lot number when prompted (*see* **Note 7**). Prime the buffer when prompted to prevent air spikes.
2. Anti-resistin sdAb-Fc fusion is diluted to 15 ug/mL in 700 uL total volume of 10 mM NaOAc pH 4.5 buffer, and it will be used for amine coupling via NHS/EDC. With the T200, this process is carried out within the Biacore Control software using the Immobilization Wizard. Thaw one aliquot of NHS and EDC for each flow cell that is being immobilized just prior to use. Place the anti-resistin, NaOH, NHS, EDC, and ethanolamine solutions in the locations and with the volumes as indicated by the Immobilization Wizard sample layout template. Run the Immobilization Wizard with a 3000 RU

immobilization target over all four individual flow cells on the CM-5 sensor chip. The wizard injects a 1:1 mixture of NHS/EDC over the carboxy-methyl dextran surface at a flow rate of 10 uL/min and 420 s, which activates free-carboxyl groups for covalent coupling with free-amines of the lysines or N-terminus on the protein of interest. The anti-resistin antibody (or other tag-specific antibody) is injected until a response of about 3000 RUs are achieved. Finally, 1 M ethanolamine is injected at a flow rate of 10 uL/min and 420 s to quench any remaining free NHS/EDC-activated carboxyl groups. The final immobilization level should be within 3000 +/− 300 RUs of anti-resistin on each flow cell. FC1 will be used as a blank reference surface, FC2, FC3, and FC4 will be used to capture recombinant S-protein (*see* **Note 8**).

3. Normalize the sensor chip. Once the indirect capture surface has been created, the sensor chip should be normalized using the Normalize function within the Tools menu of the Biacore Control software. This will prepare the chip for the upcoming Ace2 binding analysis. If the sensor chip is removed from the instrument prior to the run, it will have to be re-normalized. This will require a 120 uL aliquot of the 70% glycerol solution.

3.2 Generating the Ace2 Binding Sensorgrams

1. Repeats to include in the run. For statistical analysis, the experiment is normally repeated a minimum three times, and with each time the S-protein is captured it is considered a new surface. Use separate dilutions of Ace2 with each repeat, so a triplicate run performed on the same day will have three sets of Ace2 dilutions. For careful analyses, individual replicates will be carried out in separate runs on different days. Also, a good verification to show that the interaction is independent of surface density is to capture the increasing surface density of S-protein and perform a global fit of all of the sets of referenced sensorgrams. This will generate one set of kinetic values derived from a simultaneous fit of multiple sensorgrams collected from different S-protein capture levels (Fig. 2). This is accomplished using the multiple R_{max} function in the T200 Evaluation software. The goal is for the 1:1 binding model to follow closely to each of the sensorgrams with increasing R_{max}, showing that the interaction is not dependent on the density of captured S-protein.

2. Single-cycle kinetics method. This method will first use an indirect capture of the spike protein onto the SPR surface, followed by an Ace2 concentration series or PBST blanks to generate the referenced binding sensorgrams (Fig. 3). Within the Control software, the Method File contains the information to design each experiment. For each injection cycle consisting of S-protein capture and flowing of Ace2, there is a

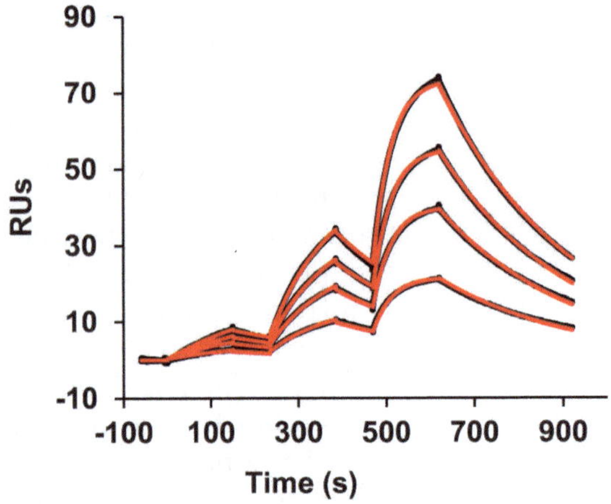

Fig. 2 Multiple Rmax sensorgrams of the same Ace2–S-protein interaction showing that the Ace2–S-protein interaction is not modified by surface density

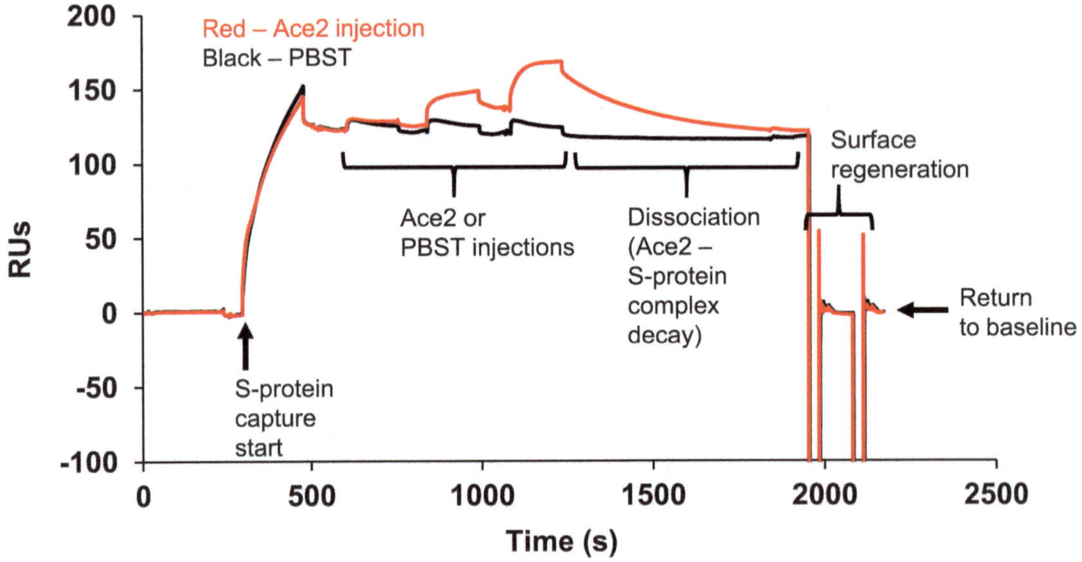

Fig. 3 Components of the SPR analysis for the Ace2–S-protein interaction using indirect capture onto a covalently-immobilized anti-resistin surface. For each Ace2 injection cycle (red), it is preceded by a buffer-only injection (black), that is critical for correct referencing

matching PBST blank injection in place of the Ace2 concentration series that needs to be added for double referencing (Fig. 4) (*see* **Note 9**). Depending on the number of samples, one to three S-proteins can be captured sequentially onto the SPR surface over FC2, FC3, and FC4 which are then used to interact with the same flowing Ace2 sample. Therefore, three different S-protein variants can be analyzed with the same Ace2 injection to increase the assay throughput. Method variables

Injection cycle	S-protein Capture			Ace2 repeat #	SCK concentrations (nM)		
	FC2	FC3	FC4		1	2	3
1	A	B	C	1	0	0	0
2	A	B	C	1	0	0	0
3	A	B	C	1	8	40	200
4	C	A	B	2	0	0	0
5	C	A	B	2	8	40	200
6	B	C	A	3	0	0	0
7	B	C	A	3	8	40	200

Fig. 4 Typical injection cycle setup for analyzing three S-protein variants (**a**, **b**, **c**) with maximum throughput utilizing all three flow cells (FC2, FC3, FC4) on the Biacore T200. Note the first injection cycle is used to help stabilize the baseline and generally not used during analysis

(outlined below) include the S-protein samples, Ace2 samples, buffer solutions, and regeneration solutions which, once entered, will be positioned by the Biacore Control Software in a plate map with the required minimum volumes calculated based on the flow rate, injection time and type of vial used.

3. Surface conditioning and buffer blank variables. Within the Conditioning portion of the Method File, set the regeneration injection of 10 mM Gly pH 1.5 at flow 30 uL/min for 30 s, repeat over three injection cycles total. Setup one to three mock buffer blank injections using the parameters indicated for the S-protein capture and Ace2 binding below to help stabilize the baseline (*see* **Note 10**).
4. S-protein capture variables. Dilute S-protein to 10 ug/mL in PBST. Set the capture injection for 120 s at flow 10 uL/min and define the FC for each capture used within the experimental variables in the Sample portion of the Method File. The capture surface is always subsequent to the reference, so if FC1 is the anti-Fc-only reference spot, FC2, FC3, and/or FC4 will be used as the flow cell(s) for capture. It is important a capture does not occur over the reference surface, otherwise the reference cell will also show Ace2 binding and the final referenced sensorgram will show no binding or negative binding. Depending on the experimental design with the T200, different flow cell configurations can be selected for referencing. Either FC1 or FC3 can be used as a blank reference surface. Therefore, referencing can be setup pairwise with FC2-1 or FC4-3; or for maximum throughput, three flow cells can be captured and referenced to FC1 in one injection cycle (FC2-1, FC3-1, FC4-1) (*see* **Note 11**).
5. Ace2 injection variables. Whereas each S-protein under analysis is captured over one specific flow cell only, one injection of

Ace2 is flown sequentially over both the reference and capture surfaces. Set three to five concentrations of Ace2 to be used for single-cycle kinetics (SCK) analysis (*see* **Note 12**). For optimum experimental design, use a maximum concentration of Ace2 that will be five- to tenfold above the expected K_D, and also have the lowest concentration below the K_D. Usually, the initial experimental setup will take one or two rounds of optimization to identify an appropriate Ace2 concentration range. Here, we selected a three-point concentration series with a fivefold dilution with a high concentration of 200 nM as the expected K_D will be in the 10–30 nM range. The sensorgrams should generate a visible amount of curvature in the highest concentrations for reliable fits to be generated when the 1:1 binding model is applied within the T200 Biacore Evaluation software (*see* **Note 13**). Set the flow rate at 50 uL/minute and a contact time of 150 s with a dissociation of 300–600 s, this sets a usable volume requirement for 96-well plates on the Biacore T200, a faster flow rate and longer association phase can be used in conjunction with larger-volume tubes in the sample rack as required.

6. Regeneration variable. Two 30 s pulses of 10 mM glycine pH 1.5 will be injected over the capture surface to remove any remaining S-protein–Ace2 complex and prepare for subsequent capture–Ace2 injection cycles (*see* **Note 14**).
7. Running the experiment and data files. Once the sample is setup according to the sample layout in the Method File, the user clicks Next through to a prompt to save the file at which point the total run time is indicated along with the volume running buffer required to complete the run. Once a file name is assigned to the run, it is saved and the run is initiated and the instrument performs the experiment as laid out in the Method File. With the Biacore T200, the run files are saved as *.BLR files which are then read by the analysis software.
8. Storing the sensor chip for reuse. After the run, the sensor chip can be slid out of its outer plastic housing and stored at 4 °C in a 50 mL Falcon tube containing PBST for use in future experiments. Ensure the buffer is above the level of the gold sensor surface. Do not touch the gold surface and handle with care (*see* **Note 15**).

3.3 Data Analysis

1. Preparation of the sensorgrams. Start the Biacore T200 Evaluation Software and open the *.BLR file of interest. The first thing to do is to examine the sensorgrams for any binding artifacts over the non-referenced control surface (the anti-Fc antibody-only surface). With the T200 this will be either FC1 (if FC2-1, 3-1, or 4-1 is used for referencing) or FC3 (if FC4-3 is used for referencing). Within the software, open all of the

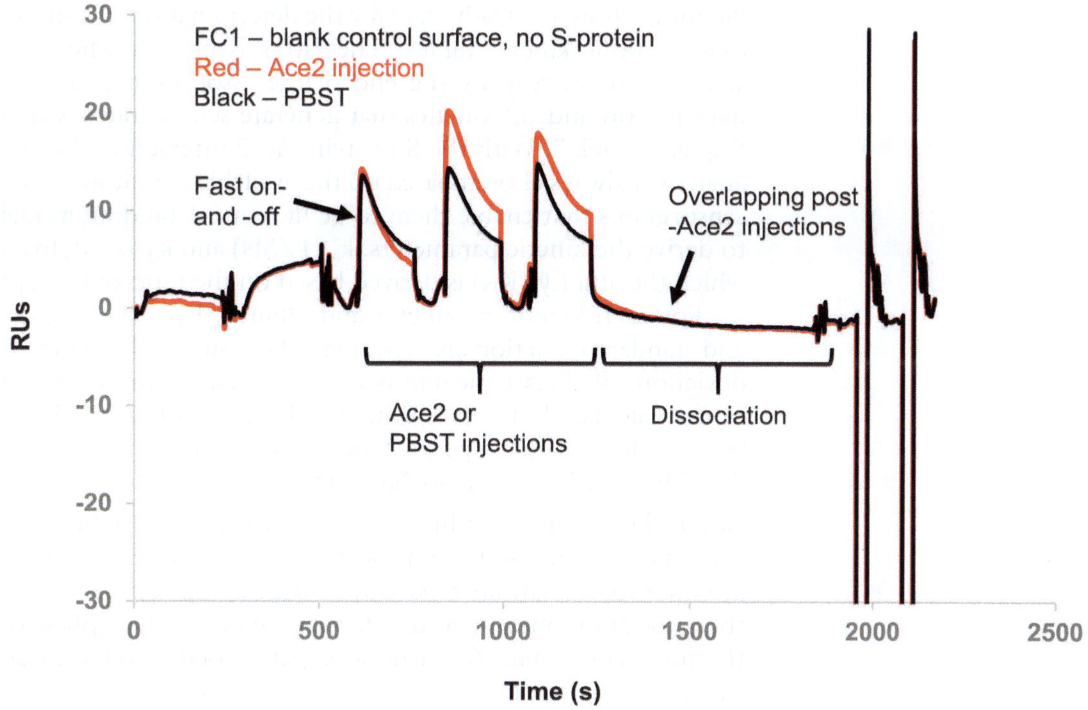

Fig. 5 A typical FC1—control surface sensorgram

sensorgrams and normalize all of them to the initial baseline before the first capture. Verify with each Ace2 injection that the sensorgrams are rapidly returning to baseline over the control surface (Fig. 5). Any increase to the control surface baseline relative to the start of the injections indicates that the Ace2 is non-specifically binding to the SPR surface. Next, check the surfaces with immobilized S-protein and ensure that post-injection of the regenerator (in this case 10 mM Glycine pH 1.5), the sensorgrams return to the pre-capture of S-protein baseline to ensure removal of all of the bound S-protein–Ace2 complex to prepare for the next injection cycle. This means that captured S-protein from the current cycle will not contaminate subsequent ones. Then check for sensorgram artifacts such as spikes caused by air or buffer mismatch with referencing. Before analysis, these can be removed in the Biacore Evaluation software.

2. Sensorgram analysis. Once these sensorgram checks are performed, the sensorgrams can be analyzed for binding. Within the software, select a single sample for analysis, or batch mode to analyze all of the binding sensorgrams at once. The latter mode can be used if the data is fairly clean and free of spikes that may need to be removed before fitting to the 1:1 binding model. Next, the type of analysis is needed to be selected, either

kinetic analysis, or steady state for the determination of affinity-only without kinetic values. The latter is chosen when the sensorgrams plateau by the end of the association phase, or have fast on-and-off kinetics that generate sensorgrams with a "square profile". With the S-protein–Ace2 interaction, kinetic analysis is always chosen based on the available curvature in the sensorgrams that enable them to be fit to a 1:1 binding model to derive the kinetic parameters, k_a (1/Ms) and k_d (1/s), from which the affinity (K_D) is derived based on the ratio of the k_d/k_a. For each kinetic parameter and affinity (K_D), the average and standard deviation are reported. The result of the standard deviation will dictate the number of significant figures to report in the calculated affinity constant [21]. For instance, 25.1E-9 ± 2E-10 nM (average ± standard deviation) compared to 25.12E-9 ± 2E-11 nM (*see* **Note 16**).

3. Residual plots. The residual plots show the difference between the experimental model and the 1:1 binding model, pay close attention to deviations especially at the beginning and end of the association phase, and the start of the dissociation phase of the interaction (Fig. 6). A good experimental setup with an expected 1:1 binding interaction should not show any large deviations in these regions with scatter around zero. This is a better gauge of the overall fit of the model to the experimental sensorgram as opposed to the commonly-used Chi^2 value (the average squared residual) which changes based on the overall R_{max}. Unexpected deviations in the residual plot often indicate that the purity of one or two of the binding components is insufficient and is likely partially inactive or aggregating, leading to non-specific binding and/or biphasic binding artifacts.
4. Use of capture levels, R_{max} and theoretical R_{max}. While affinity (K_D) is a good CQA for tracking lot-to-lot variation, capture levels of S-protein in conjunction with the R_{max} of the Ace2 binding sensorgram can also be used to extract extra information out of the sensorgrams that is useful for comparing between different S-protein productions or variants. We have seen a marked difference in capture levels between S-protein derived from different variants, as well as the same S-protein at different pH values, which suggests that the S-protein undergoes conformational changes that manifest itself in the capture levels. Additionally useful is comparing the experimental R_{max} to the theoretical R_{max} (ThR_{max}) based on the capture levels of the S-protein. The R_{max} value is the maximum response relative to the normalized baseline calculated for the sensorgrams based on the kinetic parameters derived from the 1:1 fit (*see* **Note 17**). The ThR_{max} is the maximum response if all of the captured S-protein is 100% active. This would mean for each

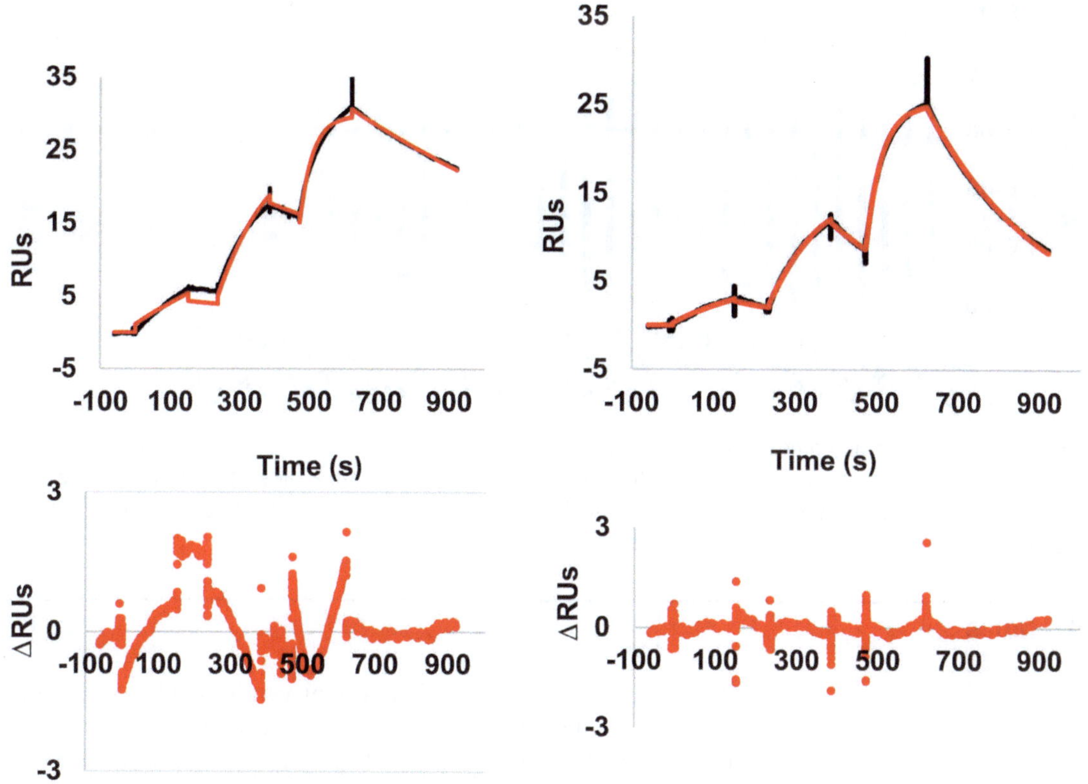

Fig. 6 Sensorgrams demonstrating a good fit of S-protein from the SARS-Cov2 Wuhan variant (**right**) versus a fit with an Omicron variant that shows some minor deviations between the experimental data and the 1:1 binding model (**left**)

captured S-protein trimer, there is full occupancy of the three S-protein RBD domains by Ace2. The ratio Rmax/ThRmax can be used to track the relative activity of the captured S-protein and can be used to benchmark between batches for the relative quality of their purification or to compare relative activity between variants. Because of the accessibility of the RBD domain between variants (refs), the R_{max} compared to the ThR_{max} can vary quite significantly. The ThRmax can be calculated with the formula: $(MW_{Ace2}$*valency/$MW_{S\text{-}protein})$ * (capture level ligand) where the MW of S-protein is for the trimer and the valency = 3 for three potential interactions between Ace2 and each S-protein trimer. This ratio will normalize for capture levels and enable for comparison between experiments (Fig. 7).

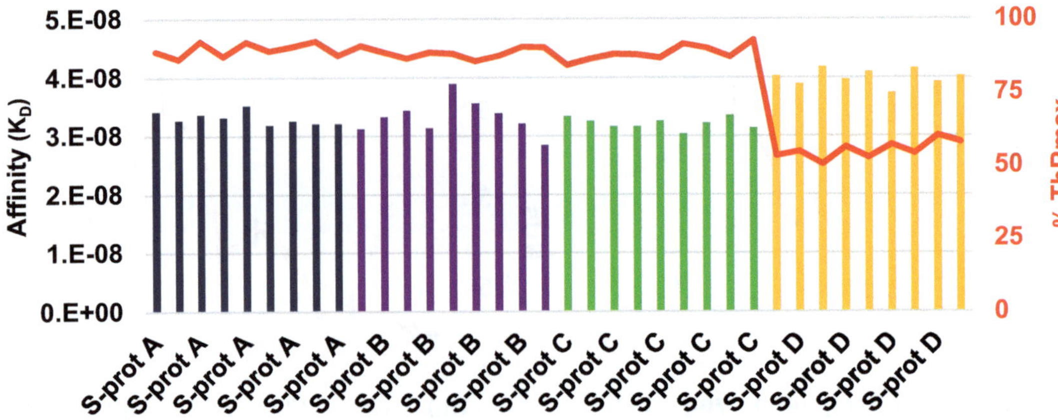

Fig. 7 Bar graph showing repeats of affinity analysis from four S-protein productions with the calculated % theoretical R_{max} overlaid in red. The S-protein production D shows both weaker binding and a lower % theoretical Rmax compared to the first three productions, indicating fewer of the captured S-protein subunits are contributing to productive Ace2 interactions

4 Notes

1. There are many label-free instrument vendors. These methods should be applicable for other SPR instruments, as well as other label-free technologies such as biolayer interferometry used in the Octet system (Sartorious Inc.) and grated-coupling interferometry used in the Creoptix WAVEcore (Malvern Panalytical Inc.).
2. The Cytiva Biacore sensor chips have a carboxymethyl polymer surface [7] that serves to generate a three-dimensional matrix for immobilization and protein–protein interactions, and aid in non-specific binding of the protein ligand and analyte with the SPR surface. The surface has a partial negative charge which is why most immobilizations will use a low pH in order for the ligand to generate a partial negative charge to interact with the activated surface.
3. If the running buffer is made from directly phosphate salts, the final 10 or 20× buffer stock should be sterile filtered to remove any insoluble particulate and dust that may have accumulated on the glassware and would be a risk for contaminating the internal fluidics of the Biacore. Generally, almost any running buffer is amenable for SPR, usually one is selected that lacks Tris buffer which can interfere with the amine coupling of capture antibody as detailed in the Method section.
4. We use the Cytiva amine coupling kit BR100050 which comes supplied with pre-weighed NHS and EDC, and 1 M ethanolamine solution.

5. This is an internally produced sdAb that was fused at the N-term of a human IgG1 Fc. This will be employed to capture the recombinant S-protein that is trimerized with the human resistin domain at its C-terminus. It is important to use a capture surface that will orient the S-protein in a manner that will allow it to productively interact with the Ace2 protein. Other purification tags on recombinant proteins such as a His-tag or Strep-tag are also frequently used to capture ligand onto the SPR surface. The anti-His antibody available from Cytiva (Cat # 2923460) works very well in our hands and it is easier to use than the available NTA sensor chip (Cytiva, Cat # 28994951). Regardless, it is crucial that the tag that is used to capture the ligand (in this case S-protein) is also not present on the analyte (in this case the monomeric Ace2), so careful experimental planning is required.
6. We use monomeric Ace2 produced at the NRC. It is good to have a working analyte stock that is more than 20× over the anticipated high concentration injection in order to minimize mismatch with the running buffer. Excipients in the analyte such as glycerol, or high concentrations of salt, can cause large refractive index changes which generate large shifts in the sensorgrams relative to the running buffer baseline that can be difficult to reference out completely. This is a common problem with commercial analyte samples.
7. It is important to create a unique name of the sensor chip and use the Immobilization Wizard when creating a sensor chip surface so the control software will embed the immobilization information (ligand and RUs) and experiment run date in the final *.BLR result files.
8. We generally don't find making high-density surfaces between 8000 and 10,000 RUs results in significantly more capture of the target ligand. Higher-density surfaces often result in worse capture and poor reproducibility between sequential S-protein capture levels which makes for worse baseline referencing.
9. Double referencing is essential for quality SPR interactions. There are two references as the name suggests, the first reference is the subtraction of the blank capture surface response (e.g., subtraction of FC1 binding signal, if FC2, FC3, or FC4 are being used for capture of S-protein) from the Ace2 binding sensorgram. This will account for any buffer bulk-shifts and non-specific binding (ideally there will be none) to the anti-resistin-only capture surface (FC1) relative to the Ace2 binding over captured S-protein on FC2, FC3, or FC4. The second is the subtraction of the blank-referenced PBST sensorgram from the blank-referenced Ace2 binding sensorgram. This is the double-reference and accounts for baseline drift in the instrument. It is best practice to have one matching PBST injection

for every Ace2 injection; however, if there are technical problems with the PBST injection, it is possible to select other blank injections.

10. It is essential that the freshly prepared anti-resistin (or other selected antibody for capture) surfaces undergo multiple regeneration and buffer blank cycles in order to remove any non-specifically bound capture antibody, and to stabilize both capture levels at the sensorgram baseline. There may be a small drop in the baseline after the first one or two cycles of acid regenerator, afterward, the baseline should remain steady over repeated binding and regeneration cycles.

11. The flow rate is set low to conserve the S-protein sample, higher flow rates won't improve the subsequent Ace2 binding. Conformation changes between S-protein variants can lead to a range of capture levels. Ultimately, the goal is to generate Ace2 binding with an Rmax below 100 RUs. If the S-protein capture levels give too high Ace2 binding, reduce the concentration or capture time accordingly. It is also good practice to vary the capture order so that in repeat Ace2 injections, the S-protein constructs are captured over different flow cells whenever possible to remove the possibility of any bias towards a particular flow cell.

12. Single-cycle kinetics is proprietary to Cytiva Inc. and is available for use with the Biacore T200. Single-cycle kinetics (SCK) uses sequential, increasing injections of Ace2 with only one dissociation for every injection cycle. Traditional multi-cycle kinetics (MCK) is where there is one analyte concentration per injection cycle, with the major disadvantages being that it takes much longer as each concentration has a long dissociation and would require a fresh S-protein capture with each Ace2 concentration so the S-protein consumption is much higher. To reduce the run length with MCK, it is possible within the T200 Method file to set the dissociation as a run variable and reduce the dissociation time of all but the highest concentrations of Ace2. In this manner, the dissociation rate will only be determined from the one long dissociation which is more analogous to SCK. Multi-cycle kinetics is useful when the test samples of interest have a very wide range of affinities and a broad concentration range of analyte needs to be used to accommodate the varied affinities. Using multi-cycle kinetics allows for a subset of concentrations to be used in a fit and discard those that are outside the useful concentration range.

13. More concentration points often result in less curvature and result in poorer fits to a 1:1 binding model. Other considerations are that a two- or threefold dilution compared to a five- or tenfold one will have a larger sample requirement, and more concentration points will mean the runs will take longer.

14. Observe that the sensorgram baseline returns to the pre-capture baseline. If the post-regeneration sensorgram does not return to baseline, additional regeneration time can be added or the regeneration solution can be changed. However, if the sensorgram does not return to baseline, this usually indicates non-specific aggregation of the analyte or ligand is occurring on the SPR surface. At this point, troubleshooting of the ligand and analyte is required to determine which binding partner is aggregating on the SPR surface. A good quality control is the routine use of analytical SEC to assess the purity of the analyte (protein that is flowing) and ligand (protein that is captured). Usually, the purity of the analyte is more crucial, and oftentimes a successful SPR run can even be performed using impure ligand, such as antibody in expression media. In common use with an antibody capture surface, regeneration with 10 mM glycine pH 1.5 is nearly always sufficient. Deviating from this regeneration solution is usually indicative of problems with one of the components of the interaction under analysis.
15. Antibody capture surfaces are very robust and amenable to several redockings and should last over many experiments if purified analyte and ligands are used. Monitor the capture levels, and discard if the capture levels are dropping substantially below the initial levels when new. Prior to redocking, carefully remove the sensor chip insert from the Falcon tube, gently rinse with ddH_2O avoiding a direct stream on the gold surface and dry under a stream of N_2 gas (we have used compressed air canisters in the past but they sometimes contain solvents which may contaminate the surface and the cold air may denature the immobilized protein surface). Re-insert into the chip's plastic housing and redock in the instrument. Identify the chip with its originally designated name used during the immobilization procedure when prompted. This ensures the chip immobilization information will be embedded within the *.BLR run files.
16. The Biacore Evaluation software fits the sensorgrams starting with a set of default parameters that are used in iterative fits until a minimum is found in the sum of the squared residuals. If the 1:1 interaction is setup correctly, a solution is almost always found that is appropriate. Rarely, the software can get stuck in a minima that gives an obviously incorrect solution with a bad fit. If this occurs, check carefully for fast air spikes that do not resolve well on the computer monitor and may add one or two data points far outside the rest of the binding sensorgram. It is also possible to alter the default parameters to aid with fits that are stuck with a poor solution. For instance, the off-rate can be adjusted to one that more closely approximates the sensorgram

to help the software to find the correct fit. It is important to verify that the final calculated rate constants are meaningful, as a fit with an association rate at 10^9 or affinity values at 10^{-15} are clearly not biologically relevant and outside the range of the instrument. It is often tempting when sensorgrams don't fit to a 1:1 binding model, that other more complex models, such as bivalent analyte, are chosen to "fit" the sensorgram. They only fit because there are more parameters that can be changed to generate a model that approximates the experimental sensorgrams. This is not good practice, and do not be tempted to report the K_D from the bivalent fit as the 1:1 binding affinity.

17. The R_{max} value should be close to the maximum response of the highest concentration. If the R_{max} is much higher, this usually indicates that there are problems with the fit, usually due to a linear response in the association phase of the sensorgrams which can indicate non-specific binding, or in cases with extremely high affinity the analyte concentration will need to be increased to generate curvature in the sensorgrams.

Acknowledgments

Thanks to Yves Durocher, Section Head Mammalian Cell Expression, for initiating this project and Matthew Stuible, Team lead Cell Engineering, for S-protein and Ace2 productions, both at NRC Montreal Canada. Thanks to Martin Rossotti, NRC Ottawa, for supplying anti-resistin antibody. Funded by the NRC Pandemic Response Challenge Program.

References

1. Else H (2020) How a torrent of COVID science changed research publishing—in seven charts. Nature 588(24):553
2. Jackson CB, Farzan M, Chen B et al (2022) Mechanisms of SARS-CoV-2 entry into cells. Nat Rev Mol Cell Biol 23(1):3–20. https://doi.org/10.1038/s41580-021-00418-x
3. Li W, Moore MJ, Vasilieva N et al (2003) Angiotensin-converting enzyme 2 is a functional receptor for the SARS coronavirus. Nature 426(6965):450–454. https://doi.org/10.1038/nature02145
4. Cheng MH, Krieger JM, Banerjee A et al (2022) Impact of new variants on SARS-CoV-2 infectivity and neutralization: a molecular assessment of the alterations in the spike-host protein interactions. iScience 25(3):103939. https://doi.org/10.1016/j.isci.2022.103939
5. Kumar R, Murugan NA, Srivastava V (2022) Improved binding affinity of Omicron's spike protein for the human angiotensin-converting enzyme 2 receptor is the key behind its increased virulence. Int J Mol Sci 23(6). https://doi.org/10.3390/ijms23063409
6. Piplani S, Singh PK, Winkler DA et al (2021) In silico comparison of SARS-CoV-2 spike protein-ACE2 binding affinities across species and implications for virus origin. Sci Rep 11(1):13063. https://doi.org/10.1038/s41598-021-92388-5
7. De Crescenzo G, Boucher C, Durocher Y et al (2008) Kinetic characterization by surface Plasmon resonance-based biosensors: principle and emerging trends. Cell Mol Bioeng 1(4):204–215. https://doi.org/10.1007/s12195-008-0035-5

8. Karlsson R, Fält A (1997) Experimental design for kinetic analysis of protein-protein interactions with surface plasmon resonance biosensors. J Immunol Methods 200(1–2):121–133. https://doi.org/10.1016/s0022-1759(96)00195-0
9. McDonnell JM (2001) Surface plasmon resonance: towards an understanding of the mechanisms of biological molecular recognition. Curr Opin Chem Biol 5(5):572–577. https://doi.org/10.1016/s1367-5931(00)00251-9
10. Myszka DG (1997) Kinetic analysis of macromolecular interactions using surface plasmon resonance biosensors. Curr Opin Biotechnol 8(1):50–57. https://doi.org/10.1016/s0958-1669(97)80157-7
11. Wang W, Thiemann S, Chen Q (2022) Utility of SPR technology in biotherapeutic development: qualification for intended use. Anal Biochem 654:114804. https://doi.org/10.1016/j.ab.2022.114804
12. Colwill K, Galipeau Y, Stuible M et al (2022) A scalable serology solution for profiling humoral immune responses to SARS-CoV-2 infection and vaccination. Clin Transl Immunol 11(3):e1380. https://doi.org/10.1002/cti2.1380
13. De Crescenzo G, Grothe S, Lortie R et al (2000) Real-time kinetic studies on the interaction of transforming growth factor alpha with the epidermal growth factor receptor extracellular domain reveal a conformational change model. Biochemistry 39(31):9466–9476. https://doi.org/10.1021/bi992987r
14. Khalifa MB, Choulier L, Lortat-Jacob H et al (2001) BIACORE data processing: an evaluation of the global fitting procedure. Anal Biochem 293(2):194–203. https://doi.org/10.1006/abio.2001.5119
15. Morton TA, Myszka DG, Chaiken IM (1995) Interpreting complex binding kinetics from optical biosensors: a comparison of analysis by linearization, the integrated rate equation, and numerical integration. Anal Biochem 227(1):176–185. https://doi.org/10.1006/abio.1995.1268
16. Oostindie SC, Lazar GA, Schuurman J et al (2022) Avidity in antibody effector functions and biotherapeutic drug design. Nat Rev Drug Discov 21(10):715–735. https://doi.org/10.1038/s41573-022-00501-8
17. Erlendsson S, Teilum K (2020) Binding revisited-avidity in cellular function and signaling. Front Mol Biosci 7:615565. https://doi.org/10.3389/fmolb.2020.615565
18. Mammen M, Choi SK, Whitesides GM (1998) Polyvalent interactions in biological systems: implications for design and use of multivalent ligands and inhibitors. Angew Chem Int Ed Engl 37(20):2754–2794
19. Rossotti MA, van Faassen H, Tran AT et al (2022) Arsenal of nanobodies shows broad-spectrum neutralization against SARS-CoV-2 variants of concern in vitro and in vivo in hamster models. Commun Biol 5(1):933. https://doi.org/10.1038/s42003-022-03866-z
20. Karlsson R, Katsamba PS, Nordin H et al (2006) Analyzing a kinetic titration series using affinity biosensors. Anal Biochem 349(1):136–147. https://doi.org/10.1016/j.ab.2005.09.034
21. Rich RL, Myszka DG (2010) Grading the commercial optical biosensor literature-class of 2008: 'the mighty binders'. J Mol Recognit 23(1):1–64. https://doi.org/10.1002/jmr.1004

Chapter 6

A Biosensor Assay Based on Coiled-Coil-Mediated Human ACE2 Receptor Capture for the Analysis of Its Interactions with the SARS-CoV-2 Receptor Binding Domain

Catherine Forest-Nault, Izel Koyuturk, Jimmy Gaudreault, Alex Pelletier, Denis L'Abbé, Brian Cass, Louis Bisson, Alina Burlacu, Laurence Delafosse, Matthew Stuible, Olivier Henry, Gregory De Crescenzo, and Yves Durocher

Abstract

Surface plasmon resonance (SPR)-based biosensing enables the characterization of protein-protein interactions. Several SPR-based approaches have been designed to evaluate the binding mechanism between the angiotensin-converting enzyme 2 (ACE2) receptor and the receptor-binding domain (RBD) of the SARS-CoV-2 spike protein leading to a large range of kinetic and thermodynamic constants. This chapter describes a robust SPR assay based on the K5/E5 coiled-coil capture strategy that reduces artifacts. In this method, ACE2 receptors were produced with an E5-tag and immobilized as ligands in the SPR assay. This chapter details methods for high-yield production and purification of the studied proteins, functionalization of the sensor chip, conduction of the SPR assay, and data analysis.

Key words Surface plasmon resonance (SPR), Ligand-oriented capture, Angiotensin-converting enzyme 2 (ACE2) receptor, Receptor-binding domain (RBD), SARS-CoV-2, high yield protein production, Kinetic model, SPR data analysis

1 Introduction

Surface plasmon resonance (SPR)-based biosensing has emerged as a valuable tool for studying protein-protein interactions in real-time and without the need of any label. With the ongoing COVID-19 pandemic, there has been a surge of research efforts to elucidate the molecular basis of the interactions between the angiotensin-converting enzyme 2 (ACE2) receptor and the receptor-binding domain (RBD) of the SARS-CoV-2 spike protein using

Authors Catherine Forest-Nault and Izel Koyuturk have equally contributed to this chapter

Steven B. Bradfute (ed.), *Recombinant Glycoproteins: Methods and Protocols*, Methods in Molecular Biology, vol. 2762, https://doi.org/10.1007/978-1-0716-3666-4_6,

various SPR-based approaches [1–3]. However, the reported kinetic and thermodynamic constants for the ACE2-RBD interaction vary widely across studies. This inconsistency is most likely due to the design of the SPR-based assays, which involve several parameters that can impact the recorded data, such as the ligand capture strategy, ligand choice and density, flow rate, and experimental temperature. These parameters must be optimized to minimize potential artifacts and facilitate data analysis and interpretation [4, 5].

This chapter presents a novel SPR assay based on a coiled-coil strategy that captures the ligand in an oriented and stable manner using the E5/K5 coiled-coil peptidic system [6], which has been shown to outperform other ligand capture strategies in reducing artifacts and increasing stability of the assay steps [7, 8]. The coiled-coil strategy involves covalent immobilization of the K5 peptides to the biosensor chip and tagging one of the proteins of interest with an E5 coil peptide to enable its subsequent capture by the surface-displayed K5 coil. In the described assay, ACE2 receptors were tagged with the E5 coil peptide (ACE2-E5) and used as ligands (the capture species in the SPR terminology), while SARS-CoV-2 RBD were the analytes (the soluble species in the SPR terminology). This chapter outlines the methods for the high-yield production of ACE2-E5 receptor and SARS-CoV-2 RBD, their purification, the functionalization of the sensor chip with the K5 coil peptide, and generation of robust ACE2-RBD sensorgrams. Additionally, it provides details on data analysis models and kinetics interpretation.

2 Materials

2.1 Cell Thawing

1. Glyco-engineered $CHO^{BRI/55E1}$ frozen cell vials from a research cell bank (stored in liquid nitrogen vapor phase).
2. L-Glutamine: 200 mM stock solution in water.
3. Culture medium mix: 75% (v/v) of BalanCD Transfectory CHO, 25% of HyClone HyCell TransFx-C with 4 mM L-Glutamine.
4. Disposable Erlenmeyer plastic shake flasks.
5. Humidified incubator controlled at 37 °C with 5% CO_2.
6. Orbital shaker set at 120 rpm, 19–25 mm throw.

2.2 Plasmid DNA Preparation (Maxiprep)

1. LB agarose plate with proper antibiotic (e.g., 50 μg/mL ampicillin).
2. Maxiprep DNA kit.
3. CircleGrow™ medium (*see* **Note 1**).

4. 250 mL Erlenmeyer shaker flask.
5. Orbital shaker plate in a non-humidified incubator controlled at 37 °C.
6. Centrifuge for 50 mL polypropylene tubes (running at 10,000 × *g*).
7. Competent *E. coli* (DH5alpha strain).
8. Equipment for agarose gel electrophoresis.
9. UV spectrophotometer.

2.3 Polyethylenimine MAX Solution

1. PEI MAX powder.
2. Milli-Q® water.
3. 1 N sodium hydroxide solution.
4. Magnetic stir plate.
5. Graduated cylinder.
6. pH meter.
7. 0.22 μm vacuum filtration unit.
8. Sterile polypropylene tubes.

2.4 BalanCD CHO Feed 4

1. BalanCD CHO Feed 4 powder.
2. Sodium Bicarbonate.
3. Glucose.
4. Kolliphor P188.
5. Milli-Q® water.
6. pH meter.
7. Balance.
8. Osmometer.
9. Magnetic stirring plate & magnetic stirring bar.
10. 0.22 μm vacuum filtration unit.

2.5 Transfection of CHO Cells with PEI MAX

1. Different glyco-engineered $CHO^{BRI/55E1}$ maintenance cells in shake flasks.
2. Culture medium mix: 75% (v/v) of BalanCD Transfectory CHO, 25% of HyClone HyCell TransFx-C with 4 mM L-Glutamine.
3. 1 mg/mL PEI MAX solution.
4. Purified plasmid DNA of interest.
5. BalanCD CHO Feed 4 stock solution.
6. N,N-Dimethylacetamide.
7. Anti-Clumping Supplement.
8. Glucose: 2 M stock solution in water.

2.6 Purification of His-Tagged Proteins

1. Nickel Sepharose® excel resin.
2. Milli-Q® water.
3. Equilibration Buffer: 50 mM sodium phosphate, 300 mM NaCl, pH 7.8.
4. Wash Buffer: 50 mM sodium phosphate, 300 mM NaCl, 10 mM imidazole, pH 7.8.
5. Elution Buffer: 50 mM sodium phosphate, 300 mM NaCl, 300 mM imidazole, pH 7.8.
6. SDS-PAGE gel electrophoresis system.

2.7 Purification of Strep-Tagged Proteins

1. Strep-Tactin® XT Superflow® high-capacity resin.
2. Milli-Q® water.
3. Equilibration/Wash Buffer: 100 mM Tris, 150 mM NaCl, pH 8.
4. Elution Buffer: BXT Elution Buffer.
5. NAP-25 desalting columns.
6. Phosphate Buffered Saline (PBS).
7. SDS-PAGE gel electrophoresis system.

2.8 Purification of FLAG-Tagged Proteins

1. Anti-FLAG M2 resin.
2. FLAG peptide.
3. Phosphate Buffered Saline (PBS).
4. Elution buffer: 0.1 mg/mL FLAG peptide dissolved in PBS.
5. Regeneration buffer: 0.1 M Glycine-HCl pH 3.5.
6. Storage: 50% glycerol, 50% DPBS, 0.02% sodium azide.

2.9 SEC Purification

1. HiPrep 16/60 Superdex 75 pg or HiPrep 26/60 Superdex 75 pg.
2. Amicon® Ultra-15 Ultracel 10 K centrifugal filter unit.
3. Milli-Q® water.
4. Running buffer: DPBS.
5. Cleaning buffer: 500 mM sodium hydroxide.
6. Storage buffer: 20% ethanol in water.

2.10 Functionalization of SPR Biosensor Chip

1. Series S Sensor chip CM5.
2. Amine coupling solution: 0.05 M N-hydroxysuccinimide (NHS), 0.2 M N-ethyl-N′-(3-diethylaminopropyl) carbodiimide hydrochloride (EDC) (*see* **Note 2**).

3. Thiol coupling solution: 0.04 M 2-(2-pyridinyldithio) ethaneamine hydrochloride (PDEA), 0.05 M Borate buffer pH 8,5. Store at 4 °C.
4. Deactivation solution (amine coupling): 1 M Ethanolamine pH 8,5. Store at 4 °C.
5. Ligand capture solution: 0.02 M Cysteine-tagged K5 peptides synthesized by the peptide facility at the University of Colorado (CGG-[KVSALKE]5) [9] and freshly dissolved in 100 mM acetic acid, pH 4.0 (*see* **Note 3**).
6. Deactivation solution (thiol coupling): 5 mM L-Cysteine, 0.05 M NaCl, 0.1 M Acetate buffer pH 4. Store at 4 °C.
7. Running buffer: HBS-EP+ (10 mM HEPES, 0,15 M NaCl, 3 mM ethylenediaminetetraacetic acid (EDTA), and 0,05% [v/v] surfactant P20, pH 7,4).
8. Regeneration buffer: 6 M Guanidine hydrochloride. Weigh 28.66 g in a 50 mL falcon. Add 45 mL of Milli-Q® water. Vortex until completely dissolved. Make up to 50 mL with Milli-Q® water and filter through a 0.22 μm syringe filter. Store at room temperature.
9. Biacore T100.

2.11 SPR Assay for the ACE2-RBD Interaction Monitoring

1. Running buffer: HBS-EP+ as in the previous section.
2. Regeneration buffer: 6 M Guanidine hydrochloride as in the previous section.
3. Ligand: ACE2-E5 purified proteins diluted at 1 μg/mL in Running buffer.
4. Analyte: RBD purified proteins diluted at several concentration in Running buffer.
5. Biacore T100.

3 Methods

3.1 Cell Thawing and Maintenance

1. Prepare a disposable 125 mL plastic shake flask filled with 29 mL of chemically-defined Culture medium mix with 4 mM L-glutamine and incubate the flask at 37 °C with 5% CO_2 on an orbital shaker with agitation for at least 1 h prior to thawing the cells.
2. Retrieve the appropriate cryogenic vial, containing 10×10^6 cells in 1 mL, stored in vapor-phase liquid nitrogen and place it in dry ice.
3. Following medium equilibration, transfer 9 mL of the equilibrated culture medium from the previously prepared 125 mL shake flask at **step 1** into a disposable 15 mL conical tube.

4. Quickly thaw cells by swirling the cryovial in a 37 °C water bath and immediately transfer thawed cell suspension into the pre-filled 15 mL conical tube.
5. Mix the 15 mL conical tube by inversion.
6. Centrifuge cell suspension at 250 × *g* for 5 min, decant supernatant and resuspend cell pellet with 10 mL of the equilibrated culture medium from the previously prepared 125 mL SF by pipetting up and down.
7. Transfer the cell suspension into the 125 mL shake flask to obtain a final concentration of 0.5×10^6 cells/mL.
8. Take out 1 mL of cells from the 125 mL shake flask to determine cell density and viability using an automated cell counter.
9. Incubate the shake flask at 37 °C and 5% CO_2 on an orbital shaker with agitation until next passage.
10. Subculture every 2 or 3 days to maintain cell densities between 1.5×10^6 and 2.5×10^6 viable cells/mL.
11. Dilute cultures to 0.125×10^6 viable cells/mL for long weekends (4 days).
12. Thaw a new vial every 10 weeks.

3.2 Plasmid DNA Preparation

The following protocol was developed in our laboratory as previously described [10]. The ACE2 construct used in this method contains a Twin-Strep-tag II-(His)6-FLAG tag on the N-terminus and an E5 coil tag was added at the ACE2 C-terminus. The RBD construct encodes a C-terminal (His)6-FLAG tag.

1. Incubate it for 16–20 h at 37 °C on an orbital shaker with agitation (250 rpm).
2. Extract plasmid DNA using a kit as recommended by the manufacturer.
3. Dissolve DNA in 1 mL of sterile TE (10 mM Tris, 1 mM EDTA, pH 8.0). This step should be performed in a laminar hood to ensure sterility.
4. Measure absorbance at 260 and 280 nm. The A_{260}/A_{280} ratio of purified plasmid DNA should be between 1.85 and 1.95.
5. Verify DNA integrity on 1% agarose gel electrophoresis.

3.3 Linear PEI-MAX (1 mg/mL Solution)

The following protocol was developed in our laboratory as previously described [10].

1. Weigh 500 mg PEI in 500 mL glass beaker and dissolve it with 450 mL Milli-Q® water by stirring (5–15 mins).
2. Adjust pH to 6.9–7.1 using 1 N NaOH dropwise.
3. Transfer solution to a 500 mL cylinder and adjust final volume to 500 mL with Milli-Q® water.

4. Filter and sterilize solution through a 0.22 μm membrane.
5. Aliquot to desired volumes and store at 4 °C.

3.4 BalanCD CHO Feed 4 Preparation (0.8× Concentration)

1. Weigh 1180 g of room temperature Milli-Q water in an appropriately sized depyrogenated vessel with magnetic stirring bar.
2. Add the entire bottle of (112.2 g) of BalanCD CHO Feed 4 powder (room temperature) to water, then turn on agitation to moderate speed.
3. Add glucose if desired (*see* **Note 4**). 0.8× feed solution already contains 16 g/L.
4. Add extra KolliphorP188 if desired (*see* **Note 5**).
5. Cover the container with aluminum foil and mix until majority of powder is dissolved.
6. Add 2.75 g of sodium bicarbonate to the solution.
7. Mix for 1.5–2 h until solution is clear.
8. Measure osmolality of the solution (acceptable range: 830–880 mOsm/kg).
9. Measure pH (acceptable range: 7.0–7.4).
10. Measure glucose concentration (acceptable range for 0.8×: 80 ± 10 mM).
11. Sterilize by membrane filtration through a 0.2 μm filter into a sterile vessel.
12. Store solution at 2–8 °C (protected from direct light) up to 9 weeks.

3.5 Transfection of CHO Cells

The following high-density transfection protocol was developed in our laboratory for transient expression in $CHO^{BRI/55E1}$ cell line [11]. It is based on a published method for high-density transfection of CHO-3E7 cells [12] with several modifications. All manipulation must be done in a sterile environment, under a certified laminar flow hood.

Two days before the transfection, cells are seeded at 1×10^6 cells/mL using culture medium mix to reach the optimal cell density of ~8×10^6 cells/mL at the time of transfection. This avoids the need of centrifugation step which could have a detrimental effect on transfection efficiency if performed shortly prior to transfection.

1. Warm the culture medium to 25–37 °C and take out DNA from −20 °C freezer to thaw.
2. From a previously seeded flask, determine the cell density and viability. On the day of the transfection, cells should be at 8×10^6 cells/mL with >99% viability.

3. Right before transfection, dilute cells to 6×10^6 cells/mL with a volume of fresh culture medium mix equal to 25% of the final culture volume and add dimethylacetamide directly into the cell suspension at a final concentration of 0.083% (v/v).
4. Incubate the culture flask at 37 °C and 5% CO_2 under agitation (120 rpm) between 30 min and 1 h.
5. Dilute plasmid DNA at 28 μg/mL in a volume of culture medium mix equal to 5% of the final culture volume (*see* **Note 6**).
6. Dilute PEI MAX at 200 μg/mL in a volume of culture medium mix equal to 5% of the final culture volume.
7. Add diluted PEI MAX to the diluted plasmid DNA at equal volume to achieve polyplex formation (*see* **Note 7**).
8. Vortex 4 s and incubate the transfection mix for 7 min at room temperature for polyplex to form.
9. Following incubation, transfer the transfection mix using a serologic pipette into previously seeded culture flask (**step 4**) and swirl.
10. Incubate 20–24 h at 37 °C, and 5% CO_2 on an orbital shaker with agitation (120 rpm).
11. At 24 h-post transfection, add Anti-Clumping Supplement (1: 500 dilution) as well as BalanCD CHO Feed 4 (2.5% v/v) and shift temperature to 32 °C, 5% CO_2.
12. Incubate cells on an orbital shaker with shaking until next Feed day.
13. On the Feed day (5 days post-transfection), analyze the viable cell density along with cell viability of the cell suspension using an automated cell counter.
14. On the Feed day, analyze the cell suspension using a calibrated metabolites analyzer (*see* **Note 8**).
15. Finally, on the Feed day, add additional Feed 4 (5%) to the cell suspension based on the initial cell suspension volume.
16. Return culture flask to the incubator at 32 °C, 5% CO_2 with agitation.
17. Harvest cells at 6–7 days post-transfection by centrifuging at $3300 \times g$ for 30 min at room temperature.
18. Filter supernatant using 0.22 μm filter units using the appropriate filter size according to the production volume (*see* **Note 9**).

3.6 Purification of Secreted His-Tagged Proteins

This section describes an immobilized metal affinity chromatography (IMAC) purification technique for polyhistidine-tagged proteins using gravity column packed with 5 mL Ni Sepharose™ excel resin.

1. Equilibrate column with 8 column-volumes (CV) of Equilibration buffer.
2. Load the filtered supernatant at 3 mL/min onto the gravity column. Save the flow-through for further analysis.
3. Wash the column with 5 CV of Wash buffer at 5 mL/min. Save the wash fraction for further analysis.
4. Elute the protein of interest with 3 CV of Elution buffer. Collect 1 mL of elution fractions and pool those containing the protein of interest.
5. Prepare samples for SDS-PAGE and/or Western blot (e.g., 18 μL sample +6 μL loading buffer (4×) with or without DTT).
6. Heat at 70 °C for 10 min.
7. Load samples on SDS-PAGE gel for analysis (Coomassie Blue staining and/or western blotting).

3.7 Purification of Secreted Strep-Tagged Proteins

The following protocol is intended for gravity flow column-based protein purification with 5 mL Strep-Tactin® XT Superflow® high-capacity resin. The manufacturer recommends using a column purification instead of batch applications to allow an efficient purification with Strep-Tactin®. We also strongly recommend performing the protein capture step using a column.

1. Equilibrate column with 10 CV of Equilibration/Wash buffer.
2. Load the IMAC-purified proteins at 2 mL/min onto the gravity column. Save the flow-through for further analysis.
3. Wash the column with 5 CV of Equilibration/Wash buffer at 5 mL/min. Save the wash fraction for further analysis.
4. Elute the protein of interest with 6 × 0.5 CV of Elution buffer and collect the eluate in 0.5 CV fractions.
5. Pool fractions containing the protein of interest.
6. Load the pooled eluted protein fractions on NAP-25 desalting columns (GE Healthcare) equilibrated with 2 CV of PBS.
7. Analyze protein purification results by SDS-PAGE followed by Coomassie staining.
8. Filter the desalted protein using a 0.22 μm syringe filter. Aliquot the purified protein in sterile screw-cap 2 mL polypropylene tubes and store them at −80 °C.

3.8 Purification of FLAG-Tagged Proteins

The following protocol is intended for gravity flow column-based protein purification with 24 mL Anti-Flag M2 resin. The column capacity is ≈1 mg protein/mL of resin. This section is optional (*see* **Note 10**).

1. Remove storage buffer with 5 CV DPBS.
2. Regenerate column with 2 CV Regeneration buffer.
3. Neutralize/equilibrate column with 5 CV (120 mL) Equilibration buffer.
4. Load sample, adjust flow to ~1 mL/min by clamping outlet tubing. Save the flow-through for further analysis.
5. Wash with 5 CV (120 mL) Equilibration buffer. Save the wash fraction for further analysis.
6. Elute the protein of interest with 8 × 0.5 CV of Elution buffer and collect the eluate in 0.5 CV fractions.
7. Analyze protein purification results by SDS-PAGE followed by Coomassie staining.
8. Pool fractions containing the protein of interest.
9. Filter the desalted protein using a 0.22 μm syringe filter. Aliquot the purified protein in sterile screw-cap 2 mL polypropylene tubes and store them at −80 °C or proceed directly for SEC purification.

3.9 SEC Purification

The following protocol is intended for purification of ACE2 proteins and RBD monomers using size exclusion chromatography (SEC) on a chromatography system. Figure 1 presents examples of SDS-PAGE and SEC pics of purified ACE2 and RBD proteins.

1. Choose column size for purification. Refer to Table 1 for the maximum sample volume. Note that samples will be concentrated up to a protein concentration of 10 mg/mL before loading. It is recommended to start column preparation the day before the purification to clean and equilibrate overnight (**steps 2–4**).
2. Remove storage buffer with 1 CV water.
3. Clean column with 1 CV CIP buffer.
4. Neutralize/equilibrate column with 3 CV Running buffer.
5. Concentrate sample to appropriate volume for the selected column using an Amicon® Ultra-15 Ultracel 10 K centrifugal filter unit.
6. Load sample using sample loop or sample pump.
7. Elute with Running buffer. It is not necessary to collect fractions for the first 0.3 CV.
8. Collect fractions for the next 0.8 CV.
9. Analyze protein purification results by SDS-PAGE followed by Coomassie staining.
10. Pool fractions containing the protein of interest.

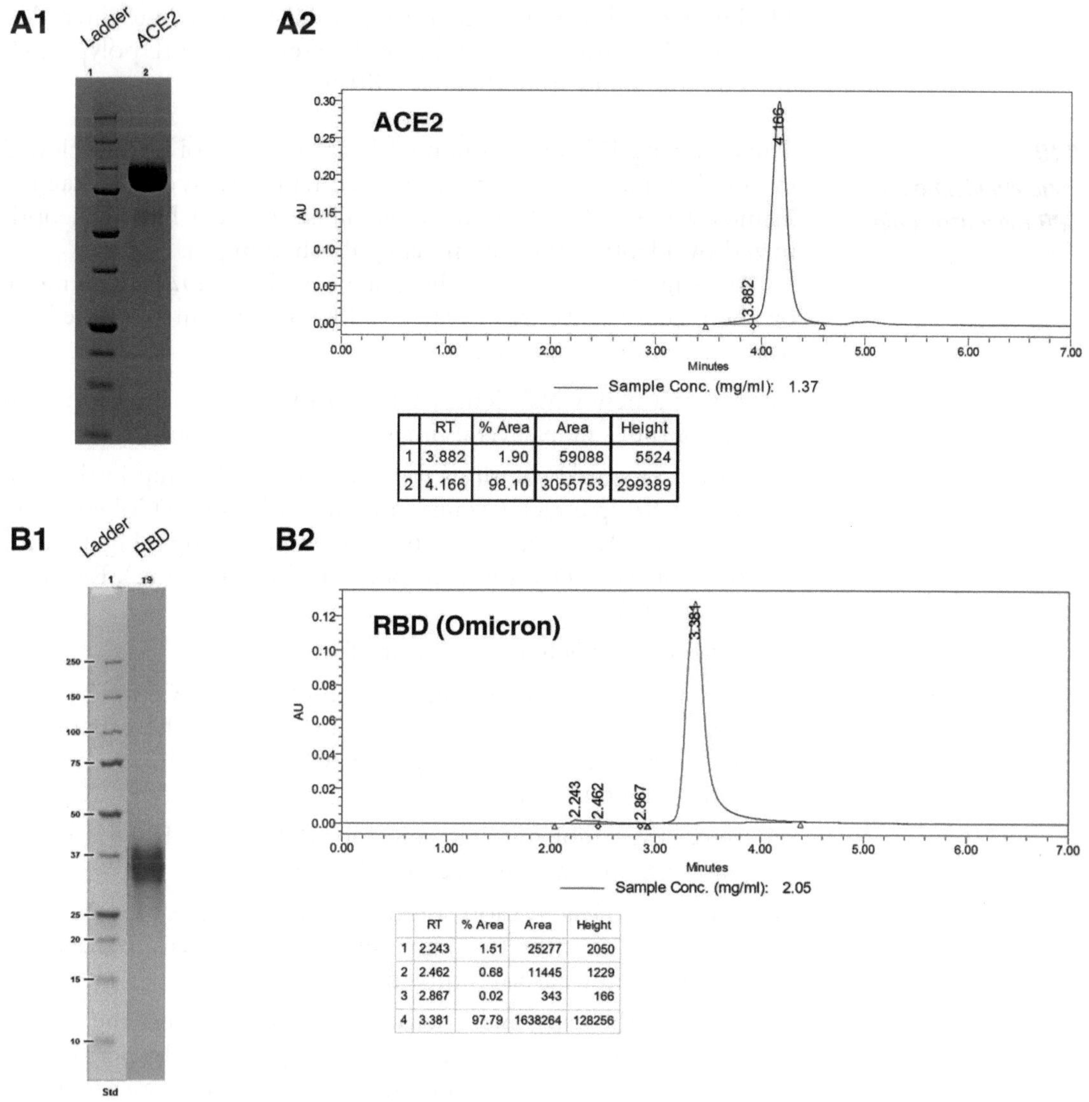

Fig. 1 Example of SDS-PAGE and SEC-UPLC pics corresponding to ACE2-E5 (Panels A1 and A2 respectively) and RBD variant Omicron B.1.1.529 (Panels B1 and B2 respectively)

Table 1
Size exclusion column selection and running conditions

Column	Maximum sample volume	Flow rate during cleaning/ equilibration	Flow rate during purification	Fraction size
HiPrep 16/60 Superdex 75 pg	5 mL	0.8 mL/min	1.0 mL/min	0.5 mL
HiPrep 26/60 Superdex 75 pg	12 mL	2.2 mL/min	2.5 mL/min	1.0 mL

11. Filter the desalted protein using a 0.22 μm syringe filter. Aliquot the purified protein in sterile screw-cap 2 mL polypropylene tubes and store them at −80 °C.

3.10 Functionalization of SPR Biosensor Chip

The following K5 covalent immobilization protocol was developed in our laboratory [6] for the subsequent display of E5-tagged ligands via coiled-coil interactions. This protocol has been optimized by adapting the parameters published in [7].

We suggest performing these steps in the *Manual Run* mode on the Biacore Software instead to allow better control of the level of reagent immobilization.

1. Insert a new CM5 Sensor Chip in the Biacore biosensor and prime the system several times with Running buffer.
2. Install prepared solutions to functionalize the chip in the biosensor Sample rack (Amine coupling solution, Thiol coupling solution, Deactivation solution (amine coupling), Ligand capture solution, Deactivation solution (thiol coupling), Regeneration buffer.
3. Set the analysis temperature at 25 °C.
4. Inject the Amine coupling solution (NHS/EDC) for 240 s at 5 μL/min over two sensor chip surfaces, i.e., the experiment and reference flow cells (e.g., 2 and 1, or 4 and 3) to activate carboxymethylated dextran.
5. Inject the Thiol coupling solution (PDEA) for 480 s at 5 μL/min over the two same sensor chip surfaces.
6. Inject the Deactivation solution (Ethanolamine) for 240 s at 5 μL/min over the two same sensor chip surfaces to inactivate the remaining carboxyl moieties.
7. Only on the experiment sensor chip surface, inject the Ligand capture solution (Cysteine-tagged K5 peptides) by performing pulses of 15 s at 10 μL/min until the signal increases to about 1300–1500 RUs (*see* **Note 11**).
8. Inject the Regeneration buffer for 20 s at 100 μL/min to remove K5 peptides that did not covalently bind to the surface.
9. Inject the Deactivation solution (L-Cysteine/NaCl) for 240 s at 5 μL/min to inactivate the remaining thiol moieties on both sensor chip surfaces.
10. Inject 3 pulses of Regeneration buffer (15 s each at 100 μL/min) on both sensor chip surfaces and verify the final level of K5 peptide immobilized on the experiment surface.
11. Prime the sensor chip with Running buffer several times before starting the kinetic experiments.

3.11 SPR Assay for ACE2-RBD Interaction Analysis

The following protocol is intended for SPR-based analysis of ACE2 and SARS-CoV-2 RBD interaction using a coiled-coil ligand-oriented capture strategy. In this method, ACE2-E5 receptors are stably captured on the K5 coil surface in an oriented manner and the RBD proteins are injected as analytes at various concentrations as depicted in Fig. 2.

1. Prime the K5 coil functionalized sensor chip with Running buffer.
2. Prepare 5 dilutions of RBD proteins in Running buffer, ranging from 0.5 to 100 nM (*see* **Note 12**).
3. Install the tubes corresponding to the ACE2-E5 receptors (diluted at 1 μg/mL), RBD dilutions, blank samples (Running buffer), and Regeneration buffer in the biosensor Sample Rack.
4. Set data collection rate at 10 Hz for each experiment cycle.
5. Set the desired experimental temperature (*see* **Note 13**).
6. Start each cycle with the injection of ACE2-E5 receptors on the experiment surface at 10 μL/min until about 60 RUs of ACE2-E5 receptors are captured (*see* **Note 14**).
7. On both experiment and reference surfaces, inject an analyte sample at a 50 μL/min long enough to reach a plateau (e.g., 460 s). This is followed by the injection of Running buffer long enough to bring the signal back to zero or to monitor signal decrease (e.g., 1800 s) (*see* **Note 15**).
8. Regenerate the surface with three 15 s pulses of Regeneration buffer at 100 μL/min at the end of each cycle.
9. Repeat each cycle (ligand injection of ACE2-E5, sample injection of RBD, regeneration, **steps 6–8**) at least 2 times for each concentration of RBD including the blank sample.
10. Repeat each experiment [9] independently at least 3 times with fresh ACE2-E5 and RBD dilutions.

3.12 SPR Data Analysis

The following protocol is intended for sensorgrams analysis to extract kinetic and thermodynamic constants relevant to the ACE2-RBD interaction.

1. Using the Biacore Evaluation Software, subtract reference surface data from experiment surface data and then subtract blank cycle. This process is called double-referencing [13].
2. As sensorgrams corresponding to multiple RBD concentrations were recorded, perform global fitting with the 1:1 kinetic model via the Biacore Evaluation Software. Examples of sensorgrams recorded at 10 and 25 °C fitted with a 1:1 kinetic model are shown in Fig. 3. Once the fit is satisfactory, both kinetic (k_{on} and k_{off}) and thermodynamic (K_D) constants can be identified (*see* **Note 16**).

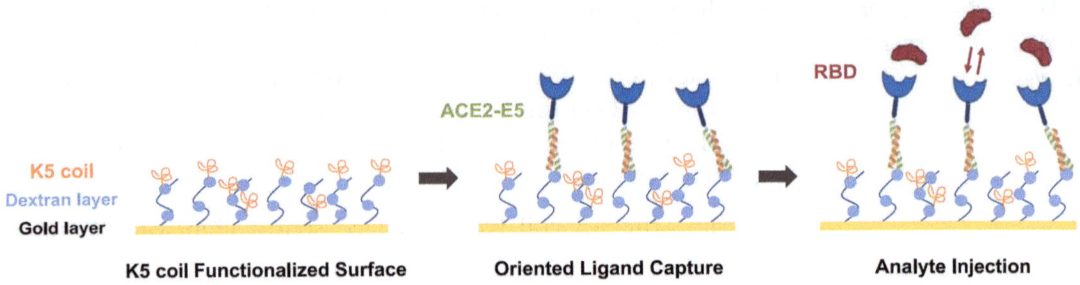

Fig. 2 SPR-based assay for the monitoring of ACE2-RBD interactions

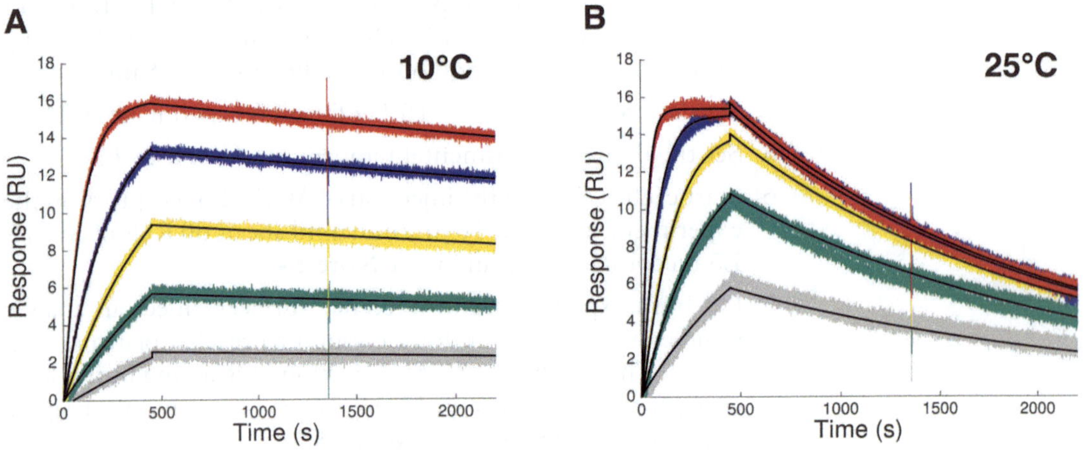

Fig. 3 Example of double-referenced SPR sensorgrams corresponding to the interaction of tethered ACE2-E5 and RBD (Omicron B.1.1.529), recorded at 10 °C (Panel A) and 25 °C (Panel B). RBD was injected at concentrations ranging from 1 to 25 nM. The sensorgrams were globally fitted with a 1:1 Langmuir binding model (solid black lines)

4 Notes

1. This rich media should generate >1 mg of plasmid DNA per 50 mL culture when using high-copy-number plasmids.
2. Amine coupling kit also available by Cytiva (1-ethyl-3-(3-dimethylaminopropyl)carbodiimide hydrochloride (EDC), N-hydroxysuccinimide (NHS), 1 M ethanolamine-HCl pH 8.5).
3. Cysteine-tagged K5 peptides dilution should be prepared rapidly and kept at −80 °C to limit oxidation of thiol groups.
4. 0.8× feed solution already contains 16 g/L.
5. Feed 4 already contains 0.1% w/v poloxamer surfactant. Final Kolliphor P188 concentration can be increased to 0.2% w/v, or

higher, by adding 1 g Kolliphor P188 per 1 L solution, depending on the desired final concentration.

6. When performing high-density transient transfection, plasmid DNA consisted of a mixture of 85% of pTT™-RBD or ACE2 constructs, 10% pTT™-Bcl-XL (anti-apoptotic effector) and 5% pTT™-GFP. Cells co-transfected with Bcl-xL shows reduced levels of apoptosis, increased specific productivity, and an overall increase in product yield. Moreover, adding 5% of a GFP-plasmid in the transfection mixture allows a visual confirmation of transfection efficiency using a fluorescence microscope or could also provide quantitative analysis if using flow cytometry, without significantly altering the expression of the gene of interest.
7. Final volume of DNA-PEI polyplexes should be 10% of the final culture volume and never process more than 6 transfections mix at the same time.
8. Glucose concentration should be equal or higher than 10 mM at all times during transient expression.
9. Filtered supernatants can be stored at 4 °C for short-term storage.
10. The previous purification step of Strep-tagged proteins should yield high purity. In this case, it is recommended to go directly to the SEC purification step to limit the number of steps and increase protein yield. If another step of purification is needed for different RBD constructs, we recommend this protocol using the FLAG tag.
11. We suggest thawing the Cysteine-tagged K5 peptides at the last minute and adding the sample to the biosensor Sample rack just before its injection on the sensor chip to limit oxidation of thiol groups. Prepare a larger volume than needed to limit manipulations and account for the dead volume needed in the biosensor sample tube for the injection needle.
12. Prepare independent dilutions from a stock solution to limit error propagation. Avoid performing serial dilutions. Choose the range of concentrations between 0.1 K_D and 10 K_D based on the theoretical affinity constant defining the interaction.
13. The experimental temperature can influence the quality of the fit to a 1:1 kinetic model [14]. Interactions between SARS-CoV-2 RBD variants and ACE2 receptors can be evaluated at lower temperatures such as 10 °C to slow down the interaction and allow for a better analysis of the binding, as shown in Fig. 2 [14].
14. Keeping a low and stable level of ligand for each cycle limits artifacts such as avidity and rebinding. We suggest conducting

a test to measure the time of injection of ligand required to obtain the same desired capture level for each cycle.

15. We suggest that the experiment is prepared in an automated method on the Biacore Software with set flow parameters and time of injections. It will limit manipulations and give an indication of the volume needed for each sample, as calculated by the biosensor software. A high flow rate decreases mass transport limitation artifacts. The time of sample and Running buffer injection can be adjusted to capture kinetics and reach plateau.
16. In this case, as the stoichiometry for ACE2-RBD interaction is 1:1, the kinetic model chosen to characterize the sensorgram data should follow a 1:1 Langmuir model. If the fit is suboptimal, verify possible artifacts such as aggregates, mass transport limitation, avidity and rebinding, and test different injection times and experimental temperatures before using another kinetic model. In such an event, the choice of another kinetic model should be based on sound biological interpretation of the mechanism of the analyte-ligand interaction for the identified parameters to be meaningful.

References

1. Barton MI, MacGowan SA, Kutuzov MA et al (2021) Effects of common mutations in the SARS-CoV-2 spike RBD and its ligand, the human ACE2 receptor on binding affinity and kinetics. elife 10:e70658. https://doi.org/10.7554/eLife.70658
2. Laffeber C, de Koning K, Kanaar R et al (2021) Experimental evidence for enhanced receptor binding by rapidly spreading SARS-CoV-2 variants. J Mol Biol 433(15):167058. https://doi.org/10.1016/j.jmb.2021.167058
3. Liu H, Zhang Q, Wei P et al (2021) The basis of a more contagious 501Y.V1 variant of SARS-CoV-2. Cell Res 31(6):720–722. https://doi.org/10.1038/s41422-021-00496-8
4. Crescenzo GD, Boucher C, Durocher Y et al (2008) Kinetic characterization by surface Plasmon resonance-based biosensors: principle and emerging trends. Cell Mol Bioeng 1(4):204–215. https://doi.org/10.1007/s12195-008-0035-5
5. Forest-Nault C, Gaudreault J, Henry O et al (2021) On the use of surface Plasmon resonance biosensing to understand IgG-FcγR interactions. Int J Mol Sci 22. https://doi.org/10.3390/ijms22126616
6. De Crescenzo G, Litowski JR, Hodges RS et al (2003) Real-time monitoring of the interactions of two-stranded de novo designed coiled-coils: effect of chain length on the kinetic and thermodynamic constants of binding. Biochemistry 42(6):1754–1763. https://doi.org/10.1021/bi0268450
7. Murschel F, Liberelle B, St-Laurent G et al (2013) Coiled-coil-mediated grafting of bioactive vascular endothelial growth factor. Acta Biomater 9(6):6806–6813. https://doi.org/10.1016/j.actbio.2013.02.032
8. Cambay F, Henry O, Durocher Y et al (2019) Impact of N-glycosylation on Fcγ receptor/IgG interactions: unravelling differences with an enhanced surface plasmon resonance biosensor assay based on coiled-coil interactions. MAbs 11(3):435–452. https://doi.org/10.1080/19420862.2019.1581017
9. Litowski JR, Hodges RS (2001) Designing heterodimeric two-stranded α-helical coiled-coils: the effect of chain length on protein folding, stability and specificity. J Pept Res 58(6):477–492. https://doi.org/10.1034/j.1399-3011.2001.10972.x
10. L'Abbé D, Bisson L, Gervais C et al (2018) Transient gene expression in suspension HEK293-EBNA1 cells. In: Hacker DL (ed) Recombinant protein expression in mammalian cells: methods and protocols. Springer, New York, pp 1–16

11. Poulain A, Perret S, Malenfant F et al (2017) Rapid protein production from stable CHO cell pools using plasmid vector and the cumate gene-switch. J Biotechnol 255:16–27. https://doi.org/10.1016/j.jbiotec.2017.06.009
12. Stuible M, Burlacu A, Perret S et al (2018) Optimization of a high-cell-density polyethylenimine transfection method for rapid protein production in CHO-EBNA1 cells. J Biotechnol 281:39–47. https://doi.org/10.1016/j.jbiotec.2018.06.307
13. Rich RL, Myszka DG (2000) Advances in surface plasmon resonance biosensor analysis. Curr Opin Biotechnol 11(1):54–61. https://doi.org/10.1016/S0958-1669(99)00054-3
14. Forest-Nault C, Koyuturk I, Gaudreault J et al (2022) Impact of the temperature on the interactions between common variants of the SARS-CoV-2 receptor binding domain and the human ACE2. Sci Rep 12(1):11520. https://doi.org/10.1038/s41598-022-15215-5

Part II

Protozoal and Nematode Glycoproteins and Polysaccharide Conjugates

Chapter 7

Production and Purification of Plasmodium Circumsporozoite Protein in *Lactococcus lactis*

Mohammad Naghizadeh, Susheel K. Singh, Jordan Plieskatt, Ebenezer Addo Ofori, and Michael Theisen

Abstract

Malaria is a vector-borne disease caused by *Plasmodium* parasites of which *Plasmodium falciparum* contributed to an estimated 247 million cases worldwide in 2021 (WHO malaria report 2022). The *P. falciparum* Circumsporozoite protein (*Pf*CSP) covers the surface of the sporozoite which is critical to cell invasion in the human host. *Pf*CSP is the leading pre-erythrocytic vaccine candidate and forms the basis of the RTS'S (Mosquirix®) malaria vaccine. However, high-yield production of full-length *Pf*CSP with proper folding has been challenging. Here, we describe expression and purification of full-length *Pf*CSP (containing 4 NVDP and 38 NANP repeats) with proper conformation by a simple three-step procedure in the *Lactococcus lactis* expression system.

Key words Circumsporozoite protein (CSP), *Lactococcus lactis*, Cloning, Recombinant protein expression, Purification, malaria, *Plasmodium falciparum*

1 Introduction

From a manufacturing perspective, selecting the optimal expression system for the production of recombinant proteins is pivotal for correct protein folding and cost-effective vaccine manufacturing, especially in low-income countries.

Lactococcus lactis (*L. lactis*) has been described as a safe, Gram-positive, non-pathogenic microorganism for the production of recombinant proteins at lab-scale [1–8]. The *L. lactis* expression system provides an efficient low-cost production system for heterogenous proteins because (1) it does not produce endotoxins and extracellular proteases, (2) it does not perform unwanted glycosylation, and (3) it can secret the recombinant protein into the culture supernatant, which facilitates upstream and downstream processing. Accordingly, we have used the *L. lactis* expression system for the manufacturing of multiple malaria vaccine candidates under

Steven B. Bradfute (ed.), *Recombinant Glycoproteins: Methods and Protocols*, Methods in Molecular Biology, vol. 2762,
https://doi.org/10.1007/978-1-0716-3666-4_7,

current Good Manufacturing Practices (cGMP) [1, 8, 9]. Here, we describe lab-scale production of the full-length *P. falciparum Circumsporozoite* protein (*Pf*CSP) in the *L. lactis* expression system.

*Pf*CSP is the most abundant protein expressed on the surface of the malaria sporozoite, and plays crucial roles in sporozoite development in the mosquito vector, motility, and host invasion [10–12]. *Pf*CSP can be divided into three domains: (1) a flexible N-terminal domain with a heparin sulfate binding site for hepatocyte attachment [13], (2) a structurally disordered central region consisting of several four-amino-acid NANP motifs that can vary among Pf. isolates and a small number of NVDP motifs [14], and (3) a structured C-terminal domain with a thrombospondin-like type I repeat (αTSR) [14]. *Pf*CSP is an attractive target for malaria vaccine development and serves as the basis for RTS'S, the first malaria vaccine to be tested in Phase 3 clinical trials (Mosquirix™) [14]. Despite multiple heterologous expression systems including *Escherichia coli* [15–17], baculovirus (Sf9) cells [18], *P. fluorescens* [18], and *Pichia pastoris* [18], full-length *Pf*CSP has been proven to be a difficult protein to express. This is possibly because of the difficulties in the formation of correctly folded protein, low yield, and stability of full-length soluble recombinant *Pf*CSP. Such challenges have often led to truncated *Pf*CSP constructs selected for manufacture [18].

In this chapter, we describe a step-by-step protocol for obtaining high yields of full-length recombinant *Pf*CSP (containing 4 cysteines, 38 NANP, and 4 NVDP repeats encompassing amino acids 26–383 of the native molecule) in *L. lactis* [18]. The central parts of the experimental procedure are presented in Fig. 1. The protocol uses a p170 promoter–pH-based and growth phase induction system for expressing the recombinant protein in *L. lactis* [18]. We describe a simple workflow using batch fermentation in a stirred bioreactor and a simple two-step purification process. The whole procedure of protein expression and purification consists of three main parts: (1) plasmid construction, (2) *L. lactis* transformation, and (3) simple chromatography-based purification.

2 Materials

Prepare all solutions using ultrapure water (18 MΩ-cm at 25 °C) and analytical-grade reagents. Store all reagents and solutions at 4 ° C unless indicated otherwise. Carefully follow material storage, labeling, and waste management regulations at your workplace.

2.1 General Supplies

1. Horizontal and vertical gel electrophoresis with appropriate power pack.
2. Polymerase chain reaction (PCR) machine (Thermocycler).

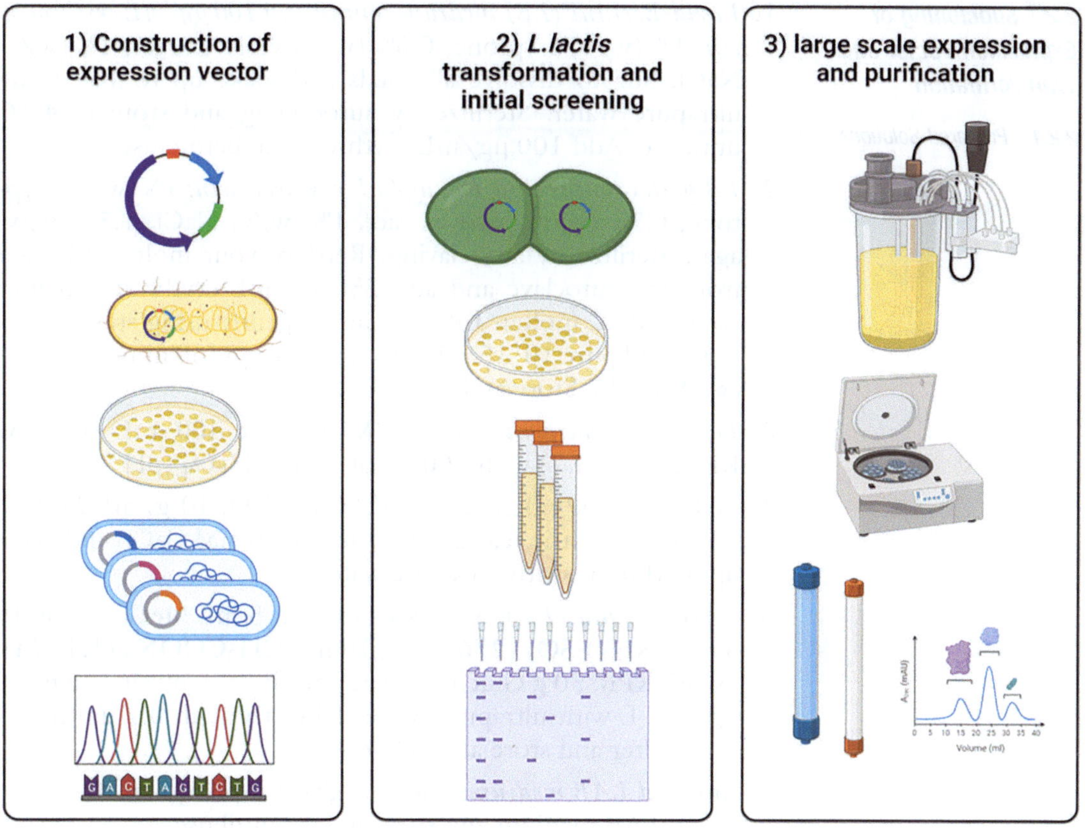

Fig. 1 Flowchart of experimental procedures

3. Immunoblotting apparatus.
4. Nanodrop/Spectrophotometer.
5. Gel-documentation system/Chemi-Doc.
6. Tabletop centrifuge and a standard centrifuge.
7. Incubator and shaker incubator.
8. Heating block.
9. Vortex.
10. Electroporation device.
11. pH meter.
12. PCR plates and microtubes.
13. Electroporation cuvettes (2 mm).
14. Freezers (−20 and −80 °C).
15. Fermenter.
16. Tangential flow filtration (TFF).
17. Protein purification system.

2.2 Subcloning of Expression Vector and Transformation

2.2.1 Prepared Solutions

1. *Luria-Bertani (LB) medium containing 100 μg/mL erythromycin:* 1% (w/v) tryptone, 0.5% (w/v) yeast extract, 1% (w/v) NaCl. Stir to dissolve all solids and make up to 1 L using ultrapure water. Sterilize by autoclaving and store at 4 °C until use. Add 100 μg/mL erythromycin before use.
2. *LB plates containing 100 μg/mL erythromycin:* 1% (w/v) tryptone, 0.5% (w/v) yeast extract, 1% (w/v) NaCl, 1.5% (w/v) agar. Sterilize by autoclaving. Retrieve your molten LB agar from the autoclave and add 250 μg/mL erythromycin into your ~60 °C molten LB agar, mix it gently using sterile technique and pure 10–15 mL into sterile plates (60 mm × 15 mm) (*see* **Note 1**). Store at 4 °C.
3. *Potassium phosphate buffer (KPB)*: mix 400 mL of 1 M KH_2PO_4 (pH 7) with 600 ml of 1 M $K_2H_PO_4$ (pH 7).
4. *MeCit (×1000):* 0.28 g/mL $FeSO_4{\cdot}7H_2O$, 10 g/mL $MgCl_2$, 2.94 g/mL Citric acid·$2H_2O$, 1 mL of 50 mM $CaCl_2$ and make up to 100 mL with ultrapure water.
5. *LAB medium (1 L):* 35 g yeast extract, 5 mL MeCit (×1000), 5 mL $(NH_4)_2SO_4$ (276 mM), 10 mL CH_3COONa (1.47 M), 20 mL KPB, 50 g Glucose, 10 μg/mL Erythromycin and make up to 1 L with ultrapure water. Filter through 0.2 μm membrane filter and store at 4 °C until use.
6. *Buffered LAB medium*: Dissolve glycerol-phosphate 4% (w/v) in 1 L LAB medium and store at 4 °C until use.
7. *Buffered LAB agar plates*: Add 15 g/L agar to the buffered LAB medium. Sterilize by autoclaving. Add 1 μg/mL erythromycin into your ~60 °C molten buffered LAB agar, mix it gently using sterile technique and pour 10–15 mL into sterile plates (60 mm × 15 mm). Store at 4 °C until use.

2.2.2 Other Reagents and Materials

1. Expression vector pSS1.
2. *Pf*CSP gene codon optimized for *L. lactis*.
3. Chemical Competent Cells (*Escherichia coli* DH5-alpha).
4. *Lactococcus lactis (L. lactis)* MG1363.
5. Restriction enzymes (*BglII, SalI*) with their respective recommended buffers.
6. High-fidelity Taq DNA polymerase.
7. T4 DNA Ligase with the respective recommended buffer.
8. TEMPase Hot Start DNA Polymerase.
9. DNA gel extraction Kit.
10. DNA Cleanup Kit.
11. Plasmid DNA Miniprep Kit.
12. Sodium dodecyl sulfate–polyacrylamide (SDS) Gel staining solution: Coomassie Brilliant Blue (Bio-Rad).

2.3 Small-Scale Expression and Cell Bank

1. Buffered LAB medium (*see* Subheading 2.1).
2. Sterile 50% (w/v) Glucose in ultrapure water.
3. Sterile 50% Glycerol (vegetable based) in ultrapure water.
4. SDS Gel staining solution: Coomassie Brilliant Blue (Bio-Rad).

2.4 Fermentation

1. LAB medium (*see* Subheading 2.1).
2. 2 M NaOH: 80 g NaOH in 1 L ultrapure water.
3. Sterile 50% (w/v) Glucose in ultrapure water.
4. A benchtop autoclavable fermenter/bioreactor.
5. Large-capacity bioprocessing centrifuge.
6. QuixStand Benchtop system (hollow fiber cartridge with cutoff at 30 kDa; surface area, 650 cm^2; GE Healthcare).
7. 500 mL Bottle Top Vacuum Filter (0.2 μm Pore 33.2cm^2 Nylon Membrane).

2.5 Protein Purification

1. 5 mL Ni^{++}–nitrilotriacetic acid column (HisTrap HP; GE Healthcare).
2. 5 mL cation exchange column (HiTrap SP HP column; GE Healthcare).
3. HisTrap binding buffer: 20 mM Hepes, 50 mM NaCl, 15 mM imidazole, and 5% Glucose (pH7).
4. HisTrap elution buffer: 20 mM Hepes, 50 mM NaCl, 700 mM imidazole, and 5% Glucose (pH7).
5. HiTrap SP HP-binding buffer: 20 mM HEPES, 1 mM EDTA, and 5% Glucose (pH7).
6. HiTrap SP HP elution buffer: 20 mM HEPES, 1 mM EDTA, 1 M NaCl, and 5% Glucose (pH7).
7. Vivaspin concentration device.

3 Methods

3.1 Subcloning of Expression Vector for the Recombinant Protein and Transformation into L. Lactis

3.1.1 Subcloning

1. Subclone a codon-optimized *Pf*CSP$_{26–383}$ (containing 4 NDVP and 38 NANP repeats) DNA fragment (NCBI reference sequence XM_001351086.1) (see [4]) into the pSS1 vector (Fig. 1) for protein expression (*see* **Note 1**).
2. Transform the ligated plasmid into competent *E. coli* (*see* **Note 2**), and plate on LB agar plates.
3. Incubate overnight at 37 °C and select positive transformants. Colony PCR or other simple verification methods can be used to screen for positive insert, for instance digestion with restriction enzymes.

4. Inoculate 3–5 positive transformants into 5 mL LB medium and culture 12–16 h with agitation (150 rpm) at 37 °C.
5. Extract plasmids via Plasmid DNA Miniprep kit and verify the inserted DNA fragment by sequencing (*see* **Note 3**).

3.1.2 Transformation

1. Thaw a frozen competent cell on ice (*see* **Note 4**).
2. Mix 2–4 μL of the plasmid DNA with 40 μL of the competent cells.
3. Transfer the mixture to an ice-chilled electroporation cuvette.
4. Pre-set an electroporation device: Voltage 2 kV, Capacitance 25 μF, Resistance 200 Ω, and time constant 4–5 ms.
5. Pulse once and immediately add 0.96 mL of ice-cold buffered LAB medium to the cuvette.
6. Transfer the mixture into a 15 mL tube and incubate for 2 h at 30 °C.
7. Centrifuge the 15 mL tube at 4000 × *g* for 10 min at room temperature.
8. Remove supernatant and resuspend pelleted bacterial cells with 200 μL LAB medium.
9. Plate the cell suspension on the buffered LAB agar plate.
10. Incubate the plate at 30 °C for up to 48 h.

3.2 Small-Scale Culture for Protein Production and Cell Banking

1. Inoculate 10–20 positive transformants into 5 mL buffered LAB medium in a 15 mL centrifuge tubes.
2. Incubate at 30 °C overnight.
3. Centrifuge the tubes at 4000 × *g* for 10 min at 4 °C, and transfer supernatants into new 15 mL tubes.
4. Remove 20 μL supernatant and mix with 4 μL of 6× SDS-PAGE loading buffer in another 1.5 mL tube.
5. Heat the sample for 10 min at 95 °C.
6. Prepare a 10% gel for SDS-PAGE and load the sample.
7. Run gel in 1× SDS-PAGE running buffer at a constant voltage of 170 V for 55 min.
8. Stain gel with 10 mL Instant Blue Stain Reagent for 30 min at room temperature.
9. Wash the gel with double-distilled water until the background of the gel turns transparent.
10. Identify whether the target protein is expressed by comparing the sample with control, as shown in Fig. 2.
11. Select the corresponding colony whose culture medium showed most prominent protein band with least host protein bands on the SDS-PAGE gel (*see* **Note 5**).

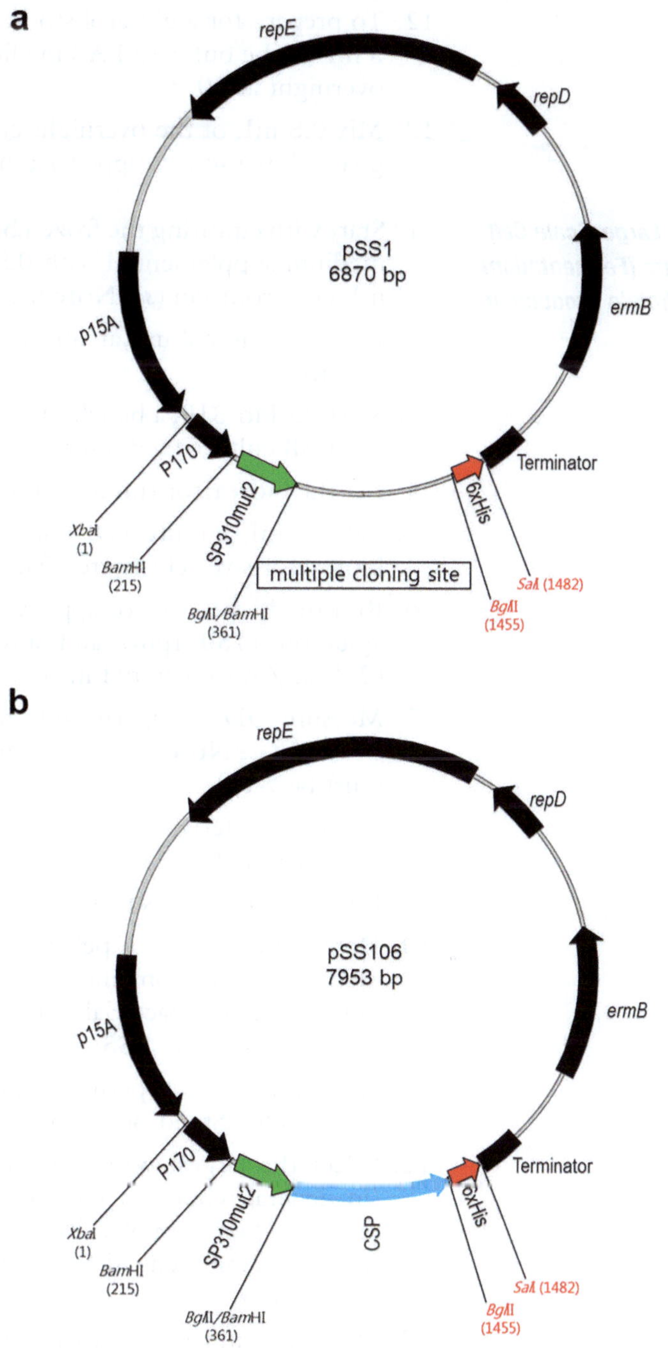

Fig. 2 *Lactococcus lactis* vector map and screening for small-scale expression of *Pf*CSP. (**a**) Schematic representation of pSS1 vector map showing structural features including open reading frames. p. 170: lactate-inducible promoter; SP310mut2: secretion signal; 6×His (Histidine Tag); terminator: transcriptional terminator; ermB: Erythromycin resistance gene; repE; repD: *L. lactis* replicon and p15A: *E. coli* replicon. (**b**) pSS106: pSS1 with CSP$_{26-383}$ as a fusion protein

12. To prepare for a glycerol stock, culture the selected colony in 5 mL of the buffered LAB medium with appropriate antibiotic overnight at 30 °C.
13. Mix 0.5 mL of the overnight culture with 0.5 mL sterile 50% glycerol in a screw-capped 1.5 mL tube, and store at −80 °C.

3.3 Large-Scale Cell Culture (Fermentation) for Protein Production

1. Start with culturing the frozen bacteria in a 5 mL buffered LAB medium supplemented with 0.5% (w/v) glucose and 10 μg/mL erythromycin (*see* **Note 6**).
2. Incubate the culture at 30 °C until the OD_{600nm} of ~0.6 is reached.
3. Start BioFlo-310, a benchtop, autoclavable bioreactor and add the LAB culture medium to the bioreactor's vessel.
4. Let the bioreactor run for 1 h to stabilize the system.
5. Take 1 mL culture (OD_{600nm} of ~0.6) and inject into the bioreactor's vessel (*see* **step 2**).
6. Run the bioreactor for approx. 14–18 h at 30 °C with gentle agitation (150 rpm) and a constant 50% glucose intake (200 μL/min) with pH maintained at 6.5 by 2 M NaOH.
7. Measure OD of supernatant and harvest the bioreactor after 14–18 h (*see* **Note 7**) (in an optimal fermentation the OD_{600nm} must be >13).
8. Remove bacterial cell pellet by centrifuging at 9000 × *g*, for 25 min at 4 °C.
9. If the supernatant is turbid, repeat the centrifugation step.
10. Determine weight of pellet by subtracting the weight of an identical, empty container from the weight of the one containing the pelleted bacterial cell (for an optimal fermentation the cell pellet must be >35 g).
11. Concentrate the supernatant (1400 mL) to approx. 200 mL using a QuixStand Benchtop system at 4 °C.
12. Select the largest pore size that retains the target molecule (in general, choose a membrane 2× smaller than the target protein). For example, 30 kDa NMWC ultrafiltration membrane is recommended for CSP (~60 kDa) concentration and diafiltration.
13. Diafiltrate and replaced the sample buffer with 1 L HisTrap binding buffer.
14. After buffer exchange remove sample from QuixStand and filter with 500 mL Bottle Top Vacuum Filter (0.2 μm Pore 33.2cm^2 Nylon Membrane).
15. Store at 4 °C until purification.

3.4 Protein Purification: Affinity Chromatography

1. Connect sample and buffer inlet and outlet tubing in an AKTAxpress pilot system (GE Healthcare) or other protein purification system.
2. Attach a 5 mL HisTrap HP (GE Healthcare) column to the AKTA system (*see* **Note 8**).
3. Perform a system wash to fill the system with the buffers and the sample.
4. Wash the HisTrap column with at least 5 column volumes (CV) filter-sterilized, double-distilled water and equilibrate with 5 CV of the HisTrap binding buffer.
5. Load the sample onto the column using 4 mL flow rate to allow sufficient binding of the target protein to the column.
6. Flush unbound samples from the column with extra HisTrap binding buffer (5 CV) and collect in a flow-through collection flask to be used as the "washing" sample for SDS-PAGE (*see* **Note 9**).
7. Collect the flow-through and the washing fraction and use 20 μL as the "HisTrap binding flow-through" sample for SDS-PAGE.
8. Then elute the protein with a linear gradient of 0–100% the elution buffer (8 CV).
9. Maintain the same flow rate (4 mL/min) and collect 2 mL eluate per fraction.
10. Select the elution fractions based on UV (280 nm) signal and analyze the fractions for impurities by SDS-PAGE.
11. Identify and pool the fractions containing the target protein (*see* Fig. 3, Peak 2).
12. Proceed to a polishing step using a cation exchange chromatography (HiTrap SP HP column; GE Healthcare) to further remove host cell proteins (*see* **Note 10**).
13. Elute the bound protein from the column using a linear gradient elution with 8 CV of the HiTrap SP HP elution buffer.
14. Identify elution fractions containing the recombinant protein and assess the purity and recovery of the target protein by the SDS-PAGE gel analysis according to the UV profile and combine all pure fractions for the proceeding concentration step, as shown in Figs. 4 and 5.
15. Concentrate the protein fractions using a Vivaspin centrifugal filter or any other centrifugal filter with 10 kDa cut-off and measure the final protein concentration using a NanoDrop.
16. Aliquot protein into small volumes, and store at −80 °C.

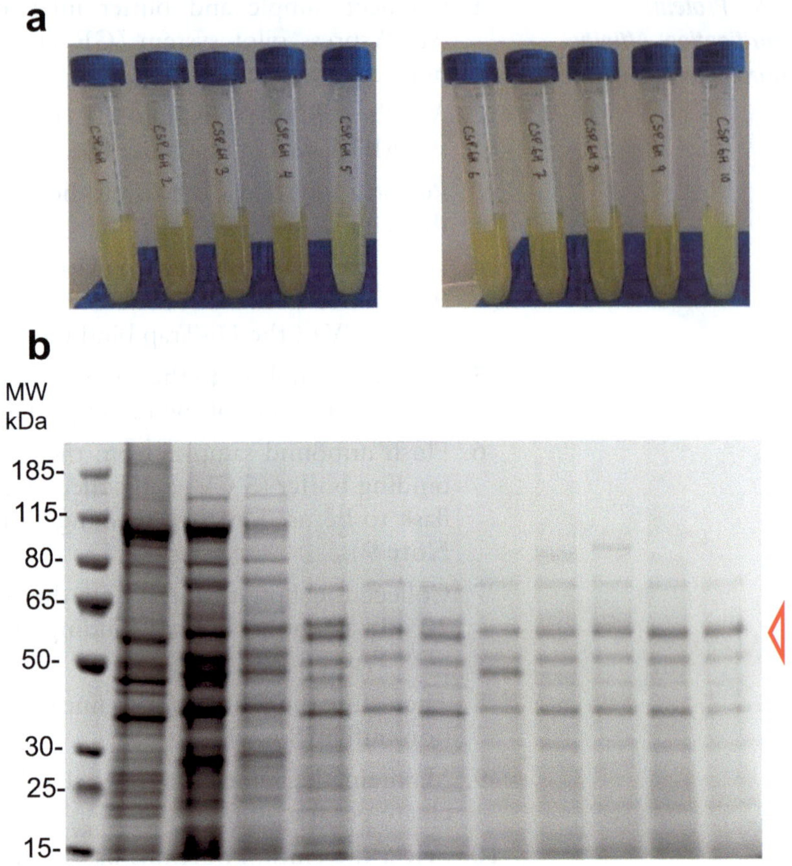

Fig. 3 Screening for a colony with the highest protein expressing in *L. lactis*. (**a**) Small-scale culture (5 mL) for the protein expression. (**b**) Coomassie blue-stained 4–12.5% SDS gel of culture supernatants (20 μL) of *L. lactis* strains expressing CSP26-383mat 30 °C. The supernatants were loaded in each lane with (+) or without (−) DTT. The sizes (kDa) of the molecular mass markers are indicated

4 Notes

1. The pSS1 vector (−6× His) is modified from pSM1013. The linearized vector is around ~6.7 kb, with a *Bgl*II restriction site in front of the open reading frame immediately downstream of the secretion signal-peptide sequence (SP310mut2) as well as a *Sal*I restriction site after the open reading frame. Several subcloning strategies can be chosen, such as double digestion using restriction enzymes assembly. pSM1013: *L. lactis* high copy-number secretion plasmid based on the pAMβ1 *L. lactis* replicon and p15A: *E. coli* replicon. The plasmid contains multiple cloning sites for insertion of a gene into unique restriction site with an inducible promoter (*SP310mut2*) (Fig. 1).

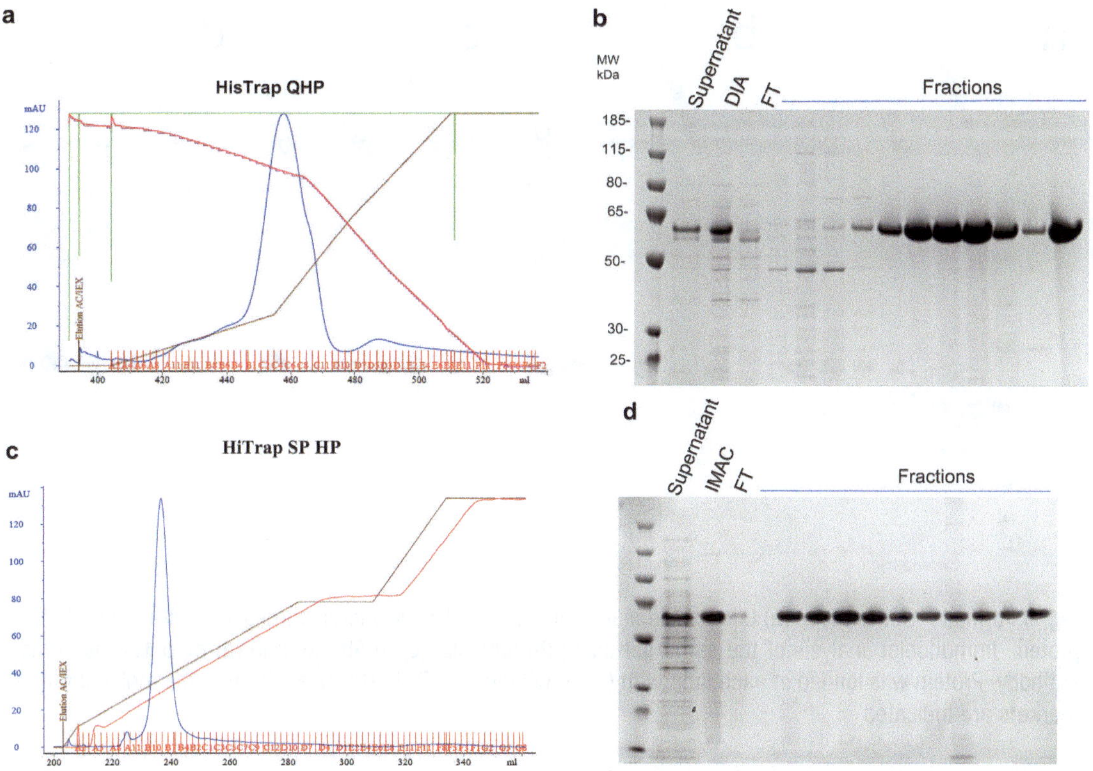

Fig. 4 Purification of *Pf*CSP. Elution profile of (**a**) Capture of full-length of *Pf*CSP (IMAC) and (**b**) pooled fraction analysis by Coomassie stained gel. Elution profiles of (**c**) polishing (IEC) of final product and (**d**) pooled fraction analysis by Coomassie stained gel. UV absorbance (Blue line), Gradient (Brown line), and conductivity (Red line). Pooled fractions from each step (5 or 10 μL) were analyzed by Coomassie blue-stained 4–12.5% polyacrylamide gels. The sizes (kDa) of the molecular mass markers are indicated

2. DH5α competent *E. coli* or any standard laboratory *E. coli* strains may be used for general cloning and subcloning applications.
3. Sequencing primer set specific for vector backbone for verification of a positive insert: Forward primer 5′-TTGCCATTTG TTAACGCT 3′ and reverse primer 5′ ATCTTTTGAAAAT TAACGT-3′.
4. The competent cells can be inoculated at the same time in a different test tube for the small-scale protein expression test as a control. It usually takes around 16 h for the culture to reach a saturation point (OD_{600} 6 to 8).
5. If the protein is not easily distinguished, immunoblotting (i.e., Western blotting) or ELISA is recommended to detect protein using anti-HIS antibody.
6. Scrape the surface of the frozen stock by a sterile loop and streak it and transfer directly into a 5 mL buffered LAB

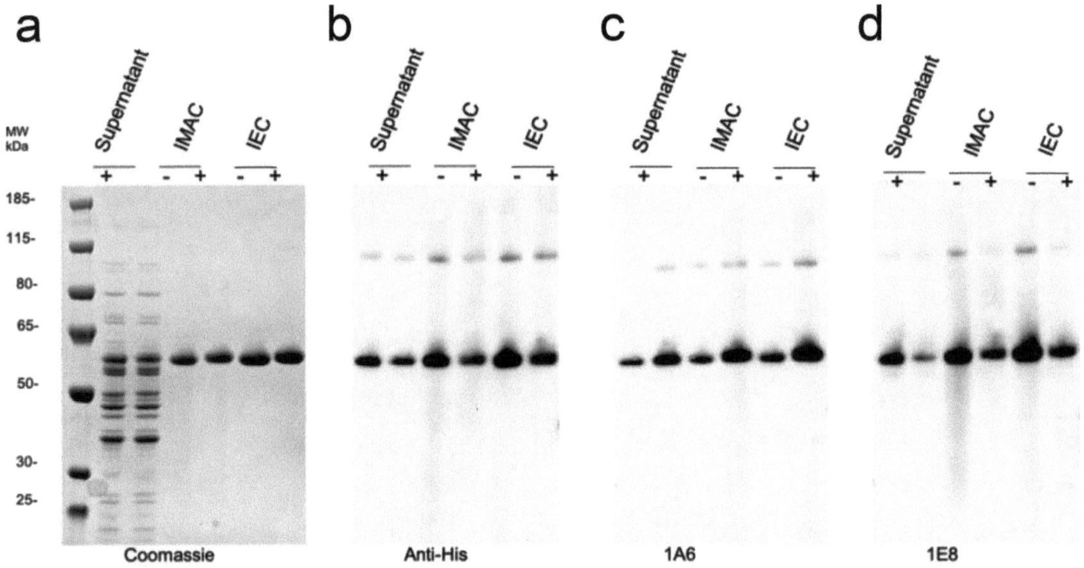

Fig. 5 Purified *Pf*CSP protein. (**a**) Coomassie blue-stained 4–12.5% polyacrylamide gel of full-length of *Pf*CSP protein. Immunoblot analysis of the same gel using (**b**) anti-His, (**c**), mAb1A6, and (**d**) mAb 1E8 as primary antibody. Protein was loaded in each lane with (+) or without (–) DTT. The sizes (kDa) of the molecular mass markers are indicated

medium with 0.5% (w/v) glucose and erythromycin 10 μg/mL. Avoid excessive thawing of the glycerol stock.

7. It is very important to harvest fermenter no later than 18 h. This is because over-fermentation increases the host cell proteins, debris, and genomic DNA in the supernatant due to cell death. Usually, 2 h after 2 M NaOH intake has stopped, is time to stop fermentation.
8. The HisTrap column size can be chosen based on availability and estimated recombinant protein expression. In this protocol, a 5 mL HisTrap column with a total binding capacity of at least 40 mg His tag protein per ml resin was used.
9. Low concentration of imidazole (5–20 mM) in the binding buffers helps reduce nonspecific binding of host proteins containing non-contiguous histidine residues on the transition metal.
10. Check the conductivity of the pooled fractions by a conductivity Meter. If the conductivity of the pooled fractions is above 10 mS/cm, dilute the pooled fractions with HiTrap SP HP binding buffer until it is below 10 mS/cm.

References

1. Singh SK, Plieskatt J, Chourasia BK et al (2021) Preclinical development of a Pfs230-Pfs48/45 chimeric malaria transmission-blocking vaccine. NPJ Vaccines 6:120. https://doi.org/10.1038/s41541-021-00383-8
2. Singh SK, Plieskatt J, Chourasia BK et al (2020) A reproducible and scalable process for manufacturing a Pfs48/45 based plasmodium falciparum transmission-blocking vaccine. Front Immunol 11:606266. https://doi.org/10.3389/fimmu.2020.606266
3. Garcia-Senosiain A, Kana IH, Singh SK et al (2020) Peripheral Merozoite surface proteins are targets of naturally acquired immunity against malaria in both India and Ghana. Infect Immun 88. https://doi.org/10.1128/IAI.00778-19
4. Chourasia BK, Deshmukh A, Kaur I et al (2020) Plasmodium falciparum Clag9-associated PfRhopH complex is involved in Merozoite binding to human erythrocytes. Infect Immun 88. https://doi.org/10.1128/IAI.00504-19
5. Singh SK, Thrane S, Chourasia BK et al (2019) Pfs230 and Pfs48/45 fusion proteins elicit strong transmission-blocking antibody responses against plasmodium falciparum. Front Immunol 10:1256. https://doi.org/10.3389/fimmu.2019.01256
6. Singh SK, Roeffen W, Mistarz UH et al (2017) Construct design, production, and characterization of plasmodium falciparum 48/45 R0.6C subunit protein produced in Lactococcus lactis as candidate vaccine. Microb Cell Factories 16:97. https://doi.org/10.1186/s12934-017-0710-0
7. Singh SK, Singh V (2022) Method for production of cysteine-rich proteins in Lactococcus lactis expression system. Methods Mol Biol 2406:189–203. https://doi.org/10.1007/978-1-0716-1859-2_11
8. Esen M, Kremsner PG, Schleucher R et al (2009) Safety and immunogenicity of GMZ2—a MSP3-GLURP fusion protein malaria vaccine candidate. Vaccine 27:6862–6868. https://doi.org/10.1016/j.vaccine.2009.09.011
9. Singh SK, Plieskatt J, Chourasia BK et al (2020) The plasmodium falciparum circumsporozoite protein produced in Lactococcus lactis is pure and stable. J Biol Chem 295:403–414. https://doi.org/10.1074/jbc.RA119.011268
10. Cerami C, Frevert U, Sinnis P et al (1992) The basolateral domain of the hepatocyte plasma membrane bears receptors for the circumsporozoite protein of plasmodium falciparum sporozoites. Cell 70:1021–1033. https://doi.org/10.1016/0092-8674(92)90251-7
11. Ménard R, Sultan AA, Cortes C et al (1997) Circumsporozoite protein is required for development of malaria sporozoites in mosquitoes. Nature 385:336–340. https://doi.org/10.1038/385336a0
12. Frevert U, Sinnis P, Cerami C et al (1993) Malaria circumsporozoite protein binds to heparan sulfate proteoglycans associated with the surface membrane of hepatocytes. J Exp Med 177:1287–1298. https://doi.org/10.1084/jem.177.5.1287
13. Ancsin JB, Kisilevsky R (2004) A binding site for highly sulfated heparan sulfate is identified in the N terminus of the circumsporozoite protein: significance for malarial sporozoite attachment to hepatocytes. J Biol Chem 279:21824–21832. https://doi.org/10.1074/jbc.M401979200
14. Zavala F, Tam JP, Hollingdale MR et al (1985) Rationale for development of a synthetic vaccine against *plasmodium falciparum* malaria. Science 80(228):1436–1440. https://doi.org/10.1126/science.2409595
15. Young JF, Hockmeyer WT, Gross M et al (1985) Expression of plasmodium falciparum circumsporozoite proteins in Escherichia coli for potential use in a human malaria vaccine. Science 228:958–962. https://doi.org/10.1126/science.2988125
16. Herrera R, Anderson C, Kumar K et al (2015) Reversible conformational change in the plasmodium falciparum circumsporozoite protein masks its adhesion domains. Infect Immun 83:3771–3780. https://doi.org/10.1128/IAI.02676-14
17. Plassmeyer ML, Reiter K, Shimp RLJ et al (2009) Structure of the plasmodium falciparum circumsporozoite protein, a leading malaria vaccine candidate. J Biol Chem 284:26951–26963. https://doi.org/10.1074/jbc.M109.013706
18. Kedees MH, Azzouz N, Gerold P et al (2002) Plasmodium falciparum: glycosylation status of plasmodium falciparum circumsporozoite protein expressed in the baculovirus system. Exp Parasitol 101:64–68. https://doi.org/10.1016/s0014-4894(02)00030-9

References

Chapter 8

Analysis of Caenorhabditis Protein Glycosylation

Katharina Paschinger, Jorick Vanbeselaere, and Iain B. H. Wilson

Abstract

Glycoproteins result from post-translational modification of proteins by glycans attached to certain side chains, with possible heterogeneity due to different structures being possible at the same glycosylation site. In contrast to the mammalian systems, analysis of invertebrate glycans presents a challenge in analysis as there exist unfamiliar epitopes and a high degree of structural and isomeric variation between different species—*Caenorhabditis elegans* is no exception. Simple screening using lectins and antibodies can yield hints regarding which glycan epitopes are present in wild-type and mutant strains, but detailed analysis is necessary for determining more exact glycomic information. Here, our analytical approach is to analyze N- and O-glycans involving "off-line" RP-HPLC MALDI-TOF MS/MS. Enrichment and labeling steps facilitate the analysis of single structures and provide isomeric separation. Thereby, the "simple" worm expresses over 200 N-glycan structures varying depending on culture conditions or the genetic background.

Key words Glycosylation, Mass spectrometry, "Off-line" MALDI-TOF MS/MS

Abbreviations

DTT	dithiothreitol
HRP	horseradish peroxidase
MALDI-TOF MS	matrix-assisted laser-desorption/ionization time-of-flight mass spectrometry
NPGC	non-porous graphitized carbon
PA	pyridylamino
PC	phosphorylcholine
RP-HPLC	reversed phase high-pressure liquid chromatography
SDS-PAGE	sodium dodecyl sulfate polyacrylamide gel electrophoresis

1 Introduction

Post-translational modifications of proteins lead to an immense diversity of proteoforms many times more than encoded directly

Steven B. Bradfute (ed.), *Recombinant Glycoproteins: Methods and Protocols*, Methods in Molecular Biology, vol. 2762, https://doi.org/10.1007/978-1-0716-3666-4_8,

by the 20,000 or so genes in a typical multicellular eukaryotic organism. There is no direct template for these: it is the interplay of sequence or structural motifs, the expression of the relevant transferring enzymes and the availability of the required donors which determine when and where they occur. This is especially so in the case of glycosylation, whereby 1–2% of the genome encodes enzymes or other proteins required for the transfer and remodeling of glycans; additionally, many proteins (such as lectins) recognize glycans and play key roles in development, physiology, and disease. Glycans have been likened to "analog" modulators of the "digital" genetic world, as their biosynthesis is not template-driven and there are few absolute on/off glycan switches [1]; rather, the large heterogeneity of glycans on any given glycoprotein results in a continuum of context-dependent biological responses.

The most common glycan modifications of proteins are N- or O-linked to amino- or hydroxyl-side chains of proteins, although C- and S-linkages are also known [2]. Probably N-glycans are the most studied forms and are known from bacteria, archaea, and almost all eukaryotes; in the latter case, asparagine residues are modified with an oligosaccharide via a core *N*-acetylglucosamine residue [3]. Certainly, N-glycans from mammals are quite well studied, whereas for invertebrate organisms the N-glycan structures and their functions remain rather unknown, but recent studies on N-glycosylation have proven that invertebrate organisms synthesize complicated N-glycomes, competing in terms of complexity with those of vertebrates [4–7]. In the case of *Caenorhabditis elegans*, despite 20 years of research on its protein glycosylation, many questions remain regarding this large subset of post-translational modifications in this organism. Lectin-based affinity purification was employed some two decades ago prior to identification of glycoproteins, but there was no protein-specific structural information regarding the attached glycans [8, 9], whereas deglycosylation, use of site-directed mutagenesis to abolish N-glycosylation sites, or lectin blotting have been employed to examine differences in the glycosylation of *C. elegans* DDR-2 and DMA-1 in wild-type and "glycomutant" worm strains [10, 11]. However, there is an overall lack of in-depth information regarding glycoprotein-specific glycosylation patterns in the worm—an exception is the analysis of *Haemonchus contortus* H11 glycoproteins recombinantly expressed in *C. elegans* [12].

As most of the available bioinformatics tools are based on mammalian structures, current databases have limited utility when considering invertebrate glycopeptide and N-glycan data. Clearly, glycan annotations on the basis of mass alone are insufficient and are often misleading. Orthogonal proofs are therefore necessary, including the use of specific detection reagents, MS/MS fragmentation, chemical or exoglycosidase treatments or reference to in-depth glycomic analyses from the same organism

[13, 14]. Recently, we have performed comprehensive N-glycomic analyses of *C. elegans* wild-type embryos, L4 larvae, and adults, as well as mutant strains, based on enzymatic release of the glycans [15, 16]. Also, we have applied chemical release methods to analyze mucin-type and glycosaminoglycan-like O-glycans [17] as well as organic extraction to isolate glycosphingolipids. Although we have not specifically examined *C. elegans* glycoproteins by this approach, we have previously analyzed a cestode antigen and royal jelly glycoproteins [18–20], whereby the N- and O-glycan release methods described here can be adapted to purified or enriched glycoproteins. Overall, various methods have been used to reveal that the glycome of *C. elegans* is highly complex, featuring motifs not found in other organisms to date as well as others found in parasitic helminths.

2 Materials

2.1 Equipment

1. Tight-fitting glass homogenizer (customize as required).
2. Vacuum centrifuge.
3. Micro-centrifuge.
4. Lyophilizer.
5. Mini Protean® Tetra cell and Power Pac power supply.
6. Trans blot SD semi-dry electrophoretic transfer cell.
7. Glass columns of 1 cm diameter and 50 cm length.
8. Multifunctional microtiter plate reader (e.g., Infinite M200, Tecan); black 96-well microtiter plates, e.g., Microfluor™ 1 or LumiNunc.
9. HPLC liquid chromatograph with fluorescence detector (e.g., Shimadzu Nexera); reverse phase chromatography column, e.g., Ascentis® Express RP-Amide (150 mm × 46 mm, 2.7 μm).
10. MALDI-TOF-TOF-MS: Autoflex Speed, UltrafleXtreme or Rapiflex MALDI-TOF-TOF; appropriate MALDI polished or ground steel target plate.

2.2 Reagents, Buffers, and Columns (See Notes 1 and 2)

2.2.1 Disruption of Biological Material and SDS-PAGE Sample Preparation

1. 2× SDS-PAGE reducing sample buffer containing 200 mg SDS, 154 mg DTT, 5 mL stacking gel buffer, 3.6 mL 87% glycerol (make up to 10 mL with water, then add a few crystals of bromophenol blue).

2.2.2 SDS-PAGE and Western Blotting

1. 12% SDS-PAGE gel (using 40% acrylamide stock, diluted with either stacking gel buffer with 0.5 M Tris/HCl pH 6.8 or separation gel buffer with 1.5 M Tris/HCl pH 8.8).
2. SDS-PAGE running buffer (25 mM Tris, 192 mM glycine, 0.1% SDS).
3. Colloidal Coomassie Blue staining solution: 0.02% (w/v) Coomassie Brilliant Blue G-250, 5% aluminum sulfate-(14–18)-hydrate [$Al_2(SO_4)_3.16H_2O$], ethanol 96%, phosphoric acid 85%. Weigh in 100 g of aluminum sulfate and dissolve it in 1500 mL of water; add 200 mL ethanol and mix well; add 0.4 g of Coomassie Brilliant Blue G-250 and mix well for at least 30 min; add slowly 47 mL of phosphoric acid and mix well; make up to 2000 mL with water (*see* **Note 3**).
4. Western blotting transfer buffer: 25 mM Tris, 192 mM glycine, 10% methanol.
5. SDS-PAGE protein standard ladder.
6. Nitrocellulose membrane (NT).
7. Extra thick blotting paper.
8. 0.5% (w/v) Ponceau S in 1% (v/v) acetic acid solution.
9. Membrane washing buffer: Tris buffered saline (TBS, i.e., 0.1 M Tris/HCl, pH 7.4, 0.1 M NaCl; typically made as a ten-fold concentrated stock) with 0.05% Tween.
10. Membrane blocking and antibody/lectin dilution buffer: Tris buffered saline with 0.05% Tween and 0.5% BSA.
11. Primary and secondary antibodies, lectins, or pentraxins (*see* Table 1).
12. SigmaFAST BCIP/NBT or SigmaFAST 3,3′--diaminobenzidine tetrahydrochloride tablets, dissolved in 10 and 5 mL respectively.

2.2.3 N-glycome Release and Analysis

1. Peptide:N-glycosidase F (PNGase F, recombinant from *Flavobacterium meningosepticum*)
2. Peptide:N-glycosidase A (recombinant Endo H-treated from *Oryza sativa* and expressed in *Pichia pastoris*, PNGase Ar from NEB).
3. For PNGase F: 50 mM ammonium hydrogen carbonate (pH 8; mixture of ammonium carbonate and ammonium hydrogen carbonate).
4. For PNGase A: 20 mM ammonium acetate (pH 5; acetic acid adjusted with ammonia).
5. 1–3 mL solid-phase extraction column and frits.
6. Acetonitrile, isopropanol, acetic acid, water.

Table 1
List of selected antibodies, lectins, and pentraxins for N-glycan epitope screening (Note 4)

Antibody (1st)	Dilution	Epitope [21, 22]	Suggested Supplier
Anti-HRP from rabbit, 10 mg/mL	1:10000	Core α1,3-Fuc/ core β1,2-Xyl	Sigma
Anti-PC (TEPC-15 mouse IgA), 10 mg/mL	1:200	PC-Hex(NAc)	Sigma
Antibody (2nd)			
Anti-rabbit IgG from goat conjugated with alkaline phosphatase	1:2000		Vector labs
Anti-mouse IgA from goat conjugated with alkaline phosphatase	1:10000		Sigma
Pentraxin			
C reactive protein (CRP) from human plasma ($CaCl_2$ 2.5 mM added)	1:200	PC-Hex(NAc)	MP biochemicals
Pentraxin recognition			
Anti-human C reactive protein from rabbit	1:1000		Dako
Lectin			
Biotinylated *Aleuria aurantia* lectin	1:1000	Core α1,6-Fuc/Le^x	Vector labs
Biotinylated wheat germ agglutinin	1:1000	β1,4HexNAc/ α2,3Sia	Vector labs
Lectin recognition			
Anti-biotin from goat conjugated with alkaline phosphatase	1:10000		Sigma

7. Dowex AG® 50 W-X8 200–400 mesh H^+ form (biotechnology grade; washed serially with 0.1 M NaOH, water, 0.1 M HCl, water, 1 M ammonium acetate and water) and pre-equilibrated with 2% acetic acid prior to usage; C18 material (Lichroprep); non porous graphitized carbon material (NPGC; e.g., ENVICarb™).
8. MALDI matrices: 6-aza-thiothymine (ATT; 3 mg/ml ATT dissolved in 50% ethanol); 2,5-dihydroxybenzoic acid (DHB; 10 mg/mL DHB dissolved in 50% acetonitrile with 0.1% TFA).
9. Glycan labeling: 2-aminopyridine (PA, >99% purity), sodium cyanoborohydride (95% purity), hydrochloric acid (37% HCl).
10. Gel filtration: Sephadex G-15 and G-25 medium.
11. Orcinol reagent: 200 mg orcinol in 100 mL H_2SO_4.

2.2.4 O-glycome Release and Analysis

1. Ammonium-based β-elimination solution: 16 μL of water, 16 μL of hydroxylamine, 32 μL of 1,8-diazabicyclo[5.4.0]undec-7-ene (DBU).
2. Fetuin (positive control).
3. Reagents for purification and labeling (cf. Subheadings 2.2.3 and 2.2.5, **item 10**).

2.2.5 Glycosaminoglycan Release and Analysis

1. Hydrazine monohydrate; distilled to yield anhydrous hydrazine.
2. Acetic anhydride.
3. Sodium bicarbonate.
4. Trifluoroacetic acid.
5. Reagents for purification and labeling (cf. Subheadings 2.2.3 and 2.2.5, **item 10**).

2.2.6 Glycan Data Analysis

1. Glycoworkbench (www.glycoworkbench.org).
2. FlexAnalysis Bruker software.

3 Methods

See flow chart and example data in Figs. 1 and 2.

3.1 Sample Preparation and Glycoepitope Recognition

The purification procedure of the (glyco)protein of interest depends on the biological material which can be whole organisms, cells, tissues, semi-purified proteins, or secreted (glyco)proteins in culture media or buffer.

3.1.1 Sample Preparation for Glycoprotein Analysis (See Notes 1 and 2)

1. Heat inactivate the biological material in boiling water for 10 min. For *C. elegans*, the worms are homogenized using a tight-fitting glass homogenizer.
2. Prior to SDS-PAGE, precipitate an aliquot of the samples with a fivefold excess volume of methanol, incubate at −80 °C for 1 h and centrifuge for 10 min at 4 °C, 21,000 g. Dry the protein pellet at 65 °C for several minutes to evaporate excessive methanol and re-dissolve the pellet in 20 μL SDS-PAGE sample buffer. In addition, heat treat the mixture for 10 min at 95 °C and after cooling, centrifuge again for 5 min at room temperature, 21,000 g.

3.1.2 SDS-PAGE and Western Blotting

For initial screening of the N-glycan epitopes, approx. 2 μg of proteins are subject to SDS-PAGE under reducing conditions, followed by protein transfer to a nitrocellulose membrane (Western blotting).

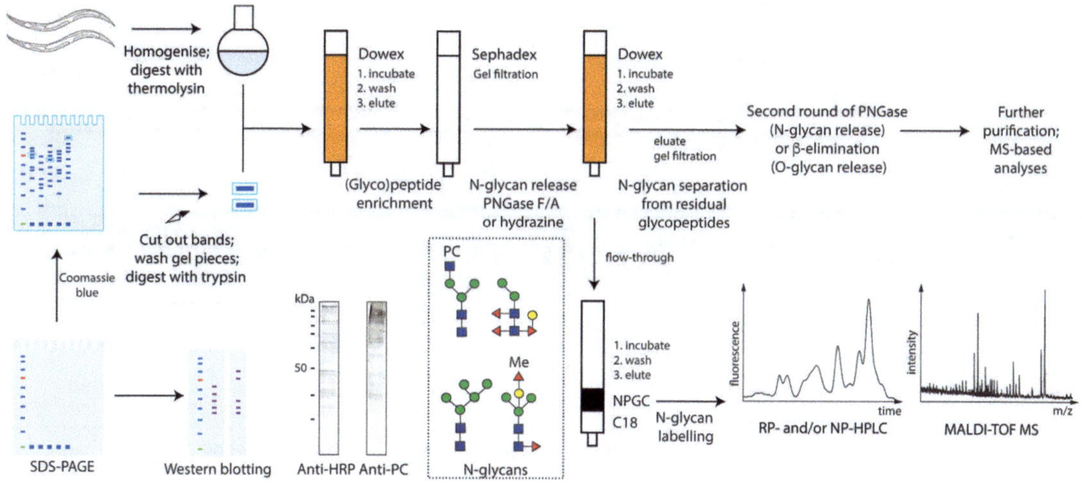

Fig. 1 A potential glycome and glycoproteomic workflow. Starting from homogenized biological material (cells, worms, etc.) or proteins separated by SDS-PAGE, there are various approaches to analyze the glycosylation. Certain glyco-epitopes can be screened using antibodies or lectins via Western blotting (e.g., anti-horseradish peroxidase or anti-phosphorylcholine antibodies), but the most reliable data comes from glycomics data. Glycans are released by an N-glycanase such as PNGase F or Ar and purified by solid phase extraction prior to labeling, HPLC, and mass spectrometry. Residual glycopeptides can be subject to another round of N- or O-glycan release prior to further purification. Other approaches are possible, including permethylation of released glycans; however, phosphorylcholine modifications are lost and natural methylation is only observed if employing perdeuteromethylation. Example N-glycan structures are shown according to the Symbol Nomenclature for Glycans, whereby circles, squares, stars, triangles, and diamonds respectively represent hexose (Man or Gal), *N*-acetylhexosamine (GlcNAc), deoxyhexose (Fuc); Me, methyl; PC, phosphorylcholine (zwitterionic modification)

1. For an initial screen for sample quality and to equalize different samples, apply 2–10 μg of protein to the SDS-PAGE and stain the gel with Coomassie Blue before attempting Western blotting.
2. Check the quality of the successful transfer by incubating the membrane with Ponceau S staining solution for 1 min. After de-staining with water (protein bands will stain red), block the membrane with Tris buffered saline containing 0.05% Tween and 0.5% BSA for 1 h at room temperature under smooth shaking.
3. Wash the membrane three times using Tris buffered saline with 0.05% Tween (washing buffer).
4. Incubate with biotinylated lectins, pentraxins or primary antibodies in blocking/dilution buffer for 60 min (*see* Table 1 and **Note 4**).
5. Wash the membrane again thrice as above and incubate with the relevant peroxidase or alkaline phosphatase-conjugated secondary antibodies in blocking/dilution buffer for 60 min.

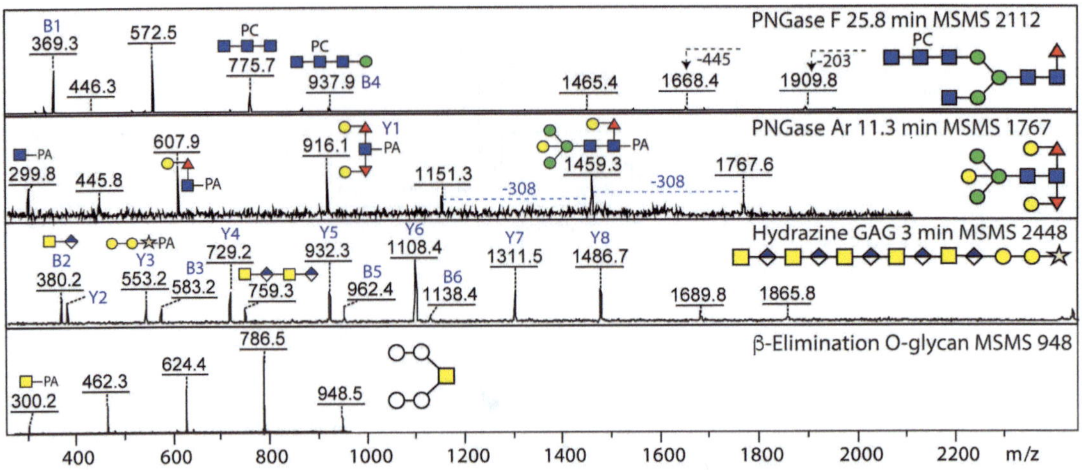

Fig. 2 Example MALDI-TOF MS/MS data for pyridylaminated *C. elegans* N-glycans, O-glycans, and glycosaminoglycan-like glycans. MS/MS data is shown for a phosphorylcholine-modified HPLC-purified N-glycan released with PNGase F, a core trifucosylated HPLC-purified N-glycan released with PNGase Ar, a non-sulfated HPLC-purified chondroitin glycosaminoglycan-like chain released by hydrazinolysis and a β-eliminated "mucin-type" O-glycan. Glycans and selected B- or Y-fragment ions are depicted according to the Symbol Nomenclature for Glycans, whereby circles, squares, stars, triangles and diamonds respectively represent hexose (Man, Gal, or undefined), *N*-acetylhexosamine (GalNAc or GlcNAc), deoxyhexose (Fuc), pentose (Xyl), or hexuronic acids (GlcA); PA, pyridylamino (fluorescent label); PC, phosphorylcholine. The pyridylamino label aids detection when purifying the glycans as well as ionization by MALDI-TOF-MS, resulting in Y-ions defining the reducing-terminus, while B-ions are more obvious for phosphorylcholine-modified or glucuronylated glycans

6. Again wash the membrane three times as above.
7. Develop the Western blots respectively for peroxidase or phosphatase conjugates with either SigmaFAST 3,3′-diaminobenzidine tetrahydrochloride or SigmaFAST BCIP/NBT (dissolve tablets first in water). Chemiluminescence or other detection methods can also be used.

3.2 Glycan Release and Analysis

3.2.1 N-glycan Release, Fractionation, and Analysis (See Notes 5–8)

1. *C. elegans* were grown in liquid culture with *E. coli* OP50 in standard S complete medium, mixed stages were harvested after cultivation at 20 °C (160 rpm) for 4–6 days and purified by sucrose density centrifugation [23, 24]. Harvested worms (2 g) were boiled, homogenized; after adjusting the pH with ammonium carbonate buffer to 8.2, $CaCl_2$ was added to a final concentration of 0.5 mM prior to addition of 2 mg thermolysin (Promega). Proteolysis was allowed to proceed for 2 h at 70 °C, prior to acidification. Other proteases can be used, but these require heat inactivation prior to further processing.
2. The peptides are enriched on a Dowex AG 50 column; first the peptides are mixed for 30 min (occasional manual stirring) in a beaker with 10 mL of the pre-conditioned cation exchange

resin in 2% acetic acid before pouring into a 10 mL glass column. The flow-through is collected and reapplied; the column is then washed with two column volumes of 2% acetic acid, prior to elution of the peptides with two column volumes of 0.5 M ammonium acetate, pH 6. 2 mL fractions are collected and assessed for the presence of carbohydrate by orcinol staining.

3. The pooled (glyco)peptides are freeze-dried, taken up in 4 mL of 0.5% acetic acid and desalted on a gel filtration column (Sephadex G25; 70 mL). The column is washed with 0.5% acetic acid; 4 mL fractions are collected and assessed for the presence of carbohydrate by orcinol staining, which can be done in solution or on thin-layer chromatography plates.
4. For deglycosylation, the freeze-dried (glyco)peptides are incubated with PNGase F and/or PNGase Ar (*see* **Note 5**).
5. Optimal conditions for the PNGase F are 50 mM ammonium carbonate, pH 8, and for PNGase Ar activity are 20 mM ammonium acetate buffer, pH 5; incubations are overnight at 37 °C with 5 U of enzyme (see also **Note 5**).
6. Purify the released N-glycans using two different columns packed with Dowex AG 50 and non-porous graphitized carbon/Lichroprep C18 (*see* **Note 6**). Wash first the Dowex AG 50 column with 2% acetic acid with 2% acetic acid. Apply the glycopeptide sample after acidifying with 10% acetic acid and collect immediately the unbound released N-glycans in the flow-through and wash fractions (three column volumes of 2% acetic acid).
7. Apply the flow-through/wash from the Lichroprep/Dowex column directly to a non-porous graphitized carbon/Lichroprep C18 column (prewashed and pre-equilibrated with first 100% acetonitrile then water). After sample application, wash the column with water and elute the neutral N-glycans with 40% acetonitrile; in cases where anionic N-glycans may occur (not in *C. elegans*, but in filarial worms), the elution is repeated with 40% acetonitrile containing 0.1% trifluoroacetic acid.
8. These fractions were subject to a further solid phase extraction step on a C18 reversed phase resin (LiChroprep) and the glycans were eluted with water and with stepwise increases in the methanol concentration (15%, 30%, 100% (v/v)).
9. Lyophilize the purified N-glycans overnight and after dissolving them in water, spot an aliquot for MALDI-TOF MS/MS analysis with 6-azathiothymine (ATT); regarding acquisition and interpretation of mass spectra, refer to **Note 7**. In comparison to peptides, higher laser power and detector gain settings are necessary to detect glycans. For a more detailed analysis, label the N-glycans by reductive amination using

2-aminopyridine and in addition subject them to HPLC and MALDI-TOF MS analysis as described below.

10. Fluorescent labeling is performed as follows: dissolve 100 mg 2-aminopyridine in 76 μL concentrated HCl and 152 μL water; add 80 μL of this solution to the dried glycan sample, prior to incubation in boiling water for 15 min. Then prepare a solution of 4.4 mg of sodium cyanoborohydride in a mixture of 9 μL of the aforementioned 2-aminopyridine solution and 13 μL water; add 4 μL of this cyanoborohydride-aminopyridine solution to the sample and continue the incubation overnight at 90 °C.
11. Removal of excess labelling reagent is performed immediately the following day by gel filtration. Dilute the sample in 1.5 mL of 0.5% acetic acid (i.e., no more than 5% of the gel filtration column volume), apply to a 30 mL Sephadex G-15 column (1 × 40 cm) equilibrated in 0.5% acetic acid, and collect 1.5 mL fractions. Transfer aliquots of fractions (80 μL) to a 96 F black plate and detect fluorescence in a microtiter plate reader (excitation/emission: 320/400 nm). Pool fluorescent glycans eluting before the excess labeling reagent and lyophilize.
12. Dissolve dried sample by washing the flask four times with 20 μL of water and transfer to a microcentrifuge tube; re-lyophilize as required and analyze an aliquot by MALDI-TOF MS.
13. Inject the major portion of sample onto an Ascentis® Express RP-Amide column pre-equilibrated with 100 mM ammonium acetate (pH 4; buffer A); elute at 0.8 mL/min using a linear gradient of 30% (v/v) MeOH (buffer B) from 0% B up to 35% B over 35 minutes (higher percentages of B generate higher pressure). The glycans are detected by fluorescence using excitation/emission wavelengths of 320/400 nm and the column is calibrated in terms of glucose units with a fluorescently labeled oligoglucose standard (partial dextran hydrolysate). Collect fractions based on fluorescence intensity and lyophilize prior to another round of MALDI-TOF MS and MS/MS to identify the glycans in the fractions (for example data, refer to Fig. 2). Alternatives to the RP-Amide column are discussed in **Note 8**.
14. Aliquots of the isolated HPLC fractions can be subject to targeted exoglycosidase digestion and chemical treatment [25]. Either α-mannosidase (jack bean), α-galactosidase (coffee bean), β-galactosidase (recombinant *Aspergillus niger* LacA prepared in house [26]) or β-hexosaminidases (recombinant *C. elegans* HEX-4 prepared in-house [27], *Streptomyces plicatus* chitinase or jack bean hexosaminidase) are used for further treatment of the samples in 25 mM ammonium acetate, pH 5

(pH 6.5 in the case of HEX-4), at 37 °C for 24 h. For removal of phosphorylcholine or α1,3-fucose residues, selected fractions are dried and incubated for 48 h at 0 °C with 3 μL 48% (v/v) hydrofluoric acid prior to evaporation in a centrifugal concentrator. The samples are diluted in water and re-evaporated, before redissolving once again. The chemically- or enzymatically-treated fractions were subject to MALDI-TOF MS and MS/MS (as above) without further purification.

3.2.2 O-glycan Release (See Note 9)

1. O-glycans can be released from *C. elegans* (extracted by methanol-chloroform precipitation of fresh lysates) via ammonium-based β-elimination [28] resulting in non-reduced O-glycans. Extracted glycoproteins are dried and incubated with a mixture of 16 μL H_2O, 16 μL hydroxylamine, and 32 μL 1,8-diazabicyclo[5.4.0]undec-7-ene at 50 °C for 30 min.
2. Released O-glycans are purified using serial small columns of Dowex AG50, C18 and non-porous graphitized carbon prior to labeling with 2-aminopyridine and MALDI-TOF MS/MS as described for N-glycans.

3.2.3 Glycosaminoglycan Release (See Note 10)

1. For chemical release, 10 mg of enriched and desalted glycopeptides (see above) are transferred into a glass reaction tube and dried overnight prior to adding 500 μL of anhydrous hydrazine (prepared from monohydrate hydrazine) and incubated at 100 °C for 5 h. Unreacted anhydrous hydrazine is removed by centrifugal evaporation.
2. Samples are cooled to 0 °C and then re-*N*-acetylated by the addition of 1 M sodium bicarbonate solution (450 μL) and acetic anhydride (21 μL) and incubated at 0 °C for 60 min.
3. The samples are then acidified by addition of 5% (v/v) trifluoroacetic acid (600 μL) to the samples and incubated at 4 °C for 60 min in order to liberate the reducing end of the glycans, followed by Dowex AG50, C18 and non-porous graphitized carbon prior to labeling with 2-aminopyridine and MALDI-TOF MS/MS as described for N-glycans [29].

4 Notes

1. The quality of water and other reagents (acetonitrile, methanol, isopropanol) used for analytical purposes should be high and free of ionic and microbial contaminants.
2. In general, contaminants should be avoided; to prevent analysis of "foreign" components from the food/nutrition source or media, the material (whole organisms or cells) should be

washed several times before the heat treatment and homogenization. After collection, the biological material should be stored at −80 °C, if not immediately homogenized. To prevent hydrolysis of the anionic or zwitterionic residues (e.g., sulfate or PC), the samples should be heat treated only in water and not in acidic buffers; however, heat inactivation is necessary to prevent degradation of the glycans by endogenous glycosidases. In the case of *C. elegans*, we have used embryos, L4 larvae (grown in liquid culture or on plates) and mixed cultures with mainly adults (grown in liquid culture) as samples for our glycan analyses [15, 16]. The procedures described here can be also adapted for proteins separated by SDS-PAGE [18–20]. For small amounts of biological samples, also a lysis buffer supplemented with protease inhibitor cocktail (Sigma) can be used prior to SDS-PAGE, while a methanol precipitation step after cell lysis prior to SDS-PAGE helps to desalt the sample and so avoid smearing upon electrophoresis.

3. Colloidal Coomassie aggregates and tiny blue dots are visible. Make sure that the staining solution is mixed well (e.g., with a magnetic mixer) before each use.

4. Results obtained from antibody or lectin binding are no structural proof of the N-glycans on the glycoprotein as their specificities are sometimes wide or not fully determined. Positive and negative controls and pull-downs to "pre-clear" endogenous biotinylated proteins, as well as Western blots with and without lectins/antibodies (i.e., just secondary reagents) or after glycosidase digestions should be considered for data interpretation. The "mini-description" of the epitopes in Table 1 is based on determination of binding of the antibodies, lectins, or pentraxins to standard ligands; these determinations are by no means exhaustive as invertebrate standards are rarely tested [21, 30]. Nevertheless, anti-horseradish peroxidase is valuable for screening of core β1,2-xylose and core α1,3-fucose [31], but the anti-xylose and anti-fucose components of the antisera are difficult to properly separate. Phosphorylcholine (PC) epitopes can be detected with either the TEPC-15 antibody or by human C-reactive protein [32].

5. PNGase F can release N-glycans from both glycoproteins and glycopeptides, whereas recombinant PNGase Ar still works best on peptides. PNGase F does not release N-glycans with core α1,3-fucose modification (but does release core α1,6-fucosylated or β1,3-mannosylated structures), while recombinant PNGase Ar can release substituted core α1,3-fucosylated glycans [7]. For protein samples, the degree of protein deglycosylation can be monitored with SDS PAGE (reduced size of the protein after deglycosylation) and Western blotting (reduced or abolished N-glycan epitope binding). If

performing serial digests on glycopeptides, first incubate with PNGase F at pH 8, then (i) acidify to pH 5 before adding PNGase Ar or (ii) purify the PNGase F-released N-glycans on Dowex AG 50 (see next step) before gel filtration of the remaining glycopeptides prior to PNGase Ar release. The result is either one combined pool of PNGase F and Ar-released glycans or separate pools; the advantage of separate pools is that low abundance α1,3-fucosylated N-glycans are more easily identified. Other PNGases have been described in recent years, but their use has not been verified with *C. elegans* N-glycans.

6. The glycopeptides should be acidified with 10% acetic acid before Dowex cation exchange chromatography. For N-glycan recovery after PNGase F or Ar release, also refer to our protocol on "Analysis of invertebrate and protist N-glycans" [25]. For O-glycosylation, there is no single universal de-O-glycosylation enzyme available; O-glycanase has a restricted substrate specificity and will not remove most extended GalNAc-Ser/Thr (mucin-type) or other O-glycan structures, therefore a chemical method is described.

7. Released N-glycans should be measured in positive and negative ion mode for the identification of potential anionic residues, such as sulfate (+80 Da), phosphate (+80 Da), glucuronic acid (+176 Da), phosphoethanolamine (+123 Da), and aminoethylphosphonate (+107 Da; +121 Da if methylated) [14]. Sialic acids are rare in invertebrates [33, 34], but are absent, e.g., from nematodes, whereas glucuronic acid has been found on N-glycans in filarial worms [35]. Invertebrate N-glycomes dramatically differ from those of mammals, so N-glycan assignments for isolated glycans or on glycopeptides should be based at least on MS/MS data analysis. Indeed, compositions based on mass alone can be misleading: for instance, a difference of 324 Da can either correspond to two hexoses or one methylaminoethylphosphonate-modified HexNAc as seen, e.g., in mollusks. Also, a difference of 176 Da may be either a methylated hexose or a glucuronic acid [14]. Nevertheless, mass differences of 146, 160, 162, or 203 can suggest the presence of fucose, methylated fucose, hexose, and *N*-acetylhexosamine residues. Various bioinformatics tools for automated glycopeptide and glycan identification [36] and the following software can be applied for glycopeptide MS (GlycoMod, GlycoX, GlycopepDB, Massy tools, and GlycoSpectrumScan) and MSMS (GlycoMiner, Protein Prospector, GlycopepID, GlycoMasterDB, etc.). As these are generally applied to mammalian glycomes and glycoproteomes, caution is required when using search engines to annotate invertebrate glycans. For publication, consider the MIRAGE guidelines for the presentation of glycomic data and descriptions of methods

[37], use of the diagrammatic Symbol Nomenclature for Glycans [38], and submission of raw spectra to databases. We have submitted mzxml files of *C. elegans* N-glycan MS/MS data to Glycopost: https://glycopost.glycosmos.org/entry/GPST000200 and https://glycopost.glycosmos.org/entry/GPST000294

8. Normal phase or non-fused core reversed-phase columns can also be used for the separation of pyridylaminated glycans [25]; normal phase separates by size and charge, while "classical" non-fused core reversed-phase columns have the disadvantage that glycans with multiple phosphorylcholine residues are highly retained and may not be efficiently eluted. On the other hand, the highest degree of variation in the phosphorylcholine-modified N-glycome was revealed using the RP-amide column [15], while "2D-HPLC" (involving re-fractionation of normal or reversed phase HPLC fractions on the other reciprocal column type) can prove valuable in separating isomeric or isobaric glycan structures.
9. There are various β-elimination procedures described in the literature, including the classical reductive method followed typically by permethylation [39] or LC-ESI-MS [40]. The non-reductive version described here has been successfully employed on *C. elegans* and allows for subsequent fluorescent labeling [15].
10. Note that hydrazine is a hazardous reagent and must only be used when applying relevant safety procedures. Although peeling reactions occur and re-*N*-acetylation must be performed, the advantages of hydrazinolysis are that more complex core-modifications of N-glycans can be released [7], in addition to non-sulfated glycosaminoglycan chains, which elute early on reversed-phase HPLC [17].

Acknowledgments

This work was supported by the Austrian Fonds zur Förderung der wissenschaftlichen Forschung (FWF; grants P32572 and P29466 to K.P and I.B.H.W.).

References

1. Varki A (2011) Evolutionary forces shaping the Golgi glycosylation machinery: why cell surface glycans are universal to living cells. Cold Spring Harb Perspect Biol 3(6):a005462. https://doi.org/10.1101/cshperspect.a005462
2. Spiro RG (2002) Protein glycosylation: nature, distribution, enzymatic formation, and disease implications of glycopeptide bonds. Glycobiology 12:43R–56R
3. Aebi M (2013) N-linked protein glycosylation in the ER. Biochim Biophys Acta 1833(11): 2430–2437. https://doi.org/10.1016/j.bbamcr.2013.04.001

4. Schiller B, Hykollari A, Yan S et al (2012) Complicated N-linked glycans in simple organisms. Biol Chem Hoppe Seyler 393:661–673
5. Eckmair B, Jin C, Abed-Navandi D et al (2016) Multi-step fractionation and mass spectrometry reveals zwitterionic and anionic modifications of the N- and O-glycans of a marine snail. Mol Cell Proteomics 15:573–597. https://doi.org/10.1074/mcp.M115.051573
6. Stanton R, Hykollari A, Eckmair B et al (2017) The underestimated N-glycomes of lepidopteran species. Biochim Biophys Acta 1861(4): 699–714. https://doi.org/10.1016/j.bbagen.2017.01.009
7. Yan S, Vanbeselaere J, Jin C et al (2018) Core richness of N-glycans of *Caenorhabditis elegans*: a case study on chemical and enzymatic release. Anal Chem 90(1):928–935. https://doi.org/10.1021/acs.analchem.7b03898
8. Hirabayashi J, Hayama K, Kaji H et al (2002) Affinity capturing and gene assignment of soluble glycoproteins produced by the nematode *Caenorhabditis elegans*. J Biochem (Tokyo) 132(1):103–114. https://doi.org/10.1093/oxfordjournals.jbchem.a003186
9. Fan X, She YM, Bagshaw RD et al (2005) Identification of the hydrophobic glycoproteins of *Caenorhabditis elegans*. Glycobiology 15(10):952–964. https://doi.org/10.1093/glycob/cwi075
10. Shimizu T, Kato Y, Sakai Y et al (2019) N-glycosylation of the Discoidin domain receptor is required for axon regeneration in *Caenorhabditis elegans*. Genetics 213(2): 491–500. https://doi.org/10.1534/genetics.119.302492
11. Rahman M, Ramirez-Suarez NJ, Diaz-Balzac CA et al (2022) Specific N-glycans regulate an extracellular adhesion complex during somatosensory dendrite patterning. EMBO Rep 23(7):e54163. https://doi.org/10.15252/embr.202154163
12. Roberts B, Antonopoulos A, Haslam SM et al (2013) Novel expression of *Haemonchus contortus* vaccine candidate aminopeptidase H11 using the free-living nematode *Caenorhabditis elegans*. Vet Res 44:111. https://doi.org/10.1186/1297-9716-44-111
13. Hykollari A, Malzl D, Yan S et al (2017) Hydrophilic interaction anion exchange for separation of multiply modified neutral and anionic Dictyostelium N-glycans. Electrophoresis 38:2175–2183. https://doi.org/10.1002/elps.201700073
14. Paschinger K, Wilson IBH (2016) Analysis of zwitterionic and anionic N-linked glycans from invertebrates and protists by mass spectrometry. Glycoconj J 33:273–283. https://doi.org/10.1007/s10719-016-9650-x
15. Paschinger K, Wöls F, Yan S et al (2023) N-glycan antennal modifications are altered in *Caenorhabditis elegans* lacking the HEX-4 N-acetylgalactosamine-specific hexosaminidase. J Biol Chem 299:103053. https://doi.org/10.1016/j.jbc.2023.103053
16. Wilson IBH, Yan S, Jin C et al (2023) Increasing complexity of the N-glycome during *Caenorhabditis* development. Mol Cell Proteomics 22:100505. https://doi.org/10.1016/j.mcpro.2023.100505
17. Vanbeselaere J, Yan S, Joachim A et al (2018) The parasitic nematode *Oesophagostomum dentatum* synthesizes unusual glycosaminoglycan-like O-glycans. Glycobiology 28(7):474–481. https://doi.org/10.1093/glycob/cwy045
18. Paschinger K, Gonzalez-Sapienza GG, Wilson IBH (2012) Mass spectrometric analysis of the immunodominant glycan epitope of *Echinococcus granulosus* antigen Ag5. Int J Parasitol 42(3):279–285. https://doi.org/10.1016/j.ijpara.2012.01.002
19. Hykollari A, Malzl D, Eckmair B et al (2018) Isomeric separation and recognition of anionic and zwitterionic N-glycans from royal jelly glycoproteins. Mol Cell Proteomics 17(11): 2177–2196. https://doi.org/10.1074/mcp.RA117.000462
20. Hykollari A, Malzl D, Wilson IBH et al (2019) Protein-specific analysis of invertebrate glycoproteins. Methods Mol Biol 1871:421–435. https://doi.org/10.1007/978-1-4939-8814-3_24
21. Iskratsch T, Braun A, Paschinger K et al (2009) Specificity analysis of lectins and antibodies using remodeled glycoproteins. Anal Biochem 386(2):133–146. https://doi.org/10.1016/j.ab.2008.12.005
22. Mikolajek H, Kolstoe SE, Pye VE et al (2011) Structural basis of ligand specificity in the human pentraxins, C-reactive protein and serum amyloid P component. J Mol Recognit 24(2):371–377. https://doi.org/10.1002/jmr.1090
23. Brenner S (1974) The genetics of *Caenorhabditis elegans*. Genetics 77(1):71–94
24. Lewis JA, Fleming JT (1995) Basic culture methods. Methods Cell Biol 48:3–29
25. Hykollari A, Paschinger K, Eckmair B et al (2017) Analysis of invertebrate and Protist N-Glycans. Methods Mol Biol 1503:167–184. https://doi.org/10.1007/978-1-4939-6493-2_13

26. Dragosits M, Pflugl S, Kurz S et al (2014) Recombinant aspergillus β-galactosidases as a robust glycomic and biotechnological tool. Appl Microbiol Biotechnol 98(8):3553–3567. https://doi.org/10.1007/s00253-013-5192-3

27. Dragosits M, Yan S, Razzazi-Fazeli E et al (2015) Enzymatic properties and subtle differences in the substrate specificity of phylogenetically distinct invertebrate N-glycan processing hexosaminidases. Glycobiology 25(4):448–464. https://doi.org/10.1093/glycob/cwu132

28. Kameyama A, Thet Tin WW, Toyoda M et al (2019) A practical method of liberating O-linked glycans from glycoproteins using hydroxylamine and an organic superbase. Biochem Biophys Res Commun 513(1):186–192. https://doi.org/10.1016/j.bbrc.2019.03.144

29. Jiménez-Castells C, Vanbeselaere J, Kohlhuber S et al (2017) Gender and developmental specific N-glycomes of the porcine parasite *Oesophagostomum dentatum*. Biochim Biophys Acta 1861(2):418–430. https://doi.org/10.1016/j.bbagen.2016.10.011

30. Purohit S, Li T, Guan W et al (2018) Multiplex glycan bead array for high throughput and high content analyses of glycan binding proteins. Nat Commun 9(1):258. https://doi.org/10.1038/s41467-017-02747-y

31. Paschinger K, Rendić D, Wilson IBH (2009) Revealing the anti-HRP epitope in *Drosophila* and *Caenorhabditis*. Glycoconj J 26:385–395

32. Paschinger K, Gonzalez-Sapienza GG, Wilson IBH (2012) Mass spectrometric analysis of the immunodominant glycan epitope of *Echinococcus granulosus* antigen Ag5. Int J Parasitol 42: 279–285

33. Aoki K, Perlman M, Lim JM et al (2007) Dynamic developmental elaboration of N-linked glycan complexity in the *Drosophila melanogaster* embryo. J Biol Chem 282:9127–9142. https://doi.org/10.1074/jbc.M606711200

34. Miyata S, Sato C, Kumita H et al (2006) Flagellasialin: a novel sulfated α2,9-linked polysialic acid glycoprotein of sea urchin sperm flagella. Glycobiology 16(12):1229–1241. https://doi.org/10.1093/glycob/cwl036

35. Martini F, Eckmair B, Neupert C et al (2019) Highly modified and immunoactive N-glycans of the canine heartworm. Nat Commun 10:75. https://doi.org/10.1038/s41467-018-07948-7

36. Tsai PL, Chen SF (2017) A brief review of bioinformatics tools for glycosylation analysis by mass spectrometry. Mass Spectrom 6: S0064. https://doi.org/10.5702/massspectrometry.S0064

37. York WS, Agravat S, Aoki-Kinoshita KF et al (2014) MIRAGE: the minimum information required for a glycomics experiment. Glycobiology 24(5):402–406. https://doi.org/10.1093/glycob/cwu018

38. Varki A, Cummings RD, Aebi M et al (2015) Symbol nomenclature for graphical representations of Glycans. Glycobiology 25(12): 1323–1324. https://doi.org/10.1093/glycob/cwv091

39. Morelle W, Faid V, Chirat F et al (2009) Analysis of N- and O-linked glycans from glycoproteins using MALDI-TOF mass spectrometry. Methods Mol Biol 534:5–21. https://doi.org/10.1007/978-1-59745-022-5_1

40. Mereiter S, Magalhaes A, Adamczyk B et al (2016) Glycomic and sialoproteomic data of gastric carcinoma cells overexpressing ST3GAL4. Data Brief 7:814–833. https://doi.org/10.1016/j.dib.2016.03.022

Chapter 9

Use of Reductive Amination to Produce Capsular Polysaccharide-Based Glycoconjugates

Federico Urbano-Munoz, Caitlyn E. Orne, Mary N. Burtnick, and Paul J. Brett

Abstract

Reductive amination is a relatively simple and convenient strategy for coupling purified polysaccharides to carrier proteins. Following their synthesis, glycoconjugates can be used to assess the protective capacity of specific microbial polysaccharides in animal models of infection and/or to produce polyclonal antiserum and monoclonal antibodies for a variety of immune assays. Here, we describe a reproducible method for chemically activating the 6-deoxyheptan capsular polysaccharide (CPS) from *Burkholderia pseudomallei* and covalently linking it to recombinant CRM197 diphtheria toxin mutant (CRM197) to produce the glycoconjugate, CPS-CRM197. Similar approaches can also be used to couple other types of polysaccharides to CRM197 with little to no modification of the protocol.

Key words *Burkholderia pseudomallei*, *Burkholderia thailandensis*, Capsular polysaccharide, Carrier protein, CRM197, Glycoconjugate, Oxidation, Reductive amination

1 Introduction

Burkholderia pseudomallei, the etiologic agent of melioidosis, is a motile, facultative intracellular, Gram-negative bacillus that can cause severe and often fatal disease in humans and animals [1]. In endemic regions, the organism can be readily isolated from environmental reservoirs such as moist soils, stagnant waters, and untreated potable water systems [2, 3]. Melioidosis can manifest as localized abscesses, visceral lesions, acute pneumonias, and fulminating septicemias [4, 5]. Because of the nonspecific clinical presentations, the lack of rapid diagnostic tests, and the intrinsic resistance of *B. pseudomallei* to commonly used antibiotics, diagnosis and treatment of melioidosis can be challenging [6, 7]. Primary routes of infection include inhalation, ingestion, and percutaneous inoculation and at present no vaccines are available for immunization against this emerging infectious disease [2, 3].

Steven B. Bradfute (ed.), *Recombinant Glycoproteins: Methods and Protocols*, Methods in Molecular Biology, vol. 2762, https://doi.org/10.1007/978-1-0716-3666-4_9,

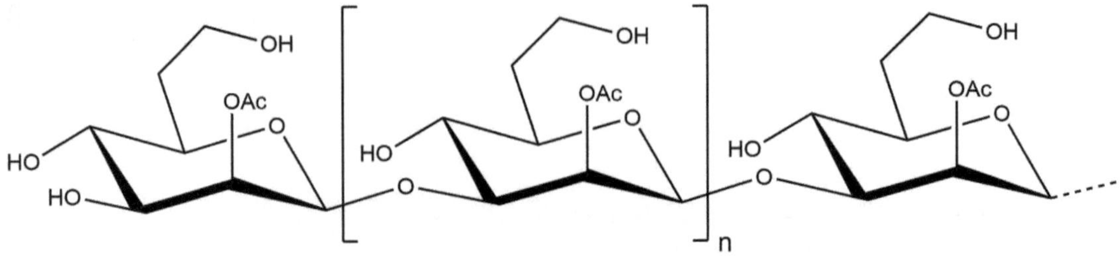

Fig. 1 Structure of the 6-deoxyheptan CPS antigen expressed by *B. pseudomallei* and CPS-expressing strains of *B. thailandensis*

Several studies have demonstrated that the 6-deoxyheptan CPS expressed by *B. pseudomallei* is both a virulence factor and a protective antigen (Fig. 1) [8–10]. Consequently, this cell surface antigen has become an important component of the subunit vaccine candidates that we are currently developing [11–13]. Immunologically, antigens can be classified as either T-cell-dependent, T-cell-independent type 1, or T-cell-independent type 2 (TI-2) [14]. Capsular polysaccharides are generally considered to be TI-2 antigens. These types of high-molecular-weight antigens are immunogenic due to their ability to cross-link multiple surface immunoglobulin molecules present on antigen-specific B cells [15], but without the involvement of T helper (Th) cells, TI-2 antigens induce poor immunological memory and little to no affinity maturation and isotype switching [16]. Additionally, without dosing at frequent intervals, antibody levels often decline. Efforts to overcome the poor immunogenicity of many clinically relevant capsular polysaccharides have led to the development of glycoconjugate vaccines [17, 18], a number of which are currently licensed for human use [19]. Covalent linkage of capsular polysaccharides to carrier proteins promotes Th-cell involvement, which improves immunological memory [20] and increases isotype switching. The affinities of the antibodies elicited by glycoconjugates also tend to be higher than those produced by polysaccharides alone [16].

Studies in our lab have shown that immunization of C57BL/6 mice with CPS-CRM197 produces high-titer IgG and opsonizing antibody responses against the CPS component of the glycoconjugate [11]. They have also demonstrated that when mice are vaccinated with the glycoconjugate material, ~70% of the animals survive lethal inhalational challenges with *B. pseudomallei*. Taken together, these studies confirm an important role for antibodies in combatting disease caused by this bacterial pathogen and the rationale for including the CPS-CRM197 in our melioidosis subunit vaccine formulations [11, 21, 22].

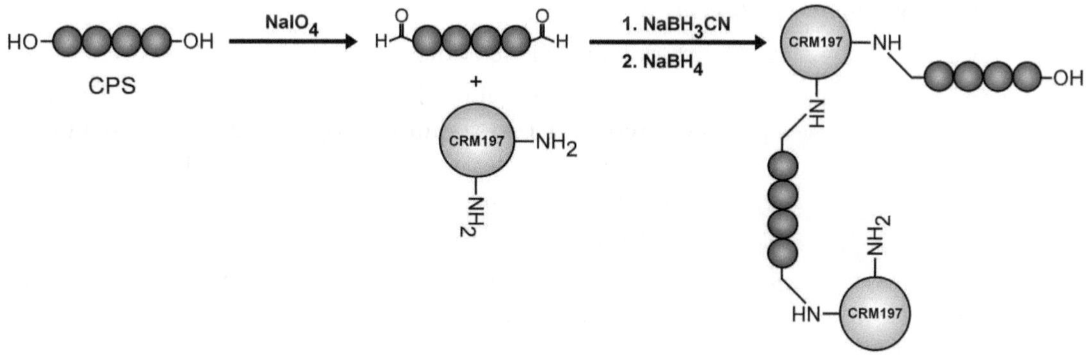

Fig. 2 Generalized scheme used for the covalent linkage of oxidized *B. pseudomallei* CPS to CRM197 via reductive amination

Herein we describe the method used in our lab to chemically couple *B. pseudomallei* CPS to CRM197. This is a relatively simple, two-step process that involves i) oxidation of the purified CPS and ii) covalent linkage of the activated CPS to surface-exposed lysine residues on the carrier protein via reductive amination (Fig. 2). Although the protocol focuses on the use of CRM197 as a carrier, alternative proteins have been explored for this purpose and can be used in its place. Overall, the approach described in this chapter results in efficient coupling of the CPS antigen to CRM197 and the production of highly immunogenic glycoconjugate material. It also provides a general guide that can be used to synthesize other glycoconjugates by covalently linking different polysaccharides to carrier proteins.

2 Materials

Prepare all of the solutions using deionized water (dH_2O) and analytical-grade reagents. All of the solutions should be prepared and stored at room temperature unless otherwise stated.

2.1 Antigens

1. CPS: The 6-deoxyheptan CPS antigen used in this protocol was purified from *Burkholderia pseudomallei* RR2683 (select agent excluded strain: Δ*purM*, Δ*rmlD*; adenine and thiamine auxotroph, O-polysaccharide deficient strain) using a modified hot-aqueous phenol procedure as previously described [11, 23, 24]. *Burkholderia thailandensis* BT2683 (Δ*rmlD*; O-polysaccharide deficient strain) can also be used as a source of the same 6-deoxyheptan CPS antigen [12] (*see* **Note 1**).
2. CRM197: The carrier protein used in this protocol (*see* **Note 2**).

2.2 Activation of CPS

1. Analytical balance.
2. 15 mL polypropylene conical tubes.
3. Phosphate buffered saline (PBS): BupH™ Phosphate-Buffered Saline. Prepare 0.1 M sodium phosphate, 0.15 M NaCl buffer at pH 7.2 solution per the manufacturer's instructions. Filter sterilize using 0.22 μm membrane.
4. Vortex mixer.
5. Sodium *meta*-periodate ($NaIO_4$): Pierce™ Sodium *meta*-periodate powder (*see* **Note 3**).
6. 4 Dram amber vial with cap.
7. Stir flea (small stir bar) for use in amber vial.
8. Magnetic stir plate.
9. Glycerol.
10. Slide-A-Lyzer G3 Dialysis Cassette: 5–15 mL, 3500 MWCO.
11. 4 L beaker, stir bar, and stir plate.
12. 10 cc syringe and 0.45 μm MCE syringe filter (3.3 cm).
13. Supplies for shell freezing: lyophilization flask, metal pan, dry ice, 200 proof ethanol, and a lyophilizer (*see* **Note 4**).

2.3 Preparation of CRM197

1. Plastic beakers, 1 L and 4 L sizes.
2. Sized stir bars for use in 1 L and 4 L beakers.
3. Magnetic stir plate.
4. Borate buffer (BB): Use 20× Borate Buffer to prepare a 100 mM BB solution. Filter sterilize using 0.22 μm membrane.
5. Slide-A-Lyzer G3 Dialysis Cassette: 1–3 mL, 3500 MWCO.
6. 5 cc syringe and 0.45 μm MCE syringe filter (3.3 cm).

2.4 Conjugation Reaction

1. Analytical balance.
2. 15 and 50 mL polypropylene conical tubes.
3. BB: Prepare 100 mM BB as described above.
4. Vortex mixer.
5. 4 Dram amber vial with cap.
6. Stir flea (small stir bar) for use in amber vial.
7. Sodium Cyanoborohydride ($NaBH_3CN$): AminoLink™ Reductant. Prepare a 1 M stock in BB immediately prior to use (*see* **Note 5**).
8. Static incubator set at 37 °C containing a magnetic stir plate.
9. Sodium dodecyl sulfate polyacrylamide gel electrophoresis (SDS-PAGE) gel electrophoresis supplies: 4–12% Bis-Tris Bolt Gels, MES running buffer, 2× SDS-PAGE sample buffer, electrophoresis apparatus, power pack, Simply Blue Safe Stain (*see* **Note 6**).

10. Sodium borohydride ($NaBH_4$). Prepare a 1 M stock in BB immediately prior to use (*see* **Note 7**).
11. Slide-A-Lyzer G3 Dialysis Cassette: 10–30 mL, 3500 MWCO.
12. High-speed tabletop centrifuge that accommodates 50 mL conical tubes.
13. 30 cc syringe and 0.45 μm MCE syringe filter (3.3 cm).
14. Supplies for shell freezing: lyophilization flask, metal pan, dry ice, 200 proof ethanol, and a lyophilizer (*see* **Note 4**).
15. BCA Protein Assay kit.

3 Methods

Perform steps as described below at room temperature unless otherwise specified. The weights and volumes of materials listed in this protocol can be scaled up or down as needed.

3.1 Activation of CPS

1. Weigh 20 mg of purified CPS and add to a 15 mL conical tube. Solubilize the CPS at 5 mg/mL in PBS. The CPS can be mixed using a vortex.
2. Add ~24 mg $NaIO_4$ to an amber vial containing a stir flea. Add the CPS solution to the amber vial and gently stir the reaction mixture for 40 min (a precipitate may appear during this step) (*see* **Note 3**).
3. Quench the reaction by adding 100 μL of glycerol to the vial (*see* **Note 8**). Allow the reaction mixture to stir for at least 15 min once the glycerol has been added. Raise the reaction volume to ~6 mL with dH_2O.
4. Transfer the reaction mixture to a 5–15 mL Slide-A-Lyzer Dialysis Cassette. Dialyze the material against 4 × 4 L of dH_2O. Protect the oxidized CPS from bright light.
5. Remove the dialysate from the cassette, sterilize with a syringe filter and lyophilize to dryness (*see* **Note 4**). Store the lyophilized material at −20 °C until required for use.

3.2 Preparation of CRM197

1. Thaw a 1 mL aliquot of CRM197 (10 mg/mL stock) and add to a 1–3 mL Slide-A-Lyzer Dialysis Cassette. Dialyze against 3 × 1 L of 100 mM BB. Remove the dialysate from the cassette, adjust the volume to 2 mL with 100 mM BB, sterilize with a syringe filter (this will yield a 5 mg/mL CRM197 working stock). Use this material immediately.

3.3 Conjugation of CPS to CRM197

1. Warm the lyophilized CPS to room temperature. Weigh 16 mg of material and solubilize in 3.2 mL of 100 mM BB (this will yield a 5 mg/mL CPS working stock). The CPS can be resolubilized using a vortex.

2. Add the CPS working stock to an amber vial containing a stir bar followed by the CRM197 working stock (this will yield a w/w ratio of 1.6:1 of CPS to CRM197) (*see* **Note 9**). Gently stir the reaction mixture on a stir plate for 5 min and remove a 10 μL sample for SDS-PAGE analysis (*see* **Notes 10** and **11**).
3. Add 125 μL of the 1 M $NaBH_3CN$ stock to the reaction mixture and incubate with gentle stirring for 3 h at room temperature (*see* **Note 5**). Move the reaction mixture to a 37 °C incubator with a stir plate and incubate with gentle stirring for up to 7 days in the dark (*see* **Note 12**). Remove a 10 μL sample for SDS-PAGE analysis on days 1, 3, 5, and 7 (*see* **Note 11**) (Fig. 3).
4. After 7 days, remove the vial from the incubator and add 125 μL of the 1 M $NaBH_4$ stock dropwise to the reaction mixture. Gently stir for 1–2 h at room temperature with the lid of the vial loosened to vent the reaction (*see* **Note 13**). Raise the reaction volume to ~10 mL with dH_2O.
5. Transfer the reaction mixture to a 10–30 mL Slide-A-Lyzer Dialysis Cassette. Dialyze the material against 4 × 4 L of dH_2O.
6. Remove the dialysate from the cassette, raise the reaction volume to ~20 mL with dH_2O and centrifuge for 5 min at 7200 × *g* in a 50 mL conical tube to remove any precipitate (*see* **Note 14**). Collect the supernatant, sterilize with a syringe filter, and lyophilize to dryness (*see* **Note 4**).
7. Weigh the lyophilized glycoconjugate material and store at −20 °C until required for use.

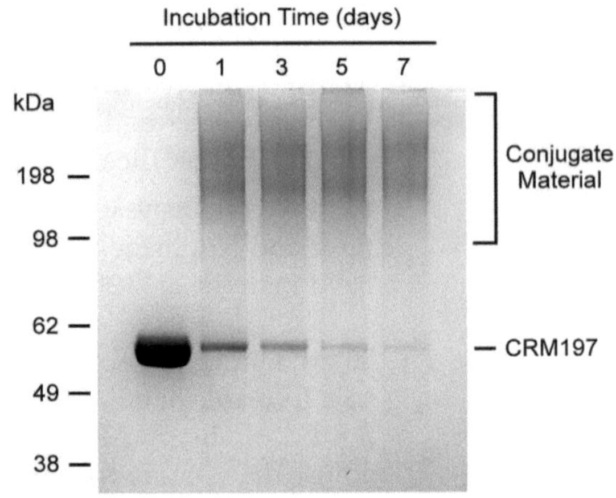

Fig. 3 Physical analysis of CPS-CRM197. SDS-PAGE and Simply Blue Safe Stain were used to assess the covalent linkage of *B. pseudomallei* CPS to CRM197. Samples were removed from the reaction mixture on days 0, 1, 3, 5, and 7 (day 0 represents the unconjugated control). ~6 μg of protein was loaded per lane on a 4–12% Bis-Tris Bolt gel. The positions of the molecular weight standards (kDa) are indicated on the left

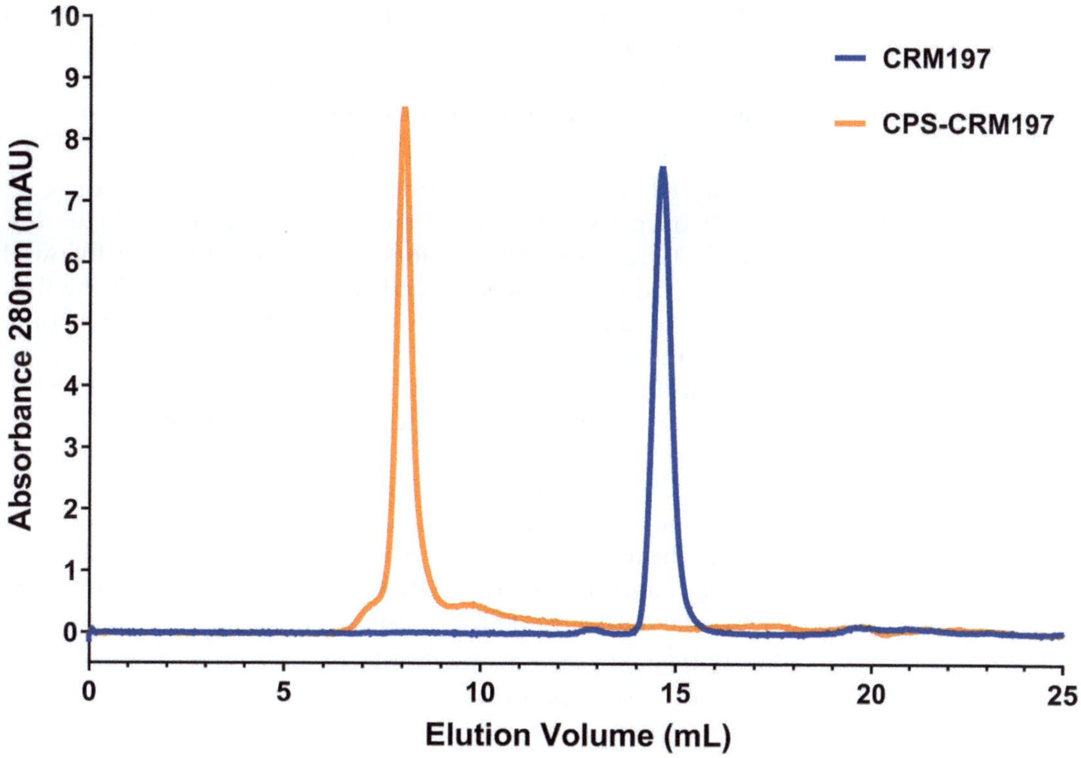

Fig. 4 SEC analysis of CPS-CRM197. FPLC was used to characterize the elution profile of the glycoconjugate material following 7 days of incubation at 37 °C. 40 μg of CRM197 (as a conjugate) or unconjugated CRM197 (control) in BupH PBS were loaded onto a Superdex 200 Increase 10/300 GL high-resolution gel filtration column and eluted using BupH PBS at a flow rate of 0.75 mL/min. CRM197 was detected at 280 nm

8. Resuspend the lyophilized glycoconjugate material in PBS at a final concentration of 2 mg/mL and assay for protein content using a BCA Protein Assay kit (*see* **Note 15**). Aliquot and store the solubilized material at −20 °C until required for use.
9. Analyze the glycoconjugate material using Fast Protein Liquid Chromatography (FPLC) system (*see* **Note 16**) (Fig. 4).

4 Notes

1. Any polysaccharide that has a reducing terminal sugar or that can be oxidized with $NaIO_4$ can be coupled to a carrier protein via reductive amination.
2. Glycoconjugates can be prepared using proteins other than CRM197. Other commonly used carriers include ovalbumin (OVA), bovine serum albumin (BSA), cationized bovine serum albumin (cBSA), keyhole limpet hemocyanin (KLH), tetanus toxoid (TT), and exotoxin A (ExoA).

3. $NaIO_4$ is a strong oxidizer and should be used in a chemical fume hood. Wear appropriate protective eyewear, gloves, and clothing to prevent eye and skin exposure. Stir gently after adding to the CPS solution to prevent foaming.
4. For lyophilization of the polysaccharide or conjugate, add the dialyzed material to a container suitable for lyophilization (e.g., Labconco fast-freeze flask) and shell freeze. To facilitate this, add dry ice to an appropriately sized metal container followed by the addition of absolute ethanol (200 proof). Tilt the flask into the dry ice/ethanol mixture and slowly rotate the bottle until the material is frozen in a shell around the bottle. Freezing of small volumes (<10 mL) is achieved by placing the sample in a 50 mL conical tube and resting it at an angle in a −80 °C freezer until frozen. For lyophilization of the frozen material, change out the cap of the conical tube to a cap that has been punctured with an 18-gauge needle and place the conical tube with the modified lid into a container suitable for lyophilization (e.g., Labconco fast-freeze flask). Place the flask with the frozen sample onto a lyophilizer until completely dry.
5. $NaBH_3CN$ is toxic and should be used in a chemical fume hood. Wear appropriate protective eyewear, gloves, and clothing to prevent eye and skin exposure.
6. SDS-PAGE should be conducted on the samples collected throughout the conjugation reaction. The use of a gradient gel (4–12%) is recommended. Preferred reagents and supplies are listed in the materials section.
7. $NaBH_4$ is a strong reducing agent and should be used in a chemical fume hood. Wear appropriate protective eyewear, gloves, and clothing to prevent eye and skin exposure.
8. Glycerol is added to the reaction mixture to exhaust any unreacted $NaIO_4$.
9. This reaction can be scaled up or down depending on the amount of glycoconjugate to be synthesized. For optimal results, the ratio of CPS to CRM197 (1.6:1 w/w) should remain the same. Ratios will need to be optimized, however, when glycoconjugates are synthesized with different polysaccharides and carrier proteins.
10. Stir the reaction mixture at ~100 rpm on a stir plate.
11. To assess conjugation, samples should be collected at various time points after the reaction has been set up for analysis by SDS-PAGE. Following the addition of the CRM197 to the CPS solution, allow the reaction to stir gently for approximately 5 min, then take a sample for SDS-PAGE analysis. This sample will serve as the "day 0" and be representative of unconjugated CRM197. Mix the sample with an equal volume

of 2× SDS-PAGE Sample Buffer plus β-mercaptoethanol (β-me) and store at −20 °C until required for use. Take additional samples from the reaction mixture on days 1, 3, 5, and 7 during the incubation at 37 °C. The reaction mixture may be briefly removed from the incubator to obtain the samples. Mix each sample with an equal volume of 2X SDS-PAGE Sample Buffer plus β-me and store at −20 °C until required for use.

12. Optimal conjugation of *B. pseudomallei* CPS to CRM97 requires up to 7 days of incubation at 37 °C. Incubation times will need to be optimized, however, when glycoconjugates are synthesized with different polysaccharides and carrier proteins.
13. Add the 1 M $NaBH_4$ solution dropwise to the conjugation reaction while it is slowly mixing on a stir plate. Addition of the $NaBH_4$ results in the production of hydrogen gas. Mix gently to prevent foam production. Leave the cap loosened to vent the mixture while the conjugation reaction is stirring.
14. Use of a high-speed tabletop centrifuge that accommodates 50 mL conical tubes is recommended for this step.
15. Use the BCA Protein Assay kit to determine the amount of CRM197 in the 2 mg/mL CPS-CRM197 stock. Once the amount of CRM197 is determined, the remaining mass of the conjugate material is assumed to be CPS.
16. The conjugate material can be analyzed by size exclusion chromatography (SEC) to determine the efficiency of the conjugation reactions. FPLC systems with UV monitors are useful for detecting both the polysaccharide (210 nm) and carrier protein (280 nm) components.

References

1. Cheng AC, Currie BJ (2005) Melioidosis: epidemiology, pathophysiology, and management. Clin Microbiol Rev 18(2):383–416. https://doi.org/10.1128/CMR.18.2.383-416.2005
2. Dance DA (2000) Ecology of Burkholderia pseudomallei and the interactions between environmental Burkholderia spp. and human-animal hosts. Acta Trop 74(2–3):159–168. https://doi.org/10.1016/s0001-706x(99)00066-2
3. Wiersinga WJ, van der Poll T, White NJ et al (2006) Melioidosis: insights into the pathogenicity of Burkholderia pseudomallei. Nat Rev Microbiol 4(4):272–282. https://doi.org/10.1038/nrmicro1385
4. Currie BJ, Ward L, Cheng AC (2010) The epidemiology and clinical spectrum of melioidosis: 540 cases from the 20 year Darwin prospective study. PLoS Negl Trop Dis 4(11): e900. https://doi.org/10.1371/journal.pntd.0000900
5. Meumann EM, Cheng AC, Ward L et al (2012) Clinical features and epidemiology of melioidosis pneumonia: results from a 21-year study and review of the literature. Clin Infect Dis 54(3):362–369. https://doi.org/10.1093/cid/cir808
6. Currie BJ, Fisher DA, Howard DM et al (2000) Endemic melioidosis in tropical northern Australia: a 10-year prospective study and review of the literature. Clin Infect Dis 31(4): 981–986. https://doi.org/10.1086/318116
7. Limmathurotsakul D, Wongratanacheewin S, Teerawattanasook N et al (2010) Increasing incidence of human melioidosis in Northeast

Thailand. Am J Trop Med Hyg 82(6): 1113–1117. https://doi.org/10.4269/ajtmh.2010.10-0038

8. Atkins T, Prior R, Mack K et al (2002) Characterisation of an acapsular mutant of Burkholderia pseudomallei identified by signature tagged mutagenesis. J Med Microbiol 51(7): 539–553. https://doi.org/10.1099/0022-1317-51-7-539
9. AuCoin DP, Reed DE, Marlenee NL et al (2012) Polysaccharide specific monoclonal antibodies provide passive protection against intranasal challenge with Burkholderia pseudomallei. PLoS One 7(4):e35386. https://doi.org/10.1371/journal.pone.0035386
10. Reckseidler SL, DeShazer D, Sokol PA et al (2001) Detection of bacterial virulence genes by subtractive hybridization: identification of capsular polysaccharide of Burkholderia pseudomallei as a major virulence determinant. Infect Immun 69(1):34–44. https://doi.org/10.1128/IAI.69.1.34-44.2001
11. Burtnick MN, Shaffer TL, Ross BN et al (2018) Development of subunit vaccines that provide high-level protection and sterilizing immunity against acute inhalational Melioidosis. Infect Immun 86(1). https://doi.org/10.1128/IAI.00724-17
12. Schmidt LK, Orne CE, Shaffer TL et al (2022) Development of Melioidosis subunit vaccines using an enzymatically inactive Burkholderia pseudomallei AhpC. Infect Immun 90(8): e0022222. https://doi.org/10.1128/iai.00222-22
13. Scott AE, Burtnick MN, Stokes MG et al (2014) Burkholderia pseudomallei capsular polysaccharide conjugates provide protection against acute melioidosis. Infect Immun 82(8):3206–3213. https://doi.org/10.1128/IAI.01847-14
14. Mond JJ, Lees A, Snapper CM (1995) T cell-independent antigens type 2. Annu Rev Immunol 13:655–692. https://doi.org/10.1146/annurev.iy.13.040195.003255
15. Snapper CM, Mond JJ (1996) A model for induction of T cell-independent humoral immunity in response to polysaccharide antigens. J Immunol 157(6):2229–2233
16. Weintraub A (2003) Immunology of bacterial polysaccharide antigens. Carbohydr Res 338(23):2539–2547
17. Lockhart S (2003) Conjugate vaccines. Expert Rev Vaccines 2(5):633–648. https://doi.org/10.1586/14760584.2.5.633
18. Stein KE (1992) Thymus-independent and thymus-dependent responses to polysaccharide antigens. J Infect Dis 165(Suppl 1):S49–S52
19. Romano MR, Berti F, Rappuoli R (2022) Classical- and bioconjugate vaccines: comparison of the structural properties and immunological response. Curr Opin Immunol 78:102235. https://doi.org/10.1016/j.coi.2022.102235
20. Kelly DF, Snape MD, Clutterbuck EA et al (2006) CRM197-conjugated serogroup C meningococcal capsular polysaccharide, but not the native polysaccharide, induces persistent antigen-specific memory B cells. Blood 108(8):2642–2647. https://doi.org/10.1182/blood-2006-01-009282
21. Biryukov SS, Cote CK, Klimko CP et al (2022) Evaluation of two different vaccine platforms for immunization against melioidosis and glanders. Front Microbiol 13:965518. https://doi.org/10.3389/fmicb.2022.965518
22. Klimko CP, Shoe JL, Rill NO et al (2022) Layered and integrated medical countermeasures against Burkholderia pseudomallei infections in C57BL/6 mice. Front Microbiol 13: 965572. https://doi.org/10.3389/fmicb.2022.965572
23. Burtnick MN, Heiss C, Roberts RA et al (2012) Development of capsular polysaccharide-based glycoconjugates for immunization against melioidosis and glanders. Front Cell Infect Microbiol 2:108. https://doi.org/10.3389/fcimb.2012.00108
24. Heiss C, Burtnick MN, Wang Z et al (2012) Structural analysis of capsular polysaccharides expressed by Burkholderia mallei and Burkholderia pseudomallei. Carbohydr Res 349:90–94. https://doi.org/10.1016/j.carres.2011.12.011

Part III

Mammalian Glycoproteins

Chapter 10

Overexpression and Purification of Mitogenic and Metabolic Fibroblast Growth Factors

Phuc Phan, Shivakumar Sonnaila, Gaetane Ternier, Oshadi Edirisinghe, Patience Salvalina Okoto, and Thallapuranam Krishnaswamy Suresh Kumar

Abstract

Fibroblast growth factors (FGFs) are proteins with a vast array of biological activity, such as cell development and repair, glucose and bile acid metabolisms, and wound healing. Due to their critical and diverse physiological functions, FGFs are believed to possess potential as therapeutic agents for many diseases and conditions that warrant further investigations. Thus, a simple, cost-efficient method to purify these biologically active signaling proteins is desirable. Herein, we introduce such techniques to purify FGFs that possess either high heparin-binding affinity or low to no heparin-binding affinity. This method takes advantage of the high affinity toward heparin sulfate from paracrine FGF1 to isolate the targeted protein. It also accounts for FGF members that have low heparin affinity, such as the metabolic FGFs, by introducing poly-histidine tags in the recombinant protein in combination with the immobilized metal affinity chromatography. Subsequently, the purified FGF products are separated from the other small protein by high-speed centrifugation. Products are then subjected to other biophysical experiments like SDS-PAGE, mass spectrometry, circular dichroism, intrinsic fluorescence, isothermal titration calorimetry, differential scanning calorimetry, and biological cell activity assay to confirm that the target proteins are purified with intact native conformation and no significant change in the intrinsic characteristics and biological activities.

Key words Recombinant proteins, Heparin-affinity chromatography, Immobilized metal affinity chromatography, Metabolic FGFs, FGF1

1 Introduction

Fibroblast growth factors (FGFs) are one of the most important families of 23 structurally related proteins that are important for cellular biological processes. The human acidic fibroblast growth factor (FGF1) belongs to the FGF1 subfamily, whose signaling aids in the regulation of cellular processes such as cell proliferation, cell differentiation, cell development, and angiogenesis. FGFs elicit a cellular response by activating the specific tyrosine kinase fibroblast

Steven B. Bradfute (ed.), *Recombinant Glycoproteins: Methods and Protocols*, Methods in Molecular Biology, vol. 2762,
https://doi.org/10.1007/978-1-0716-3666-4_10,

growth factor receptors (FGFRs) with the help of negatively charged glycosaminoglycan known as heparin sulfate. Heparin not only aids in increasing the stability of the molecule but also protects the protein from proteolytic and enzymatic digestion. FGF1-bound FGFRs, in turn, induce downstream signaling through some common pathways, such as mitogen-activated protein kinase (MAPK), phosphoinositide-3-kinase/AKT, and phospholipase-C_l pathways. hFGF1 is unique from the other members of the FGF family in its ability to activate all four known tyrosine kinase fibroblast growth factor receptors (FGFRs). Therefore, FGF1 is known as a universal ligand and is recognized as a robust wound-healing agent. FGF1's characteristics can also be exercised to treat disorders such as diabetic ulcers, bed sores, coronary ischemia, etc. [17].

Although FGF1 has extraordinary properties as a wound-healing agent, its potential for practical use is limited owing to its low stability. Therefore, several studies have been working on improving the stability of hFGF1 via site-directed mutagenesis [2]. In this regard, overexpression and purification of hFGF1 are of immense importance to the pharmaceutical and biotechnological industries. As hFGF1 is a heparin-binding protein, herein, we illustrate a simple and cost-effective method to purify hFGF1 from bacterial cell lysate using heparin-affinity chromatography.

Contrasting to mitogenic FGFs, including hFGF1, metabolic FGFs possess lower heparin affinity and increased metabolic activity [18]. FGF19 is one such FGFs, along with FGF21 and FGF23. It is secreted in the small intestine. But due to its endocrine nature, FGF19 is secreted and sent to the liver, commonly through the hepatic portal, where the 21.8 kDa protein regulates bile acid synthesis. The effects of FGF19 in the regulation of lipid homeostasis in adipocytes and its ability to increase glucose uptake make it a prospective candidate in the treatments of diabetes-induced obesity. Yet, the inherent structural instability of the molecule challenges some of its pharmaceutical applications [7, 22].

The next member of metabolic FGFs is FGF21, which is known for its ability to regulate glucose uptake, rendering the molecule an attractive target to treat metabolic diseases like type-2 diabetes and obesity [9, 13]. The protein is around 22.3 kDa with 181 residues [13]. Similar to FGF19, FGF21 is unstable due to its dynamic but unstable core structure [25]. Additionally, the C-terminal of FGF21 is a poly-rich region that is prone to protease cleavage despite its critical role in asserting FGF21's activity through binding to coreceptor β-klotho [8, 10, 14, 24]. Combining both problems, FGF21 is believed to be difficult to apply as a therapeutic agent for metabolic diseases despite its potential [13].

Upon its discovery, FGF23 has been identified as a key mediator in phosphate homeostasis and vitamin D metabolism in the body. FGF23 is often produced by osteocytes and osteoblasts, especially in the human body [18]. The mature FGF23 with

227 residues is released into circulation [16]. The main physiological functions of FGF23 are exerted through the activation of FGFRs prominently, isoforms of FGFR1, FGFR3, and FGFR4 [23]. The interactions between FGF23 and its specific coreceptor, α-klotho, with FGFR activate downstream signaling pathways through MAPK and extracellular signal-regulated kinase (ERK) pathway [11]. Alterations in FGF23 levels in humans have been associated with several types of bone diseases, such as rickets, X-linked hypophosphatemia (XLH), and tumor-induced osteomalacia (TIO), among others [20, 21]. A further elevated level of FGF23 is associated with hyperphosphatemic familial tumoral calcinosis and chronic kidney disease [3]. Therapeutic modulation of FGF23 overexpression or disrupting α-klotho-FGF23-FGFR interactions prevent higher FGF23 levels in the body and pave better clinical management of patients suffering from FGF23-related disorders [5, 15].

Since metabolic FGFs possess low heparin binding activity, heparin affinity chromatography cannot be utilized to purify metabolic FGFs the same way as mitogenic FGFs. Instead, each member of the FGF19 subfamily is tagged with poly-histidine to aid in the purification process. The fusion of peptide affinity tags into the protein of interest in order to perform affinity chromatography is widely employed in biochemical studies [4]. Since fused peptide tags have an affinity for metal ions, such incorporation allows for immobilized metal affinity chromatography (IMAC) to purify the proteins of interest, the metabolic FGFs in this case. Poly-histidine tags are commonly used for IMAC due to their strong interactions between transition metal ions (e.g., Co^{2+}, Ni^{2+}, Cu^{2+}, and Zn^{2+}) which are immobilized on a matrix (e.g., sepharose) [4, 6]. Here, imidazole rings in histidine residues act as electron donors and form coordination bonds with the transition ion, and immobilize the poly-histidine tagged protein. The immobilized protein can then be eluted by increasing the ionic strength of the column buffer, including imidazole in the column buffer, or reducing pH [19].

2 Materials

All solutions are made using ultrapure (deionized) water and analytical-grade reagents. All solutions and reagents are stored at 4 °C unless otherwise stated.

2.1 Buffers

1. **1× Phosphate Buffered Saline (1× PBS):** 150 mM Sodium Chloride, 10 mM Disodium Hydrogen Phosphate, 1 mM Sodium Azide, 25 mM Ammonium Sulfate; pH 7.2.
2. **Equilibration/Lysis/Wash buffer:** 1× PBS, pH 7.2.

3. **Elution Buffer:** 1× PBS with a stepwise increase of sodium chloride concentrations (heparin-affinity chromatography) or 1× PBS with a stepwise increase of imidazole concentration (immobilized metal affinity chromatography).

2.2 Media and Cells

1. **Luria-Bertani (LB) Miller Broth media:** Dissolve 12.5 g of LB ready-made powder in 500 mL of water. Dispense into flasks, cover with aluminum foil, and autoclave with the liquid cycle for 30 min at 121.5 °C at 15-lb pressure.
2. **LB agar media:** Dissolve 8 g of LB agar ready-made powder in 200 mL of water in a 500-mL bottle and autoclave with the liquid cycle of 30 min at 121.5 °C at 15-lb pressure.
3. **Competent *BL21(DE3)*/star competent cells.** *E. coli* BL21-Pro cells containing HB-pET22b™.
4. **Dulbecco's Modified Eagle Medium (DMEM).**
5. **NIH/3T3 fibroblast cell line.**

2.3 Solutions

1. **Ampicillin:** 100 mg of ampicillin in 1 mL of 50% ethanol and store in −20 °C freezer.
2. **Isopropyl-1-thio-B-D-galactopyranoside (IPTG):** 238 mg of IPTG in 1 mL of water and stored at −20 °C.
3. **Glycerol stock** of *E. coli* cells transformed with the HB-pET22b™ expression vector.
4. **SDS–polyacrylamide gel electrophoresis (SDS-PAGE) set up**: 30% polyacrylamide (29% acrylamide and 1% N,N′--Methylenebisacrylamide), 1.5 M Tris–HCl (pH 6.8 for stacking gel and pH 8.8 for resolving gel), 10% ammonium persulfate, N,N,N′,N′-Tetramethyl-ethylenediamine.
5. **Loading dye:** 9.6 mL di$H_2$0, 2.4 mL 0.5 M Tris (pH ~ 6.8), 4 mL 10% SDS, 2 mL Glycerol (100%), 0.9 mL B-mercaptoethanol, 6 mg Bromophenol blue.
6. **Staining solution:** 15 mL of Ethanol, 5 mL of Glacial Acetic Acid, 30 mL of DI water, 100 mg of Coomassie Brilliant Blue–R – 250.
7. **De-staining solution:** 120 mL of Ethanol, 40 mL of Glacial Acetic Acid, 200 mL of DI water.
8. **10× Running buffer**: 30.3 g of Tris base; 144.0 g Glycine, 10.0 g of SDS dissolve and bring the total volume to 1.0 L.
9. **TBS/T solution:** 20 nM Tris pH 7.4, 0.15 M NaCl, 0.1% Tween.
10. **Towbin buffer:** 25 mM Tris-HCl, 192 mM glycine, pH 8.3 with 20% methanol (v/v) and 0.01% SDS.
11. **5% skim milk:** 5 gm dissolved in 100 mL 1× TBS-T solution 10 mL 10% Tween-20.

12. **Phenyl methyl sulfonyl fluoride (PMSF):** 34.8 mg of PMSF in 1 mL of 100% ethanol and stored in 20 °C freezer.
13. **Sodium Azide** (powder).
14. **Disodium Ethylenediaminetetraacetate ($EDTA\text{-}Na_2$) (1 M)**: dissolve 3.72 g of $EDTA\text{-}Na_2$ to 10 mL water to have 1 M solution.
15. **Fetal Bovine Serum:** sterile solution.
16. **Penicillin/Streptomycin:** sterile solution with 10,000 units of penicillin and 10 mg of streptomycin/mL.
17. **Poly-D-Lysine:** Dilute stock solution (1 mg/mL) to 20 ug/mL with water to coat wells at 100 μL volume per well.
18. **CellTiter Glo 2.0 Cell Viability Assay.**
19. **Trypan Blue Dye.**
20. **Acetone.**
21. **Ethanol.**
22. **Urea (8 M):** dissolve 480 g of crystallized ACS-grade Urea in water to make 1 L of 8 M Urea.
23. **100% Trichloro acetic acid (TCA) solution:** Dissolve 10 g of TCA crystal in 10 mL of water to make 100% solution.
24. **NaOH (2 M):** dissolve 80 g of crystallized ACS-grade Urea in water to make 1 L of 2 M NaOH.
25. **HCl (2 M):** dilute stock concentrated HCl (varied by brand) with water to achieve 2 M solution.

2.4 Equipment

1. Micropipettes.
2. 2 L Erlenmeyer flasks.
3. 1 L Erlenmeyer flasks.
4. 500 mL capped bottle.
5. Temperature-controlled incubator.
6. CO_2-controlled and temperature-controlled incubator.
7. Laminar Flow Hood.
8. A mechanical device to disrupt *E. coli* cells (e.g., an Ultrasonicator, French press, or cell homogenizer).
9. pH meter.
10. UV-vis spectrophotometer.
11. Tabletop centrifuge.
12. Vortex.
13. Nalgene™ PPCO Centrifuge Bottles.
14. Heparin-Sepharose resin.
15. Ni-Sepharose resin.

16. Columns with dimensions 1.6 cm × 20 cm can be used to pack resin with a bed volume of ~15 mL.
17. Econo UV monitor.
18. Low-flow peristaltic pump.
19. Oakridge tubes.
20. Ultrafiltration centrifugal concentrating devices with appropriate molecular-weight cut-offs.
21. Water bath.
22. Beckman Coulter Avanti centrifuge.
23. Beckman Coulter Avanti J-E Rotor JA-10.
24. Beckman Coulter Avanti J-E Rotor JA-25.
25. Sorvall™ Legend™ Micro 21 Microcentrifuge.
26. Rocker.
27. Degas machine.
28. Electrophoresis setup.
29. SDS-PAGE setup.
30. Western Blot setup.
31. Nitrocellulose membrane.
32. Isothermal titration calorimeter.
33. 2.5 mL Hamilton gas-tight syringe with a blunt-end needle long enough to reach the volume of the sample cell.
34. 2 mL all-glass syringe (e.g., Becton Dickinson 2 cc Yale syringe).
35. Precut needle supplied with the ITC instrument.
36. Differential scanning calorimeter (DSC), calibrated.
37. CD spectrophotometer (calibrated) and cells of appropriate optical path length (depending on the expected protein concentration).
38. Titrator accessory for CD spectrophotometer.
39. Fluorescence spectrophotometer (calibrated) and quartz cuvettes of appropriate optical path length (depending on the expected protein concentration).
40. Cell counter machine (Tecan Spark) and appropriate cell chips.
41. Inverted microscope.
42. 96-well cell culture plates.
43. T-75 cell culture flask.
44. Filtration unit.

3 Methods

3.1 Construction of an FGF1 Gene Containing Expression Vector and Transformation of BL21 Star E. coli *Competent Cells with the Constructed Vector*

1. Thaw the competent cell stock aliquot *BL21 star* (about 50–100 μL) by placing it on ice for about 5 min (*see* **Note 1**).
2. Pipet 5 μL of plasmid DNA (pET22b™) construct and add to the competent cells on ice and incubate for at least 30 min (*see* **Note 1**).
3. Heat-shock the incubated competent cells at 42 °C for 50 s, followed by cooling on ice for 5 min.
4. After the ice treatment, 800 μL of cold LB broth is added to the above cells and placed in a 37 °C shaker for 60 min.
5. Centrifuge the incubated cells for 10 min at 13,000 rpm.
6. Discard the supernatant and dissolve the pellet in 100 μL of LB media.
7. Spread the culture onto LB agar + ampicillin plates. Ampicillin's final concentration should be 100 μg/mL.
8. Incubate the spread plate at 37 °C for 14–16 h.

3.2 Transforming Expression Vectors Contain Metabolic FGFs Genes into BL21 Star E. coli *Competent Cells with the Constructed Vector*

1. Each metabolic FGF genes (FGF19, FGF21, and FGF23) is designed to have a poly-histidine (6xHis) tag at the N-terminal of the protein and is inserted into a plasmid DNA construct (pET22b™) (*see* **Notes 1** and 7).
2. 1 tube of competent cell, *BL21 star* (about 50–100 μL), is thawed by placing it on ice for about 5 min (*see* **Note 1**).
3. Add about 5 μL of each plasmid to individual tubes of competent cells on ice and incubate on ice for around 30 min.
4. Place competent cells in a 42 °C hot bath for 30 s, followed by cooling on ice for 3 min (heat shock treatment).
5. Add 800 μL of LB broth to the tube and place it in a 37 °C shaker at 250 rpm for 60 min.
6. Prepare LB-agar plates as instructed by the manufacturer. Add ampicillin into warm but still liquid LB agar plates to the concentration of 100 μg/mL.
7. Collect the cells by centrifuging the tubes after incubation at 13,000 rpm for 3 min.
8. Remove the supernatant and redissolve the pellet in fresh 100 μL of LB media.
9. Proceed to spread the culture onto LB agar + Ampicillin plates.
10. Incubate the spread plate at 37 °C for 14–16 h. Discrete colonies spread evenly on the plate should be expected.

3.3 Preparation of Starter Culture

1. Sterilize the 100 mL of LB broth (2.5 w/v % of LB media to water (DDI)) in an Erlenmeyer flask (500 mL) by autoclaving at 120 °C, 15 lbs pressure for 60 min.
2. Cool down the media to room temperature, then add ampicillin antibiotic to 10 mL of sterilized LB media in 50 mL tube. Ampicillin's final concentration should be 100 μg/mL or 269 μM.
3. Using a pipette tip to pick a single colony from overnight LB agar plates and add it to the falcon tube containing prepared 10 mL LB broth.
4. Incubate the culture in a shaker (~250 rpm) at 37 °C until the OD at 600 nm reaches a value of 0.6–0.8 (~12–14 h).

3.4 Preparation of Small-Scale Bacterial Culture for Protein Overexpression

1. Into a clean 50 mL tube, add fresh 10 mL of LB media and 1 mL of cell suspension from the starter culture.
2. Add 10 μL of Ampicillin into the mixture to a final concentration of 100 μg/mL.
3. Repeat **steps 1** and **2** for a second clean 50 mL tube.
4. Let both cultures grow at 37 °C until OD reaches 0.6 (for 1–2 h).
5. Add in the mixture 10% v/v of IPTG (final concentration 1 mM) into the first tube to start protein expression. Do not add IPTG to the second tube, as it is an uninduced sample.
6. Allow cells to grow for 4 h at 37 °C shaking at 220–250 rpm.
7. Collect the cells via centrifuge the cultures at 4500 rpm for 25 min.
8. Remove the supernatant and resuspend the cells in 1 mL of buffer (10 mM phosphate buffer, 100 mM NaCl, pH = 7.2).
9. Transfer each mixture to individual microcentrifuge tubes and centrifuge at 13,000 rpm for 5 min.
10. Carefully transfer out the supernatant and resuspend the pellets in 1 mL of buffer (10 mM phosphate buffer, 100 mM NaCl, pH = 7.2).
11. Lyse the cell membrane via ultrasonication (amplitude #15 Watts output) with an alternate cycle of 10 s of ON and OFF for 20 min.
12. Centrifuge the lysate at 13,000 rpm for 5 min and transfer the supernatant, which contains proteins, into a new microcentrifuge tube.
13. Wash the pellet, which contains cell debris, with 1 mL of buffer (10 mM phosphate buffer, 100 mM NaCl) three times to eliminate any residual supernatant.
14. Resuspend the pellets in 10–20 μL of 8 M urea.

15. Resuspended pellets and lysate supernatants from both cultures should then follow the sample preparation and gel running protocols in Subheading 3.8.
16. Observe any indications that the protein is expressed in a sufficient amount on SDS-PAGE results before proceeding to Subheading 3.5: Large-scale protein expression (*see* **Note 2**). The resulting gel should show a band corresponding to the molecular size of the target FGF protein.

3.5 Preparation of Large-Scale Bacterial Culture for Protein Overexpression

1. Sterilize the 500 mL of LB broth (2.5 w/v % of LB media to water (DDI)) in 2 L Erlenmeyer flasks by autoclaving at 120 °C, 15 lbs. pressure for 60 min.
2. Cool down the media to room temperature, then add ampicillin antibiotic to the sterilized LB media (final ampicillin concentration: 100 μg/mL or 269 μM).
3. Add 25 mL of the starter culture (Subheading 3.3) to the mixture.
4. Grow the culture at 37 °C and 250 rpm until the OD at 600 nm reaches ~0.6 to 0.8 (~1.5–2 h).
5. Prepare 1 M Isopropyl-1-thio-B-D-galactopyranoside (IPTG) stock solution (2.38 gm of IPTG in 10 mL DI water).
6. Add 500 μL of 1 M IPTG stock solution (final IPTG concentration to 100 μg/mL) to the cell culture to induce expression.
7. Incubate the induced culture for 4–6 h at 220–250 rpm at 37 °C.
8. Centrifuge bacterial cell culture in centrifuge bottles (1 L) at 6000 rpm for 25 min at 4 °C.
9. Decant out the supernatant and resuspend the pellet in approximately 40 mL of buffer (10 mM phosphate buffer, 100 mM NaCl, pH = 7.2) and transfer to a new 50 mL tube (one 50 mL tube can contain harvested cells from up to 1 L of LB culture).
10. Centrifuge the cell suspension again at 6000 rpm for 15 min at 4 °C and discard the supernatant.
11. Cell pellets can be stored at −20 °C for future use or be used immediately in protein purification.

3.6 FGF1 Purification Using Heparin Affinity Chromatography

Heparin affinity chromatography is an adsorption-based technique in which biomolecules are separated specifically and reversibly by heparin-immobilized solid supports. Purification of FGF1 can be carried out well using established conditions for heparin-sepharose chromatography.

3.6.1 Column Packing

Protein purification methods, especially gradient elution, require efficient packing of the resins. If packed too densely, the resins might crack, which may cause channeling and breakthrough. On the other hand, a too loosely packed column can further compress, and liquid gaps in the column can occur. Uneven flow, loss of resolution, and band broadening are expected, which causes loss of product. Correct column packing helps avoid such issues and improves both consistency and performance (*see* **Note 4**).

1. Degas 15–20 mL of a 50% slurry of Heparin-Sepharose in 20% ethanol.
2. Pour the slurry into a 1.6 cm × 20 cm Econo UV column along the walls using a glass rod and allow gravity settlement of resin.
3. Tap the column softly while packing to make sure there is no formation of air gaps/bubbles.
4. Wash the resin with equilibration buffer with a constant flow rate of 1–2 mL/min for at least 5 column volumes of buffer (~100 mL).

3.6.2 Bacterial Cell Lysis and Loading

1. Add 35–40 mL of lysis buffer to the pellet from the −20 °C freezer.
2. Let it thaw for 20 min at room temperature (*see* **Note 5**).
3. Vortex the pellet to dissolve until the cell suspension becomes turbid.
4. Place the cell suspension on an ice bath for at least 10 min.
5. Ultrasonicate (amplitude #15 W output) the resuspended cells to lyse the cell membrane with an alternate cycle of 10 s of ON and OFF for 20 min.
6. Centrifuge the lysate at 19,000 rpm for 30 min at 4 °C using Nalgene centrifuge bottles.
7. Collect the clear supernatant into a clean falcon tube.

3.6.3 Using Heparin-Affinity Chromatography to Purify FGF1

1. Dispense 10 μL of supernatant, as the "supernatant" sample, for SDS-PAGE analysis.
2. Clear supernatant loaded onto pre-equilibrated Heparin-–Sepharose column at a flow rate of 0.2–0.4 mL/min (*see* **Note 3**).
3. Pass 1× PBS buffer wash to elute the non-binding elements from the supernatant.
4. Record the absorbance for bacterial contaminant proteins by a single wavelength detector (280 nm) until the absorbance reaches baseline.
5. Collect the eluent from this fraction in a clean falcon tube and label it as unbound.

6. Continue washing with 10 mM phosphate buffer with 100 mM NaCl.
7. Record the absorbance for bacterial contaminant proteins by a single-wavelength detector (280 nm) until the absorbance reaches baseline.
8. Collect the eluent from this fraction in a clean falcon tube and label it as 100 mM NaCl fraction.
9. Continue washing with 10 mM phosphate buffer with 300 mM NaCl.
10. Record the absorbance for bacterial contaminant proteins by a single-wavelength detector (280 nm) until the absorbance reaches baseline.
11. Collect the eluent from this fraction in a clean falcon tube and label it as 300 mM NaCl fraction.
12. Repeat the above **steps 9–11** for buffer fractions 10 mM phosphate buffer with 500 mM NaCl, 10 mM phosphate buffer with 800 mM NaCl, and 10 mM phosphate buffer with 1500 mM NaCl.
13. All the fractions should be immediately placed on ice after collection (*see* **Note 6**).

3.7 Immobilized Metal Affinity Chromatography is Used to Purify Metabolic FGFs

As stated previously, IMAC is an excellent choice to purify poly-histidine tagged protein as they offer notable separation of targeted proteins (which are tagged by poly-histidine) from the complex lysate. Since metabolic FGFs do not possess heparin-binding motifs nor have high heparin affinity, they cannot be purified using heparin affinity chromatography as FGF1 (*see* Subheading 3.6.2, **step 1**). Thus, IMAC and poly-histidine tag combination can be used to purify metabolic FGFs out of lysate after the large-scale protein overexpression experiment (Subheading 3.5).

3.7.1 Column Packing

Similar to heparin-binding chromatography, IMAC also requires efficient packing of the resins to avoid loss of resolution and product as well as improve both consistency and performance (*see* **Note 4**).

1. Make a 50% slurry of Ni-Sepharose in 20% ethanol (15–20 mL) and degas.
2. Using a Econo Chromatography column, pour the mixture down a glass rod along the wall of the column.
3. Allow the column to sit and settle. Occasionally, tap the column softly while packing to get rid of air gaps/bubbles.
4. Wash the resin with ddH_2O, about 10× the column volume (~200 mL), to make sure no ethanol remains on the column.

5. Wash the resin again with buffer (10 mM phosphate buffer, 100 mM NaCl, 25 mM $(NH_4)_2SO_4$, pH = 7.2) with a constant flow rate of 1–2 mL/min for at least 5 column volumes of buffer (~100 mL).

3.7.2 Bacterial Cell Lysis and Loading

1. Thaw the pellet from the −20 °C freezer, and add 30 mL of buffer (10 mM phosphate buffer, 100 mM NaCl, 25 mM $(NH_4)_2SO_4$, pH = 7.2) to the pellet.
2. Mix the pellet in the buffer until no solid is left in the mixture (*see* **Note 5**).
3. Incubate the cell suspension on ice.
4. Ultrasonicate (amplitude #15 W output) the mixture to rupture the cell membranes with an alternate cycle of 10 s of ON and OFF for 25 min.
5. Separate the cell debris from the lysate by centrifuge at 19,000 rpm for 25 min at 4 °C using Nalgene centrifuge bottles.
6. Collect the clear supernatant into a clean falcon tube.

3.7.3 Purification of Metabolic FGFs Using IMAC

1. Remove 10 μL of the supernatant, as supernatant sample, for SDS-PAGE analysis.
2. Add the supernatant after centrifuge onto a pre-equilibrated Ni-Sepharose column at a slow flow rate of 0.6–0.8 mL/min.
3. Allow the equilibration buffer to wash through the resins and elute the non-binding elements from the supernatant.
4. Record absorbance of eluents by a single-wavelength detector at 280 nm until the absorbance reaches baseline.
5. Collect all eluants in clean tubes and label as unbound.
6. Continue washing with 10 mM phosphate buffer with 20 mM imidazole.
7. Note the absorbance again using a single-wavelength detector at 280 nm until it reaches baseline.
8. Collect the eluent from this fraction in a clean falcon tube and label it as 20 mM imidazole fraction.
9. Repeat **steps 6–8** for buffer fractions of 50 mM imidazole, 100 mM imidazole, 250 mM imidazole and 500 mM imidazole. All fractions should be included with 10 mM phosphate buffer.
10. All fractions should be placed on ice immediately after collection (*see* **Note 6**).

3.8 Denaturing (SDS) Polyacrylamide Gel Electrophoresis (PAGE)

One-dimensional gel electrophoresis helps to separate and identify proteins under denaturing conditions by the addition of 0.1% SDS. Based on the molecular size, the proteins move through a polyacrylamide gel matrix toward the anode. The following gel size of 1 mm × 14 cm × 14 cm is the best fit for this protocol. FGFs' molecular weights range from 15–30 kDa. Thus, a 12–15% acrylamide gel provides adequate separation for data interpretation. The constitution of the gel can be adjusted as needed.

3.8.1 Gel Casting

1. Assemble the glass-plate sandwich of the electrophoresis using a clean thin short plate and a clean thick supporting plate.
2. Place the glass sandwich into a supporting clamp and/or into the casting stand as instructed by the manufacturer.
3. Check for leaks by filling the sandwich with water and observe the water level within the glass sandwich. Reassemble the glass sandwich with new glass plates and/or casting stand if leakage is observed. Remove the water from the glass sandwich if no leakage is observed.
4. Prepare the resolving gel solution as suggested in Table 1. Do not add 10% ammonium persulfate or TEMED until ready.
5. Add in the mixture the following solutions: 10% ammonium persulfate and TEMED, as suggested in Table 1, and mix gently.
6. Immediately pour the separating gel solution into the glass sandwich until the height of the solution between the glass plates is ~11 cm, leaving 2–3 cm empty at the top of the short plate. (Use the solution immediately; otherwise, it will polymerize in the tube).
7. Fill the top of the gel slowly with a layer of isopropyl alcohol.
8. Let the setup stay undisturbed at room temperature for at least 20–30 min.

Table 1
Materials for making SDS-PAGE gel

Materials	Resolving gel	Stacking gel
Total volume	8 mL	5 mL
ddH_2O	2.6 mL	2.6 mL
30% Acrylamide	3.2 mL	1 mL
0.5 M Tris pH 8.8	2 mL	–
0.5 M Tris pH 6.8	–	1.25 mL
10% SDS	80 μL	50 μL
10% APS	80 μL	50 μL
TEMED	8 μL	8 μL

3.8.2 Stacking Gel

1. Prepare the stacking gel solution as directed in Table 1. Do not add 10% ammonium persulfate or TEMED until ready.
2. Pour off the isopropyl alcohol layer from the glass sandwich. Remove the remaining isopropyl alcohol with dry wipes.
3. Pour into the mixture the following solutions as suggested in Table 1 and stir gently: 10% ammonium persulfate and TEMED.
4. Immediately pour the stacking gel solution slowly into the center of the sandwich along the edge of one of the spacers until the height of the solution in the sandwich is ~1 cm from the top of the plates.
5. Immediately, insert a 1 mm Teflon comb into the layer of stacking gel solution.
6. Let the stacking solution polymerize at room temperature for roughly 10–20 min.
7. Once polymerized, the gel can be stored at 4 °C in a damped paper towel covered in aluminum foil or used immediately.

3.8.3 Sample Preparation and Loading

1. Dispense the eluted fractions from the purification or protein samples into an Eppendorf tube (100–200 μL each). The volume of samples can change depending on how concentrated or diluted the sample is (*see* **Note 10**). Please check the Note section (Subheading 4) (*see* **Note 4**) for more details.
2. Add 10% v/v from the 100% TCA stock to the sample tube (i.e., if the sample is 100 μL, add 10 μL of the TCA).
3. Mix the sample well by vortex and centrifuge the sample at 13,000 rpm for 3 min.
4. Carefully discard the supernatant and add 500 μL of 100% acetone.
5. Vortex the solution and centrifuge the sample again for 3 min at 13,000 rpm.
6. Dispose of the supernatant completely using a pipette.
7. Heat the Eppendorf tube in a heat block for 30 s.
8. Dissolve the white pellet by adding 10–15 μL of 8 M urea solution.
9. Add loading dye to the urea layer up to a total volume of 20 μL and mix.
10. Remove the Teflon comb from the stacking layer of the prepared gels without tearing the edges of the polyacrylamide wells.
11. Place the prepared gels in an electrophoresis apparatus as instructed by the manufacturer.

12. Fill the apparatus with the appropriate electrode buffer (Tris-glycine buffer).
13. Load the samples slowly into the wells (15–20 μL).
14. Add a commercial protein marker in one of the lanes (3 μL) as needed.

3.8.4 Gel Running

1. Cover the sample chamber with the apparatus cover.
2. Start the gel run by connecting the power supply to the apparatus (100 mA of constant current and 200 V of constant voltage) until the bromophenol blue front has reached the bottom of the separating gel (~60 min).
3. Stop the run and remove the gel sandwich.
4. Without damaging the gel, carefully separate the gel from the glass plates and place it in water.
5. Remove the water and stain the gel with Coomassie blue staining solution for approximately 5 min.
6. Discard the staining solution and then wash the stained gel in water.
7. Add the de-staining solution to the gel and place it on a shaker for 6–8 h at room temperature. If desired, add a couple of Kimwipes to aid the de-staining process.
8. Once the stained gels' background turned clear, remove the de-staining solution and proceed to protein detection (Fig. 1).
9. Electrode running buffer, staining solution, and de-staining solution can be reused up to 5 times.
10. Sample cane be further analyzed using mass spectrometry to confirm the molecular weight and the identity of the product (Fig. 2).

3.9 Transferring the Protein from the Gel to the Membrane/ Western Blot

The western blotting technique is a common technique that detects and characterizes certain proteins in the sample. Gel electrophoresis (SDS-PAGE) is first used to resolve the sample mixture, and then the gel undergoes blotting or electro-transfer of samples from the gel to a nitrocellulose membrane. Using a specific type of antibody, certain specific proteins are to be detected in the membrane. Thus, western blot aids in detecting proteins of interest in a mixture. The technique provides more information about the targeted protein's molecular weight, and abundance in a sample mixture (*see* **Note 8**).

3.9.1 Set Up Blotting Sandwich and Transfer

1. Perform SDS-PAGE as stated in Subheading 3.8. The resulting gel is used for blotting/transferring.

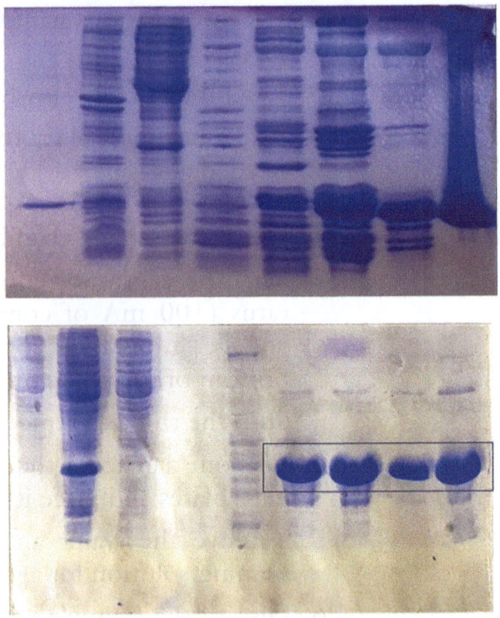

Fig. 1 SDS-PAGE of purification FGF1 (15.9 kDa) and FGF19 (21.8 kDa). (**a**) *FGF1 purification.* Lane 1: Supernatant; Lane 2: unbound; Lane 3: unbound 2; Lane 4: 100 mM NaCl & 10 mM phosphate buffer fractions; Lane 5: 300 mM & NaCl 10 mM phosphate buffer fractions; Lane 6: 500 mM & NaCl 10 mM phosphate buffer fractions; Lane 7: 800 mM & NaCl 10 mM phosphate buffer fractions; Lane 8: 1500 mM & NaCl 10 mM phosphate buffer fractions. (**b**) *FGF19 purification.* Lane 1: Pellet; Lane 2: Supernatant; Lane 3: 20 mM imidazole & 10 mM phosphate buffer fractions; Lane 4: 50 mM imidazole & 10 mM phosphate buffer fractions; Lane 5: 100 mM imidazole & 10 mM phosphate buffer fractions; Lane 6: 250 mM imidazole & 10 mM phosphate buffer fractions; Lane 7: 500 mM imidazole & 10 mM phosphate buffer fractions; Lane 8: FGF1 marker for comparison

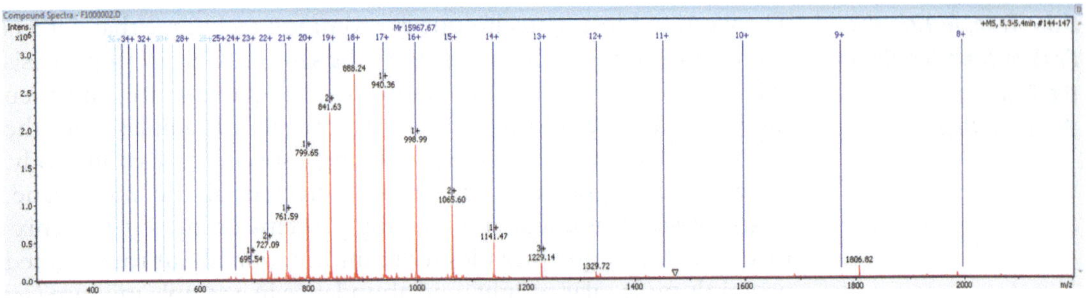

Fig. 2 Mass spectrometry of protein product after purification. Mass spectrometry confirms the molecular weight of the product to be 15.9 kDa, matching the actual FGF1 molecular weight

2. Soak the following items, sponge, western blot cassette, 2 filter papers, and transfer membrane (nitrocellulose membrane) for 5 min in Towbin solution.

3. Assemble the blotting stack as instructed by the manufacturer. In this case, the order is bottom cassette – sponge – filter paper – gel – nitrocellulose membrane – filter paper – top cassette. Please see the notes section for further details.
4. Set up the reservoir as instructed by the manufacturer.
5. Add Towbin buffer into the reservoir until full. (Usage of stirrer and ice bags in the reservoir is recommended due to the high amount of heat generated during the transfer process).
6. Adjust the run parameters to 150 V, 75 mA, then connect the reservoir to the electrophoresis setup to carry on blotting for 120 min.
7. Carefully separate the nitrocellulose membrane from the blotting stack into a clean box.
8. Soak the membrane in 5% skim milk (w/v%) in a new TBS-T solution and keep on a rocker for 30 min.
9. Wash the membrane with 5 mL of TBS-T solution.
10. Add 5 mL of 0.2% BSA in TBS-T to the membrane, followed by the addition of a primary anti-6xHistidine alkaline phosphatase-conjugated antibody (2 μL).
11. Incubate overnight at room temperature with continuous rocking.
12. Wash the membrane with TBS-T solution 3 times with 3 mL each time to remove excess antibodies.
13. Add 1 mL of NBT/BCIP solution to the membrane, then incubate for 5–10 min at room temperature on a rocker.
14. Proceed to protein detection (Fig. 3).

3.10 Isothermal Titration Calorimetry (ITC)

ITC is a common and useful technique to observe the interactions of ligand-receptor in solution. The dissociation constant K_d has an inverse relationship with the binding affinity. Thus, the more the K_d value is, the less the binding affinity the protein has for the ligand.

1. Degas the protein samples by gently stirring under a vacuum for 15 min.
2. Clean the calorimeter reaction cell with buffer.
3. Remove the buffer from the reaction cell using the 3 mL syringe.
4. Load the sample into the reaction cell using the 2.5 mL Hamilton syringe. Gently place the needle of the syringe on the bottom of the reaction cell and slowly inject the sample while moving the syringe up to dislodge any air bubbles that may otherwise become trapped in the reaction cell.
5. Stop injecting when the sample can be observed to be coming up out of the reaction cell port.

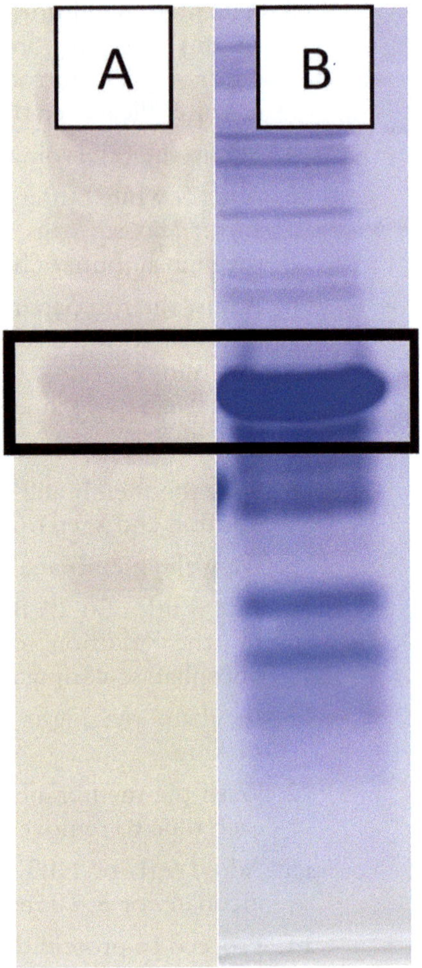

Fig. 3 Western blot of FGF21. Western blot using anti-6x histidine antibody showed that the protein product also has polyhistidine tag. Lane A: FGF21 after transferring to nitrocellulose membrane and detect via anti-6x His alkaline phosphatase-conjugated antibody. Lane B: same FGF21 sample in corresponding SDS-PAGE gel

6. Remove the excess solution from around the cell porthole but leave the sample cell filled up to the top of the reaction cell opening.
7. Load the syringe with a heparin sample (syringe capacity up to 40 μL).
8. Using the Origin Version 7.0, set the temperature of the reaction cell to 25 °C.
9. Initiate the titration according to the manufacturer's instructions.
10. Carry on the titration sequentially with 1.3 μL aliquots from the syringe to the cell (FGF1) with a 12-s interval between

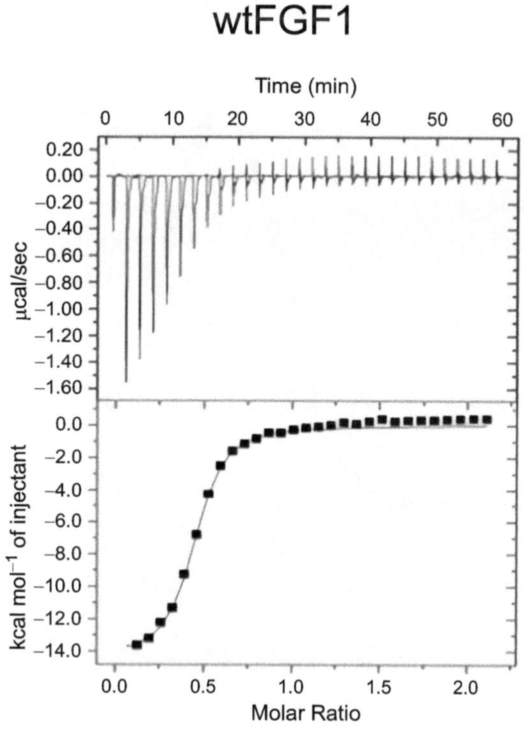

Fig. 4 Isothermal Titration Calorimetry graph of purified FGF1 product. From the data of the graph, the product of purification showed a calculated binding affinity to heparin at a value of $K_d = 1.8$ μM, which closely resembles FGF1, further confirming the product identity after purification [12]

injections to allow each point along the titration sufficient time to reach equilibrium.

11. Refer to the manufacturer's instructions for how to set the parameters in the Origin program.
12. The given concentration of FGF1 sample and heparin is in a molar ratio of 10:1 (*see* **Note 11**). Make sure you follow the same ratio to carry out the titration.
13. After completion of the ITC titrations, clean and store the calorimeter cell as recommended by the instrument manufacturer.
14. Analyze the binding data using the Origin Version 7.0 version provided by the instrument manufacturer (Fig. 4).

3.11 Differential Scanning Calorimetry

Analyzing the thermal stability of a protein in a dilute solution involves determining changes in the partial molar heat capacity of the protein at constant pressure (ΔCp). Differential scanning calorimetry (DSC) is an analytical technique that measures the molar heat capacity of samples as a function of temperature. It measures the thermal transition temperature (melting temperature; Tm) and

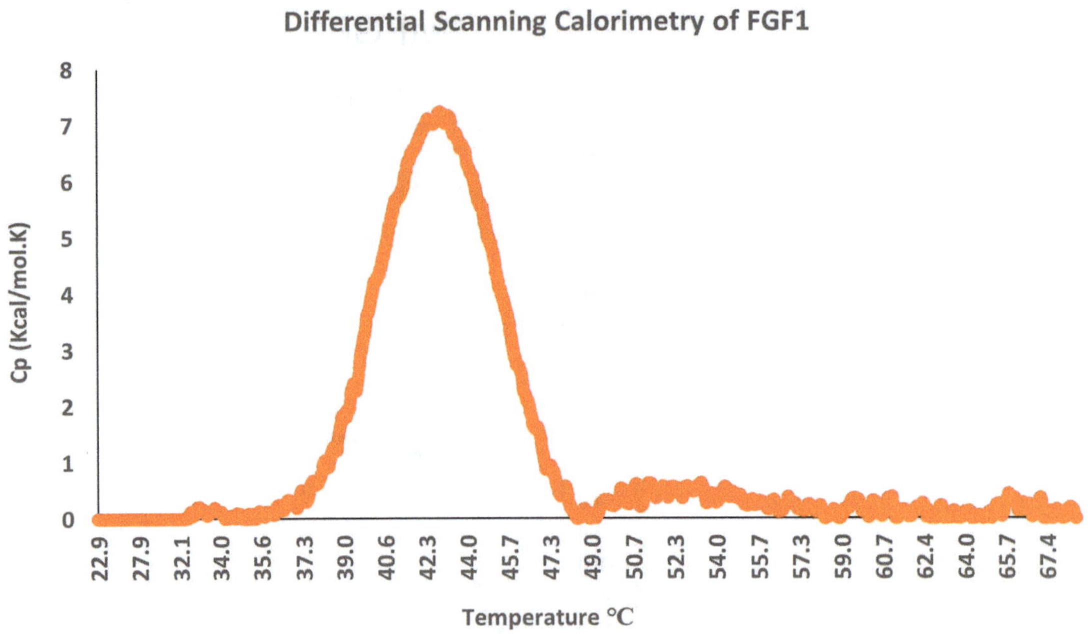

Fig. 5 Differential scanning calorimetry of FGF1. Thermograms of heat-treated wtFGF1 showing a T_m of ~43 °C, matching previously reported FGF1's T_m indicating the purified product is FGF1 [12]

the energy required to disrupt the interactions stabilizing the tertiary structure (ΔH) of proteins.

3.11.1 DSC Run

1. Start the machine and open the Microcal PEAQDSC program.
2. Clean the sample and reference cell with DI water 2–3 times (Follow the instructions manual for a cell cleaning procedure).
3. Rinse both the cells with buffer 2–3 times.
4. Fill the sample and reference cell with buffer (250 μL), then tighten the lid tightly until the program shows the cell pressure above 40 psi.
5. Set the run with 3–4 buffer runs before you run the sample.
6. Program command a run according to your convenience.
7. Program command the temperatures between 25 and 85 °C with a ramping temperature rate of 2 °C/min to buffer – buffer baseline of the instrument.
8. Replace the buffer with the protein solution (250 μL) before the sample run start.
9. After the completion of a run, analyze the results using the program (Fig. 5).
10. After completion of the experiment, clean cells as recommended by the instrument manufacturer.

3.12 Circular Dichroism

Circular dichroism (CD) spectroscopy is a very important technique in structural biology for examining the folding, structural changes, and especially secondary structures of proteins. Its usage as a quantitative/qualitative method has been based on empirical methods that use a variety of computational algorithms with reference databases composed of spectra of known protein (crystallographic) structures. These algorithms permit the determination of the secondary structure of an unknown protein. In addition, new computational and bioinformatic methods have been developed, and new reference databases have been created, which greatly improve and facilitate the analyses of the secondary conformation of proteins using CD spectra. CD spectrum can provide a complex fingerprint that can be compared with that of a specimen that has been well-authenticated as being native and functional. This constitutes an empirical but powerful means of characterization.

3.12.1 Recording a CD Spectrum

1. Set up the CD spectrometer by purging the optics with Nitrogen gas, turning on the coolant, and finally switching on the lamp.
2. Allow the machine to warm up for 15 min. Regulate the thermostat system to the 25 °C temperature.
3. Enter the settings in the required program for the scan.
4. Set wavelength. For far-UV spectra, it is usual to scan between 250 nm and 180 nm.
5. Set bandwidth as 10 nm.
6. Set averaging time or time constant, scan speed, and the number of accumulations following the instructions in the manual.
7. Clean and dry a 0.1 cm quartz cuvette for experimental purposes.
8. Fill the cuvette with buffer and place it in the cell holder.
9. Scan the baseline using the same cell and buffer with above mentioned same instrument settings.
10. Remove the buffer solution, rinse the cell with water and dry, refill the cell with the clear protein solution, and scan the sample using the same instrument settings.
11. Save the raw data at a desired location. Proceed to analyze this data to produce the CD graph by graphing ellipticity (mdeg) against wavelength (nm). In order to compare different spectra together, each dataset needs to be normalized with the sample's concentration to produce the sample's molar ellipticity as shown in Fig. 6.

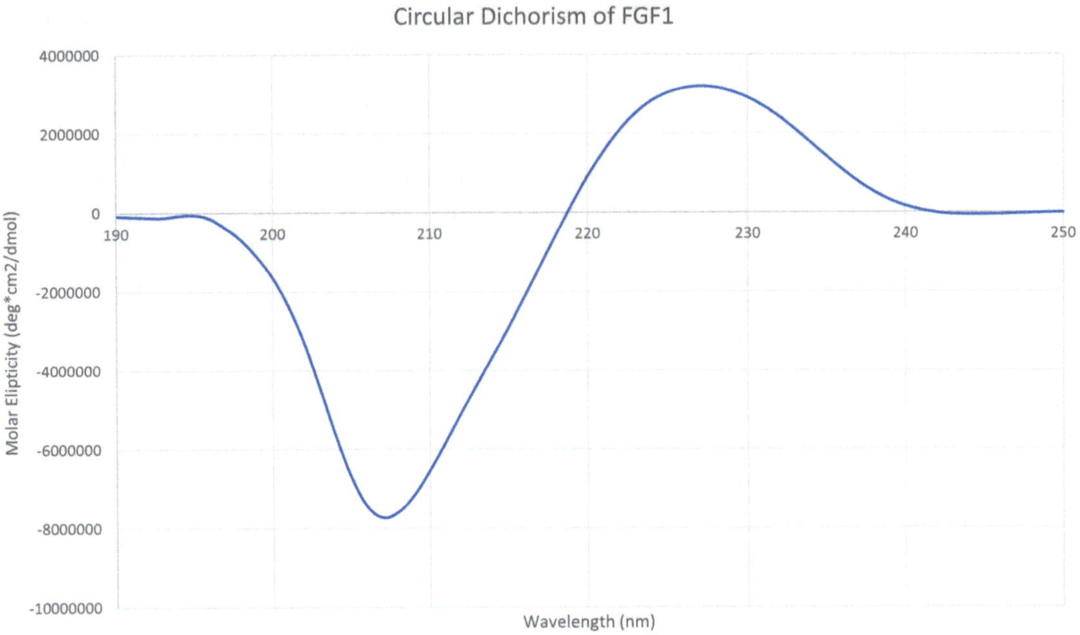

Fig. 6 Far UV CD spectra of FGF1 product. The product showed typical β-barrel conformation that is common in reported FGF1 secondary structure. It has two strong signals for β-barrel conformation: positive peak at 228 nm and negative trough at 207 nm, indicating the product maintains native conformation of FGF1 [12]

3.13 Intrinsic Fluorescence Spectroscopy

Protein samples should be kept on ice when measurements are not being taken. The concentration required ranges from 0.3 mg/mL to 0.5 mg/mL.

3.13.1 Preparation

1. Purge the fluorescence spectrophotometer with nitrogen gas for 10 min or as recommended by the manufacturer.
2. After 10 min, turn on the fluorescence spectrophotometer and lamp, and open the corresponding software on the computer.
3. A quartz fluorescence cuvette (1-cm path length, 1.4 mL) should be cleaned.

3.13.2 Settings

1. The fluorescence profile can be obtained by setting the excitation wavelength at 280 nm and recording the emission from 300 nm to 450 nm.
2. Set the number of iterations to 3 or more as needed.
3. Enable the shutter to open and close automatically.
4. Indicate the file format, name, and save location for the file.
5. Set relative fluorescence intensity range from 500 RFI to 1100 RFI or as needed.

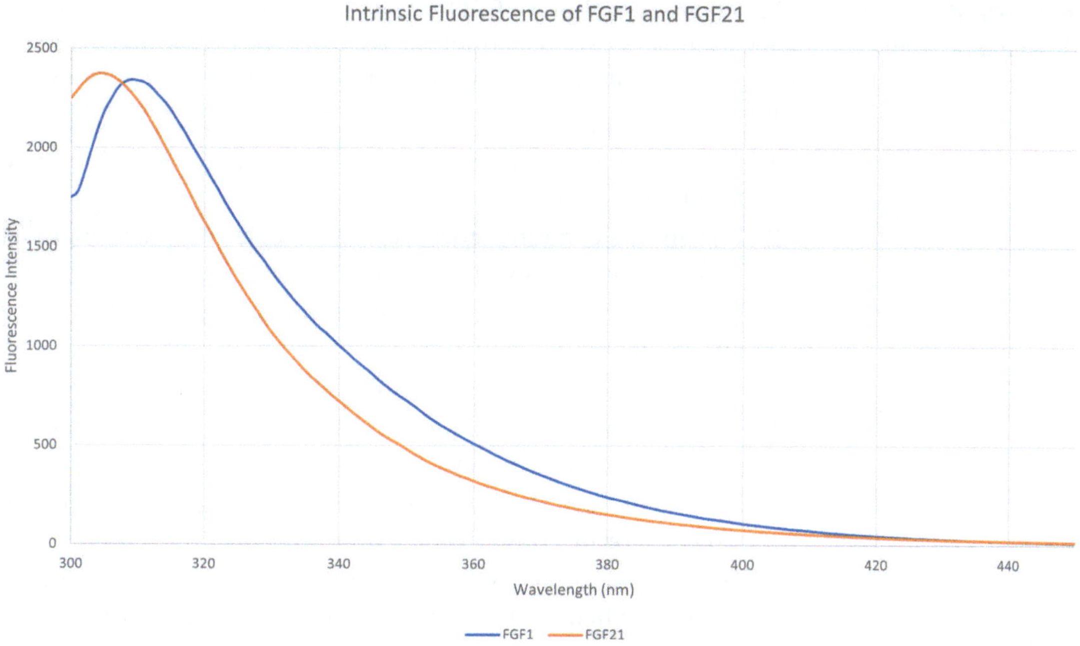

Fig. 7 Intrinsic fluorescence spectra of FGF1 and FGF21 products. The FGF1 product (in blue) showed typical tyrosine emission at 310 nm. FGF1 has a tryptophan residue in its sequence, but does not display tryptophan emission at 350 nm, which is a unique feature of FGF1 protein, indicating the product conforms to the native tertiary structure of FGF1 [12]. Similarly, the FGF21 product (in green) also showed tyrosine emission at 305 nm since FGF21 does not have tryptophan residue in its sequence

3.13.3 Measurements

1. Measure the fluorescence of an empty cuvette in the holder.
2. If a fluorescence peak is observed, the cuvette will need to be cleaned again until no fluorescence signal is obtained in the spectra.
3. Add 1 mL of the buffer (10 mM phosphate, 100 mM NaCl, pH = 7) to the cuvette and measure the fluorescence.
4. Dispose of the buffer and place 1 mL of the sample in the cuvette.
5. Measure the fluorescence of the sample.

3.13.4 Analysis

1. Open the saved spectra measurements and export them in the desired format.
2. Subtract the fluorescence measurements of the buffer from the samples' fluorescence measurements.
3. Plot the graph of the relative fluorescence intensity (RFU) versus the wavelength (nm) to obtain the intrinsic fluorescence of the targeted FGF as shown in Fig. 7.

3.14 Standard Urea Denaturation Assay

The chemical stability of FGFs can be evaluated with denaturants such as urea. A urea denaturation assay may be performed manually or using a titration machine. In both cases, the parameters are the same.

1. Turn the fluorescence machine on following the steps indicated in Subheading 3.13.
2. For intrinsic fluorescence, set the excitation wavelength at 280 nm and monitor the emission from 300 nm to 450 nm.
3. Ensure that the cuvette is clean by measuring the RFU of the empty cuvette.
4. Measure the buffer for later subtraction.
5. Place 0.5 mL of the protein sample in the cuvette and measure the fluorescence intensity.
6. Add 20 μL of 8 M urea to the protein and mix thoroughly with a pipette or by constant stirring at 300 rpm.
7. If performed manually, keep adding urea until saturation is reached (when the RFI at increased urea concentration remains stable).
8. If a titration machine is used, set the end point at 8 M of the final urea concentration in the cuvette.

Analysis

9. Subtract the buffer from each of the trials.
10. Calculate the ratio of the peaks at 305 nm and 350 nm after buffer subtraction (called ratio, R).
11. Follow the equation below:

$$\text{Fraction unfolded} = \frac{R \text{ of the first Urea conc.} - R \text{ of individual Urea conc.}}{R \text{ of the first Urea conc.} - R \text{ of the last Urea conc.}}$$

12. Plot the graph of the fraction unfolded (y-axis) vs. the concentration of urea in molar (M) at each urea addition.
13. The C_m value is given by the equation of the plot and can be approximated to be the urea concentration corresponding to the 0.5 (or 50%) of the fraction unfolded (Fig. 8).

3.15 Alternative Urea Denaturation Assay/ Sypro Orange Assay

In some cases, especially the metabolic FGFs, the protein may not have any major emission shifts and display only one maximum emission peak (~350 nm). Thus, they are unable to undergo regular Urea Denaturation Assay, which requires a fluorescent emission shift between two emission peaks. In this case, a fluorescence dye such as Sypro Orange can be used. The emission range monitored, as well as the excitation wavelength, will depend on the dye. In the following, the emission range is set based on the Sypro Orange dye:

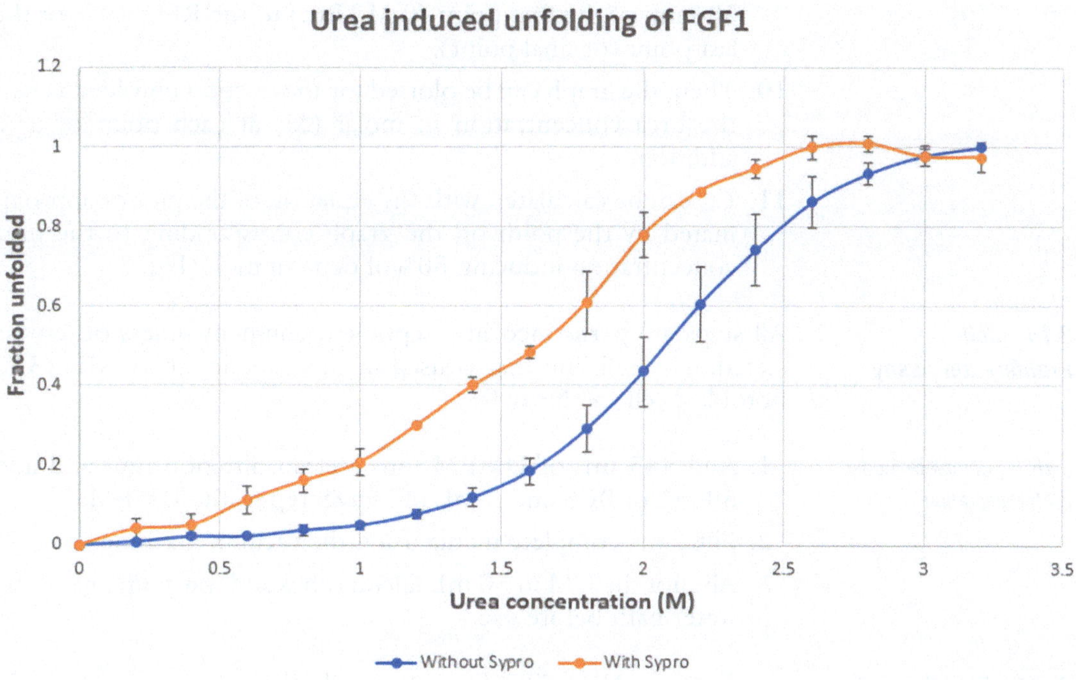

Fig. 8 Urea-induced unfolding of FGF1 product using the standard method and the alternative/SPYRO method. The standard method (in blue) showed a C_m of ~2 M, while the Spyro method reported a C_m of 1.6, resembling the reported C_m of FGF1 [12]

1. For Sypro Orange, the concentration in the final volume (after adding urea) is 10×.
2. Prepare a protein sample of 0.3–0.5 mg/mL.
3. Calculate the needed amount of Sypro Orange based on the concentration of stock Sypro Orange dye and the final total volume of samples (i.e., from a 5000× stock of Sypro Orange and an expected final volume of 1.5 mL, 3 μL of dye should be added from the beginning).
4. Place the calculated volume of Sypro Orange in the 500 μL of protein sample and mix thoroughly.
5. Set the excitation at 491 nm and monitor the emission from 510 nm to 650 nm.
6. Set the *y*-axis maximum to 500.
7. Measure the protein fluorescence emission at 0 M of urea.
8. Add 20 μL of 8 M urea at a time and measure until saturation or to the final mixture volume of 1.5 mL.

Analysis

9. When using Sypro Orange, the fraction unfolded can be defined as the ratio of fluorescence intensity value (RFU) at

586 nm of the first point (0 M Urea) to the RFU value of the last point (or final point).

10. Then, the graph can be plotted for the fraction unfolded versus the urea concentration in molar (M) at each point of urea addition.
11. C_m can be calculated with the equation of the plot or approximated by the point on the graph corresponding to the urea concentration inducing 50% of denaturation (Fig. 8).

3.16 Cell Proliferation Assay

All steps are performed in a septic environment unless otherwise stated. The cell line that is used in this experiment is NIH/3T3 fibroblast cell (*see* **Note 9**).

3.16.1 Complete Media (CM) Preparation

1. Add 445 mL of DMEM into a clean filtration unit and add 50 mL of BCS and 5 mL of Pen/Strep to the DMEM.
2. Filter the complete media using the vacuum machine.
3. Aliquot the CM in 50 mL falcon tubes and warm up in a 37 °C water bath before use.

3.16.2 Expanding Cells

1. Retrieve NIH-3T3 fibroblast cells that are stored in liquid nitrogen (vapor phase).
2. Thaw the cells in a 37 °C water bath until cells are completely thawed (~1–2 min).
3. Transfer 1 mL of cells into a sterile 15 mL falcon tube.
4. Add 10 mL of warm CM into the falcon tube with cells (add slowly and dropwise after 5 mL).
5. Centrifuge cells at 500*g* for 7 min at 37 °C.
6. Slowly get rid of the liquid, but ensure the pellet is not removed.
7. Add 1 mL of complete media to the cell pellet and carefully mix.
8. Add 15 mL of CM into a 75-cm^2 cell culture flask.
9. Transfer 1 mL of cell suspension from the previous step into the cell culture flasks.
10. Incubate the cells at 37 °C in an incubator with 5% CO_2 until 80% confluence is reached.

3.16.3 Detaching Cells

1. Thaw Trypsin-EDTA (stored at −60 °C) and PBS (stored at 4 ° C) in a 37 °C water bath.
2. Remove the CM in the cell culture flask.
3. Wash cells twice with 3 mL of PBS and discard PBS.
4. Add 3 mL of Trypsin-EDTA (ensure the base of the flask is covered).

5. Incubate at 37 °C in an incubator with 5% CO_2 for 7–8 min.
6. Check the flask under the microscope to confirm if cells have fully detached or not.
7. Transfer the detached cells in CM into a new 15 mL tube.
8. Add slowly 7 mL of CM into the tube (for a total of 10 mL).
9. Centrifuge cells at 500*g* for 7 min at 37 °C.
10. Carefully remove the supernatant without removing the pellet.
11. Add 1 mL of warm CM into the 15 mL tube and mix well.
12. Obtain the cell counts per milliliter.

3.16.4 Cell Proliferation Assay

1. Once cells reach 80% confluency, remove the CM in the flask and add 15 mL of DMEM (incomplete media) to starve the cells.
2. Incubate at 37 °C in an incubator with 5% CO_2 for 12–15 h.
3. Use white 96-well plates for luminescence (or black 96-well plates for fluorescence).
4. Pre-coat wells with PDL (final concentration – 2 μg /well, or 3–10 μg/cm^2) for 3–12 h.
5. Remove PDL after incubation and wash wells with PBS twice and discard PBS.
6. Detach the cells as described earlier and get the total cell count per mL.
7. Obtain viable cell count per mL.
8. Dilute the cell suspension to 200,000 cells/mL in CM.
9. Add 50 μL from cell suspension with 200,000 cells/mL (cell density = 10,000 cells/well) into each well.
10. Filter the protein to remove endotoxins.
11. Dilute the protein in CM up to the desired concentrations.
12. Add appropriate amounts of diluted protein into each well based on your plate template. (Maximum volume of 96-well is 200 μL. But the working volume is 100 μL).
13. Cell volume and protein volume should not exceed 50 μL each. Both volumes should add up to 100 μL per well.
14. Incubate the plate at 37 °C in the incubator with 5% CO_2 for 24 h.
15. Obtain cell counts per mL for each well.
16. Plot cell counts for each concentration to compare to the control group (0 ng/mL FGF1) and observe the increase in cell counts (Fig. 9).

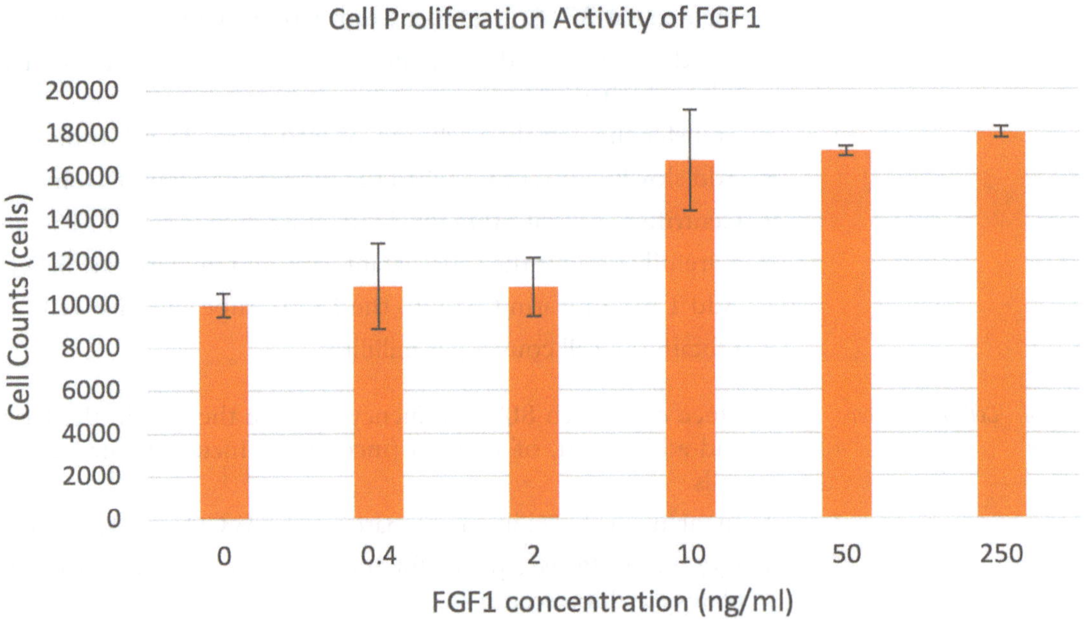

Fig. 9 Cell proliferation activity of FGF1. The FGF1 is known for its ability to induce cell proliferation through receptors. The assay above showed that the FGF1 product, after purification, was able to induce more cells to proliferate as more protein was added to the cells compared to control (where no FGF1 was added or 0 ng/mL). This result indicates the purified product also retained the biological activity of FGF1 after purification using the above method

4 Notes

1. One of the most crucial factors to be considered during cloning is the selection of the expression host(s). *Escherichia coli* is preferred for producing recombinant proteins due to their ease of handling, cost-effectiveness, rapid growth, and ability to express proteins at high levels (Subheading 3.2).
2. The protein expression yields are influenced by parameters such as growth temperature, antibiotic concentration, the time at which expression is induced, and the concentration of the inducer (IPTG). These parameters require optimization (Subheading 3.4).
3. Due to the relatively slow binding kinetics between protein targets and resins, it is important to keep the flow rate low during sample loading for maximum binding capacity (Subheading 3.6.3).
4. The efficiency of both the Heparin-Sepharose and Ni-Sepharose matrices depends on proper maintenance and storage of the column. The most suitable solvent to store the column is 20% ethanol with 1 mM sodium azide at 4 °C (Subheadings 3.6.1 and 3.7.1).

5. For metabolic FGFs which are prone to protease cleavage, it is recommended to include protease inhibitors such as PMSF during ultrasonication to prevent protein degradation (Subheadings 3.6.2 and 3.7.2, **step 2**). Alternatively, having EDTA (final concentration in lysate ~1 mM) is also recommended to inhibit metalloprotease in the lysate.
6. Some crucial parameters that must be carefully considered are additives to the storage buffer after purification (Subheadings 3.6.3 and 3.7.3). Such additives can help stabilize the products and/or extend the shelf-life, prevent degradation, etc. Each FGFs may require different storage buffers with slight changes to the common 10 mM phosphate buffer, 100 mM NaCl, 25 mM $(NH_4)_2SO_4$, pH = 7.2. Some additives to be considered are glycerol (5–10%), EDTA-Na (1 mM), β-Mercaptoethanol (1 mM), etc.
7. In this protocol, FGF19, FGF21, and FGF23 are characterized with the His-tag fused to the protein. To remove the His-tag, an Enterokinase cleavage site is recommended to be placed between the His-tag and the target protein. Enterokinase is a protease that cleaves after lysine at its cleavage site Asp-Asp-Asp-Asp-Lys leaving no extra amino acid on the target protein sequence (Subheading 3.2).
8. Western blot protocol (Subheading 3.9) differs between different manufacturers. Please adjust the assembly of the western blot cassettes, run time, voltage and current of electrophoresis, and the use of a secondary antibody as necessary. Western blot protocol in this chapter will only be applied to proteins with poly-histidine tags (metabolic FGFs) and not others, as the antibody used here is specific for only such tags.
9. Cell proliferation assay protocol (Subheading 3.16) in this book chapter is used to measure only FGF1 product after purification as it has been reported previously [1] and, thus, can be us as verification of biological activity of the purified product.
10. The sample volumes for SDS-PAGE sample preparations (Subheading 3.8.3) are often varied depending on how much proteins are present within the fractions or samples. Generally, the more proteins present in the fractions, the less volumes are needed for sample preparations. If the sample for SDS-PAGE is too concentrated, the pellet may not be able to dissolve completely in 8 M urea and/or may appear streaky on the gel after the SDS-PAGE experiment. However, if the sample is too diluted, no pellet will be visible after the TCA addition and/or may not show on the gel after the SDS-PAGE run.

11. In the ITC experiment, the minimum molar ratio between FGF1 and heparin is 1:10, but the working ratio can be varied depending on the specific experiment parameters (Subheading 3.10).

Proteins Addressed in This Book Chapter

(i) *FGF1*

(ii) *FGF19*

(iii) *FGF21*

(iv) *FGF23*

Acknowledgments

We would like to thank the AIMRC NIH COBRE (P20GM139768), Department of Energy (DE-FG02-01ER15161), University of Arkansas Honors College for financial support. T.K.S.K. is the Mildred – Cooper Chair of Bioinformatics and would like to gratefully acknowledge this endowment grant for funding this research.

References

1. Agrawal S (2020) Design and characterization of FGF-1 mutant (s) with increased stability and enhanced bioactivity. University of Arkansas
2. Agrawal S, Govind Kumar V, Gundampati RK, Moradi M, Kumar TKS (2021) Characterization of the structural forces governing the reversibility of the thermal unfolding of the human acidic fibroblast growth factor. Sci Rep 11:15579. https://doi.org/10.1038/s41598-021-95050-2
3. Benet-Pagès A, Orlik P, Strom TM, Lorenz-Depiereux B (2004) An FGF23 missense mutation causes familial tumoral calcinosis with hyperphosphatemia. Hum Mol Genet 14:385–390. https://doi.org/10.1093/hmg/ddi034
4. Bornhorst JA, Falke JJ (2000) Purification of proteins using polyhistidine affinity tags. Methods Enzymol 326:245–254. https://doi.org/10.1016/s0076-6879(00)26058-8
5. Carpenter TO et al (2018) Burosumab therapy in children with X-linked hypophosphatemia. New Engl J Med 378:1987–1998. https://doi.org/10.1056/nejmoa1714641
6. Crowe J, Döbeli H, Gentz R, Hochuli E, Stüber D, Henco K (1994) 6xHis-Ni-NTA chromatography as a superior technique in recombinant protein expression/purification. Methods Mol Biol 31:371–387. https://doi.org/10.1385/0-89603-258-2:371
7. Dolegowska K, Marchelek-Mysliwiec M, Nowosiad-Magda M, Slawinski M, Dolegowska B (2019) FGF19 subfamily members: FGF19 and FGF21. J Physiol Biochem 75:229–240. https://doi.org/10.1007/s13105-019-00675-7
8. Dunshee DR et al (2016) Fibroblast activation protein cleaves and inactivates fibroblast growth factor 21. J Biol Chem 291:5986–5996. https://doi.org/10.1074/jbc.M115.710582
9. Gaich G et al (2013) The effects of LY2405319, an FGF21 analog, in obese human subjects with type 2 diabetes. Cell Metab 18:333–340. https://doi.org/10.1016/j.cmet.2013.08.005
10. Gillum MP, Potthoff MJ (2016) FAP finds FGF21 easy to digest. Biochem J 473:1125. https://doi.org/10.1042/BCJ20160004
11. Ho BB, Bergwitz C (2021) FGF23 signalling and physiology. J Mol Endocrinol 66:R23–R32. https://doi.org/10.1530/JME-20-0178
12. Kerr R et al (2019) Design of a thrombin resistant human acidic fibroblast growth factor

(hFGF1) variant that exhibits enhanced cell proliferation activity. Biochem Biophys Res Commun 518:191–196. https://doi.org/10.1016/j.bbrc.2019.08.029

13. Kharitonenkov A et al (2013) Rational design of a fibroblast growth factor 21-based clinical candidate, LY2405319. PLoS One 8:e58575. https://doi.org/10.1371/journal.pone.0058575
14. Lee S et al (2018) Structures of β-klotho reveal a 'zip code'-like mechanism for endocrine FGF signalling. Nature 553:501–505. https://doi.org/10.1038/nature25010
15. Liu SH, Xiao Z, Mishra SK, Mitchell JC, Smith JC, Quarles LD, Petridis L (2022) Identification of small-molecule inhibitors of fibroblast growth factor 23 signaling via in silico hot spot prediction and molecular docking to α-Klotho. J Chem Inf Model 62:3627–3637. https://doi.org/10.1021/acs.jcim.2c00633
16. Lu Y, Feng JQ (2011) FGF23 in skeletal modeling and remodeling. Curr Osteoporos Rep 9: 103–108. https://doi.org/10.1007/s11914-011-0053-4
17. Ornitz DM, Itoh N (2015) The fibroblast growth factor signaling pathway. Wiley Interdiscip Rev Dev Biol 4:215. https://doi.org/10.1002/wdev.176
18. Phan P, Saikia BB, Sonnaila S, Agrawal S, Alraawi Z, Kumar TKS, Iyer S (2021) The saga of endocrine FGFs. Cells 10:2418. https://doi.org/10.3390/cells10092418
19. Porath J (1992) Immobilized metal ion affinity chromatography. Protein Expr Purif 3:263–281. https://doi.org/10.1016/1046-5928(92)90001-d
20. Shimada T et al (2001) Cloning and characterization of FGF23 as a causative factor of tumor-induced osteomalacia. Proc Natl Acad Sci U S A 98:6500–6505. https://doi.org/10.1073/pnas.101545198
21. Shimada T et al (2002) Mutant FGF-23 responsible for autosomal dominant hypophosphatemic rickets is resistant to proteolytic cleavage and causes hypophosphatemia in vivo. Endocrinology 143:3179–3182. https://doi.org/10.1210/endo.143.8.8795
22. Somm E, Jornayvaz FR (2018) Fibroblast growth factor 15/19: From basic functions to therapeutic perspectives. Endocr Rev 39:960–989. https://doi.org/10.1210/er.2018-00134
23. Suzuki Y et al (2020) FGF23 contains two distinct high-affinity binding sites enabling bivalent interactions with α-Klotho. Proc Natl Acad Sci U S A 117:31800–31807. https://doi.org/10.1073/pnas.2018554117
24. Zhen EY, Jin Z, Ackermann BL, Thomas MK, Gutierrez JA (2016) Circulating FGF21 proteolytic processing mediated by fibroblast activation protein. Biochem J 473:605–614. https://doi.org/10.1042/BJ20151085
25. Zhu L et al (2021) Dynamic folding modulation generates FGF21 variant against diabetes. EMBO Rep 22:e51352. https://doi.org/10.15252/embr.202051352

Chapter 11

Production and Purification of Antibodies in Chinese Hamster Ovary Cells

Lauren Stuart

Abstract

Antibodies are versatile biological molecules with widespread applications in research and medicine. This protocol outlines the generation of monoclonal IgG antibodies from Chinese hamster ovary cells. It includes steps for cell maintenance, transient transfection, and antibody purification via protein A affinity chromatography. The methods described are intended for the production of milligram amounts of protein but can be adapted for most small- to mid-scale applications.

Key words Recombinant protein expression, Monoclonal antibody, Immunoglobulin, Transient transfection, Protein A affinity chromatography

1 Introduction

Antibodies are naturally occurring glycoproteins produced by the immune system in response to infection or vaccination. Their ability to bind specific targets with high affinity makes them a useful scientific tool. Recombinant antibodies have been used to treat numerous medical conditions, including cancer, autoimmune diseases, and viral infections [1, 2]. They are also a key reagent in flow cytometry, western blots, and several other biological assays [3]. The most common antibody format used for pharmaceutical and research purposes is the ~150 kDa immunoglobulin G (IgG1) molecule [4].

A variety of expression systems exist for the production of antibodies in vitro. Bacterial and yeast cells are regularly used to make antibody fragments, while mammalian cells are effective for the generation of full-length IgGs [5]. Chinese hamster ovary (CHO) cells are frequently used to manufacture pharmaceutical antibodies. CHO cells are advantageous because they offer both high expression levels and advanced mechanisms of protein folding and post-translational modification [6]. Because these cells

Steven B. Bradfute (ed.), *Recombinant Glycoproteins: Methods and Protocols*, Methods in Molecular Biology, vol. 2762, https://doi.org/10.1007/978-1-0716-3666-4_11,

originate from hamsters, they are unlikely to be contaminated with human viruses, reducing the risk of production loss and improving researchers' safety [7].

Protein expression in mammalian cells is accomplished by delivering the gene of interest to the cell nucleus in a process called transfection. During stable transfection, the introduced DNA is integrated into the nuclear genome, allowing it to be expressed continuously [8]. Establishing a stable cell line requires time-consuming selection and amplification steps, making it ideal for large-scale protein production, but less useful for small-scale expression [7]. This chapter focuses on the process of transient transfection, which introduces DNA into the nucleus, but does not integrate it into the genome. The desired protein is only expressed for a limited period of time but can be generated much more rapidly [8]. Chemical-based transfection methods rely on the use of positively charged species like calcium phosphate or cationic lipids [9]. These chemicals form complexes with the negatively charged DNA to facilitate transport into the cell nucleus [8, 9].

Once a protein is expressed, it needs to be isolated from the other components of the cell mixture. Several characteristics of proteins, such as size, charge, and binding affinity, can be taken advantage of for their purification [10]. Affinity chromatography is a fast and highly selective method of purification driven by interactions between the protein of interest and an immobilized ligand [3]. The affinity-based purification of IgG antibodies can be done using a bacterial-derived ligand called protein A [11]. The constant region of the antibody binds to protein A and all non-binding species are washed away [12]. The antibody is then eluted using a low pH buffer, which disrupts the associations between the antibody and the ligand.

2 Materials

2.1 Cell Maintenance

1. Laminar flow hood.
2. CO_2 incubator with orbital shaker.
3. ExpiCHO-S™ Cells, 1×10^7 cells frozen in liquid nitrogen (*see* **Note 1**).
4. ExpiCHO™ Expression Medium.
5. Sterile Erlenmeyer shake flasks (polystyrene, vented, non-baffled).
6. 0.4% trypan blue solution.
7. Automatic cell counter or hemacytometer.

2.2 Transient Transfection

1. ExpiFectamine™ CHO Transfection Kit (*see* **Note 2**).
2. OptiPRO™ SFM.
3. Antibody plasmid DNA (*see* **Note 3**).

2.3 Harvesting Cells

1. 0.22 μM membrane filters.

2.4 Antibody Purification

1. ÄKTA™ Pure with UNICORN™ software or alternative protein purification system (*see* **Note 4**).
2. 5 mL HiTrap® MabSelect™ SuRe column (*see* **Note 5**).
3. Phosphate buffered saline (PBS): 137 mM NaCl, 2.7 mM KCl, 10 mM Na_2HPO_4, 1.8 mM KH_2PO_4, pH 7.4.
4. Pierce™ IgG Elution Buffer.
5. 0.5 N NaOH.
6. 20% (vol/vol) ethanol.
7. 1 M Tris-HCl, pH 7.8.
8. Digital pH meter or pH paper.
9. UV-Vis spectrophotometer.
10. Amicon® Ultra-15 Centrifugal Filters, 30 kDa MWCO.

3 Methods

Work in a laminar flow hood and use proper aseptic technique when preparing cells and transfection reagents. Perform all steps at room temperature unless otherwise specified.

3.1 Cell Maintenance

1. To establish a new culture, remove a vial of CHO cells from storage in liquid nitrogen and incubate in a 37 °C water bath until thawed (~2–3 min).
2. Using a serological pipette, transfer the cells to a sterile 125 mL shake flask containing 25 mL pre-warmed ExpiCHO™ Expression Medium.
3. Grow the cells in a 37 °C incubator with 8% CO_2, 80% humidity, and shaking at 125 RPM (*see* **Note 6**).
4. Monitor the culture over the next few days. Mix an aliquot of cells 1:1 with 0.4% trypan blue and use a hemocytometer or automatic cell counter to determine the cell density and viability (*see* **Note** 7). When the density reaches 4×10^6–6×10^6 cells/mL (typically 3–4 days post-thaw), passage the cells by seeding a fresh flask at 0.15×10^6–0.3×10^6 cells/mL in prewarmed ExpiCHO™ Expression Medium.
5. Continue passaging the cells every 3–4 days, scaling up the culture volume as necessary (*see* **Note 8**).

3.2 Transient Transfection

1. The day prior to transfection (Day −1), passage the cells to a density of 3×10^6–4×10^6 cells/mL and grow overnight.
2. Immediately before transfecting (Day 0), passage the cells to the desired volume at 6×10^6 cells/mL (*see* **Note 9**).
3. In a sterile tube, dilute the antibody DNA in cold OptiPRO™ SFM. See Table 1 for the recommended transfection conditions at various scales.
4. In a separate tube, dilute the ExpiFectamine™ reagent in cold OptiPRO™ SFM. Do not store the diluted ExpiFectamine™ for more than 5 min.
5. Combine the diluted DNA with the diluted ExpiFectamine™ and mix gently. Incubate at room temperature for 1–5 min.
6. Slowly add the transfection mixture to the cells (*see* **Note 10**). Grow at 37 °C with 8% CO_2, 80% humidity, and shaking at 125 RPM.
7. 18–22 h post-transfection (Day 1), supplement the cells with ExpiFectamine™ CHO Enhancer and ExpiCHO Feed. See Table 2 for the recommended volumes. Change the incubator settings to 32 °C, 5% CO_2, 80% humidity, and 127 RPM (*see* **Note 11**).
8. Five days after transfection (Day 5), supplement the cells with an additional volume of ExpiCHO Feed.

3.3 Harvesting Cells

1. After the second feed, begin monitoring the viability of the cells. The culture should be harvested when the viability drops below ~70% (typically 10–14 days after transfection).
2. Transfer the culture to an appropriately sized centrifuge container. Spin at 14,000*g* for 10 min (*see* **Note 12**).
3. Collect the supernatant and pass it through a 0.22 μM filter. Proceed immediately to purification or store overnight at 4 °C.

3.4 Antibody Purification

Filter all buffers through a 0.22 μM filter before use.

1. Equilibrate a 5 mL mAb select SuRe column with 10 column volumes (CV) of PBS at a flow rate of 5 mL/min (*see* **Note 13**).
2. Load the filtered cell culture supernatant directly onto the column at 5 mL/min (*see* **Note 14**).
3. Wash the column with 10 CV of PBS at 5 mL/min (*see* **Note 15**).
4. Perform an isocratic elution of the bound antibody using 10 CV of IgG elution buffer at 5 mL/min. Collect the eluate in 2 mL fractions.
5. Perform a column CIP: wash the column with 5 CV of PBS at 5 mL/min, then regenerate the column by washing with 5 CV

Table 1
CHO cell transfection conditions for 25–400 mL cultures [13]

		DNA mix			Expifectamine mix	
Culture volume (mL)	Flask size (mL)	Optipro SFM (mL)	Heavy chain DNA (μg)	Light chain DNA (μg)	Optipro SFM (mL)	Expifectamine (μL)
25	125	1	12	12	0.920	80
50	250	2	24	24	1.840	160
100	500	4	48	48	3.680	320
200	1000	8	96	96	7.360	640
400	2000	16	192	192	14.720	1280

Table 2
Enhancer and feed volumes for CHO cell transfection

Culture volume (mL)	Day 1 Enhancer (μL)	Days 1, 5 Feed (mL)
25	150	4
50	300	8
100	600	16
200	1200	32
400	2400	64

of 0.5 N NaOH at 5 mL/min. Wash again with 5 CV of PBS at 5 mL/min (*see* **Note 16**).

6. For column storage, wash with 5 CV of 20% ethanol at 1.8 mL/min. Keep the column at 4 °C when not in use.
7. Pool the elution fractions. Immediately neutralize the solution to pH ~7 using 1 M Tris-HCl, pH 7.8 (*see* **Note 17**).
8. Determine the protein concentration using a spectrophotometer. Blank the instrument with IgG elution buffer that has been neutralized with 1 M Tris-HCl.
9. Run the antibody on an SDS-PAGE gel to confirm its size and purity (*see* **Note 18**).
10. If necessary, the solution may be buffer exchanged or concentrated in a 30K Amicon filter. Aliquot the antibody and store at −80 °C.

4 Notes

1. Alternative CHO-S lines may be used in place of ExpiCHO™ cells. Refer to the manufacturer's recommendations for cell maintenance and transfection conditions.
2. The ExpiFectamine™ CHO Transfection Kit is designed for use with ExpiCHO-S™ cells. Alternative commercial transfection reagents may be used according to the manufacturer's recommendations.
3. The antibody DNA should be contained in an appropriate mammalian expression vector. Use transfection-grade DNA to avoid contaminating the cells.
4. There are a variety of established methods for antibody purification, including commercial spin column kits for small samples. The HiTrap® columns from Cytiva are compatible with syringes and peristaltic pumps, as well as automated liquid chromatography systems like the ÄKTA™.
5. The MabSelect™ SuRe column will only work for antibodies that bind protein A. Traditional ScFvs, Fabs, and other non-protein A binding antibodies will require alternative methods of purification. A single 5 mL column can hold up to ~175 mg of protein (~35 mg per mL of resin). If more than 175 mg is expected, two or three 5 mL columns can be connected in series.
6. The RPM values in this protocol are for incubators with a 19-mm shaking diameter. Adjust the speed accordingly for different diameters.
7. The viability of the culture should be close to 100% within a few days of thawing. Do not use the cells for transfection if the viability is below 95%.
8. Allow the thawed cells to recover for at least 2 passages before using for transfection. Discard the culture after ~30 passages.
9. The volume of the transfection will depend on how much antibody is required and how well the antibody expresses. We typically see yields between 0.2 and 1 g of protein per liter of culture, though the ExpiCHO™ system claims to generate up to 3 g/L. The transfection conditions described in Tables 1 and 2 can be scaled linearly for cultures between 25 and 400 mL but may need to be optimized for volumes outside of this range.
10. Vigorous mixing at this step can lead to reduced transformation efficiencies. Pipette gently to avoid disrupting the transfection complexes.

11. If a 32 °C incubator is not available, leave the cells at 37 °C and 8% CO_2. Supplement with 50% more feed on day 1 (e.g., 6 mL for a 25 mL culture) and do not feed on day 5. The cells will be ready to harvest earlier (~8–10 days after transfection), and the final antibody titer will be lower.
12. The sample needs to be clarified before purification to prevent clogging the column. Larger culture volumes may require longer centrifuge times.
13. Optionally, when attaching a new column to the ÄKTA™, run PBS through the column line at 0.5 mL/min to prevent air from entering the system. Before using a column for the first time, carry out the equilibration, elution, and CIP steps as described in Subheading 3.4. Although not essential, performing a "blank run" without the antibody will help wash out trace amounts of leaked ligand that could otherwise contaminate the sample.
14. The flow rate may be increased to 10 mL/min for large culture volumes.
15. If the chosen purification system has a UV monitor, continue washing until the UV signal stabilizes (i.e., a fluctuation of less than 0.2 mAU for 1 min).
16. A CIP (cleaning in place) step is recommended to sanitize the column and prevent cross-contamination between consecutive purification runs.
17. A pH of 7 can be reached with ~1/20 volume of tris per volume of protein. Add the tris slowly and mix well to avoid precipitating the protein. Cloudiness is indicative of precipitated or aggregated protein and can be resolved by passing the solution through a 0.22 μM filter.
18. Typical IgG antibodies will produce a single ~150 kDa band under non-reducing conditions and a ~50 kDa heavy and ~25 kDa light chain band under reducing conditions.

References

1. Berger M, Shankar V, Vafai A (2002) Therapeutic applications of monoclonal antibodies. Am J Med Sci 324(1):14–30
2. Lu RM, Hwang YC, Liu IJ et al (2020) Development of therapeutic antibodies for the treatment of diseases. J Biomed Sci 27:1
3. Ayyar BV, Arora S, Murphy C et al (2012) Affinity chromatography as a tool for antibody purification. Methods 56(2):116–129
4. Correia I (2010) Stability of IgG isotypes in serum. mAbs 2(3):221–232
5. Verma R, Boleti E, George AJ (1998) Antibody engineering: comparison of bacterial, yeast, insect, and mammalian expression systems. J Immunol Methods 216(1–2):165–181
6. Frenzel A, Hurst M, Schirrmann T (2013) Expression of recombinant antibodies. Front Immunol 4:217
7. Lai T, Yang Y, Ng SK (2013) Advances in mammalian cell line development technologies for recombinant protein production. Pharmaceuticals 6(5):579–603

8. Kim TK, Eberwine JH (2010) Mammalian cell transfection: the present and the future. Anal Bioanal Chem 397(8):3173–3178
9. Fus-Kujawa A, Prus P, Bajdak-Rusinek K et al (2021) An overview of methods and tools for transfection of eukaryotic cells in vitro. Front Bioeng Biotechnol 9:701031
10. Lee CH (2017) A simple outline of methods for protein isolation and purification. Endocrinol Metab (Seoul) 32(1):18–22
11. Lee HG, Kang S, Lee JS (2021) Binding characteristics of staphylococcal protein A and streptococcal protein G for fragment crystallizable portion of human immunoglobulin G. Comput Struct Biotechnol J 19:3372–3383
12. Huse K, Böhme HJ, Scholz GH (2002) Purification of antibodies by affinity chromatography. J Biochem Biophys Methods 51(3):217–231
13. Gibco (2018). ExpiCHO™ expression system user guide. https://assets.thermofisher.com/TFS-Assets%2FLSG%2Fmanuals%2FMAN0014337_expicho_expression_system_UG.pdf

Chapter 12

Mammalian Antigen Display for Pandemic Countermeasures

Andrea Quezada, Ankur Annapareddy, Kamyab Javanmardi, John Cooper, and Ilya J. Finkelstein

Abstract

Pandemic countermeasures require the rapid design of antigens for vaccines, profiling patient antibody responses, assessing antigen structure-function landscapes, and the surveillance of emerging viral lineages. Cell surface display of a viral antigen or its subdomains can facilitate these goals by coupling the phenotypes of protein variants to their DNA sequence. Screening surface-displayed proteins via flow cytometry also eliminates time-consuming protein purification steps. Prior approaches have primarily relied on yeast as a display chassis. However, yeast often cannot express large viral glycoproteins, requiring their truncation into subdomains. Here, we describe a method to design and express antigens on the surface of mammalian HEK293T cells. We discuss three use cases, including screening of stabilizing mutations, deep mutational scanning, and epitope mapping. The mammalian antigen display platform described herein will accelerate ongoing and future pandemic countermeasures.

Key words Surface display, Antigen, Spike, Hemagglutinin, Influenza, Coronavirus, Flow cytometry

1 Introduction

Pandemics are becoming more frequent due to increased encroachment into zoonotic reservoirs and climate change. By one estimate, the probability of observing extreme pandemics like COVID-19 in one's lifetime—currently 38%—may double in coming decades [1]. The rapid development of subunit vaccines is the most important pandemic countermeasure, as highlighted by the record-breaking speed of SARS-CoV-2 vaccine deployment [2]. Subunit vaccines include only the components, or antigens, that best stimulate the immune system to create a durable immunological memory of a specific pathogen. Additional pandemic countermeasures include mapping the humoral immune responses of convalescent

Andrea Quezada and Ankur Annapareddy contributed equally to this work.

Steven B. Bradfute (ed.), *Recombinant Glycoproteins: Methods and Protocols*, Methods in Molecular Biology, vol. 2762, https://doi.org/10.1007/978-1-0716-3666-4_12,

and immunized patients and tracking the emergence of new variants. Accelerating these countermeasures using modern molecular biology approaches is imperative for minimizing the global disruption from future viral threats. Antiviral subunit vaccines focus the immune system on viral entry glycoproteins. Targeting the pre-fusion conformation of these glycoproteins can produce potent neutralizing antibodies that prevent viral entry [3–10]. Structure-guided antigen design is the leading approach for developing subunit vaccine antigens [11, 12]. However, structure-guided protein engineering requires the relatively low-throughput expression, purification, and biochemical characterization of antigen candidates. Machine learning and other computational engineering approaches can accelerate antigen development by predicting stabilizing mutations [13–16]. However, these candidates must still be validated using the same laborious biochemical approaches. In short, antigen design is bottlenecked by the need to individually express, purify, and test each protein candidate. Accelerating the antigen design-build-test cycle is a central plank of future pandemic countermeasures.

A second key plank in future pandemic preparedness is the ability to map the binding sites of antigen-specific monoclonal antibodies (mAbs). Understanding how mAbs bind and neutralize viruses improves our understanding of conserved epitopes, sheds light on neutralization mechanisms, and anticipates viral escape potential. For example, understanding which conserved and rare epitopes lead to potent and broad-spectrum neutralizations may lead to the design of antigens that are protective against emerging viral lineages, such as the SARS-CoV-2 variants of concern. More broadly, such approaches can lead to the development of vaccines that protect against multiple viral clades or even entire viral families, as was recently demonstrated for pan-influenza and pan-coronavirus vaccine candidates [17–21].

Cell surface display is a high-throughput approach for antigen design and antibody epitope mapping. Tethering antigens (or their subdomains) to cells bypasses the need for biochemical purifications and can be used to pool antigen variants in a single experiment. The most common cell surface display approaches leverage the budding yeast *S. cerevisiae* [22]. *S. cerevisiae* has been used for epitope mapping and deep mutational scanning of influenza hemagglutinins and the SARS-CoV-2 spike receptor binding domain (RBD) [23–28]. These experiments provide valuable insight into the mechanisms for viral evolution and immune escape but also face several limitations. First, yeast is unable to produce many full-length antigens (e.g., the SARS-CoV-2 spike ectodomain). The humoral immune response to SARS-CoV-2 produces potent anti-spike neutralizing antibodies that target the N-terminal domain as well as the S2 stalk [29–34]. These domains are outside the RBD and cannot be addressed via yeast display. Second, antigens produced in yeast do not recapitulate mammalian

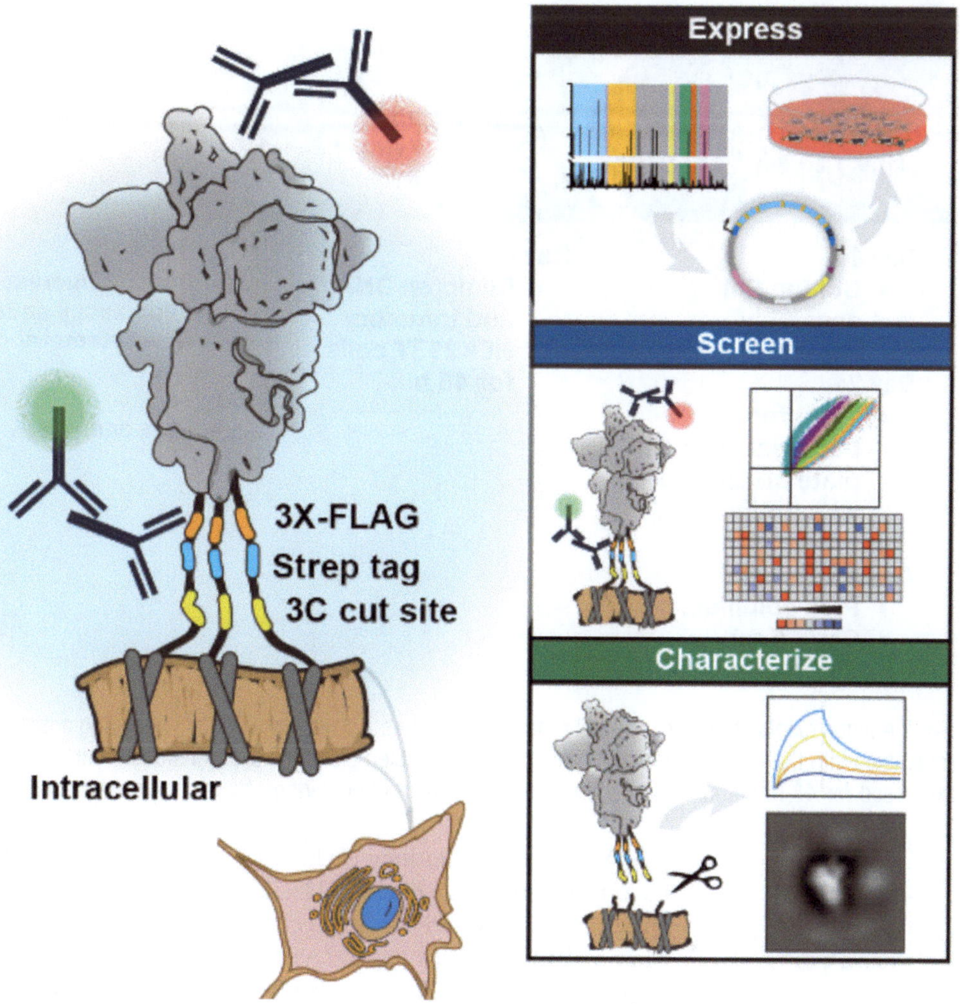

Fig. 1 Overview of the antigen display platform. The platform consists of a rapid cloning pipeline, flow cytometry, immunofluorescence microscopy, and antigen cleavage for further biophysical characterization. The antigen is tethered to the cell via a short transmembrane domain and a flexible linker. The linker encodes a 3C protease cut site which enables cleavage and purification of displayed antigens (yellow), a StrepTactin purification tag (blue), and a 3×FLAG epitope tag for immunostaining. A Golden Gate (GG) cloning pipeline facilitates high-throughput antigen cloning (right, top). Flow cytometry enables high-throughput screening (right, middle). Downstream biophysical characterizations, such as bio-layer interferometry (BLI) and negative stain electron microscopy (nsEM), can be performed on cleaved antigens (right, bottom)

glycosylation [35]. These differences may alter a protein's antigenicity toward cell receptors and antibodies [36]. To overcome these limitations, we developed a mammalian cell surface display platform that measures antigen expression and antibody binding on the surface of mammalian cells.

Mammalian antigen display is designed for phenotypic screening of full-length viral glycoproteins on the surface of mammalian cells (Fig. 1). As a proof of principle, we displayed the SARS-CoV-2 spike protein and the influenza hemagglutinin (HA) on the surface of

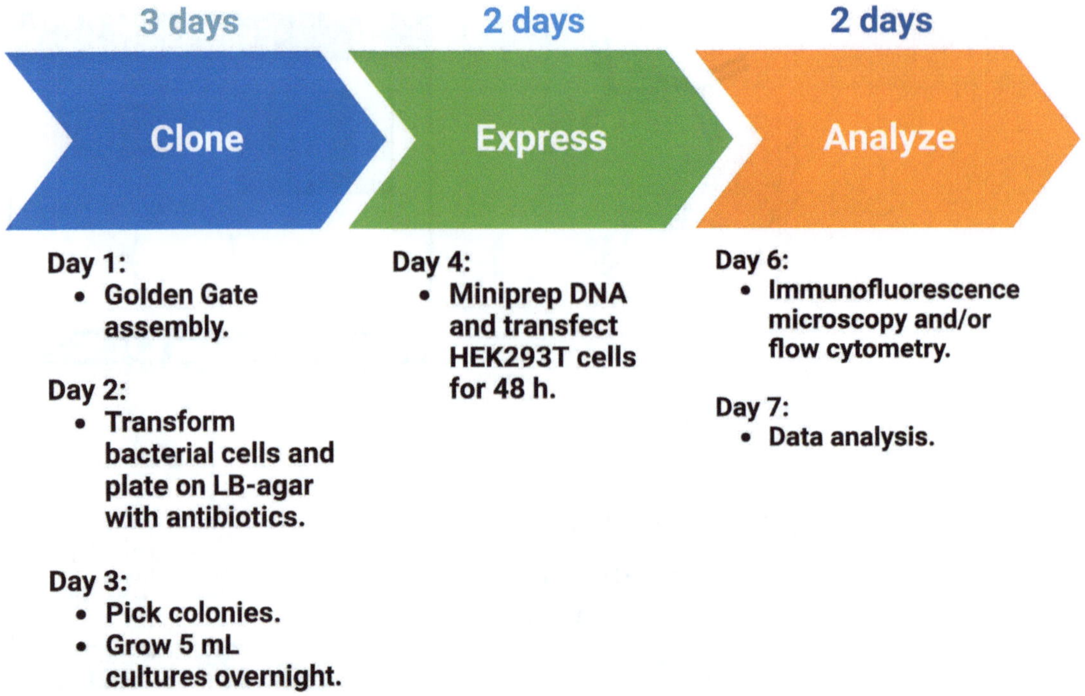

Fig. 2 Antigen display timeline. The full procedure requires 7 days. Gene blocks encoding antigen variants are assembled into the drop-out vector (3 days). The assembled plasmids are transfected into HEK293T cells and expressed for 48 h. Immunofluorescence microscopy and flow cytometry take two additional days

human embryonic kidney (HEK293T) cells [37, 38]. Although we have tested this approach in HEK293Ts, the protocols are general and can be adapted to any cell type in about a week (Fig. 2). Mammalian cell lines express viral proteins with glycosylation patterns comparable to those found from bona fide viruses [39, 40]. We focus on the SARS-CoV-2 spike ectodomain coding sequence (residues 1-1208) containing six pre-fusion stabilizing prolines and a mutated furin cleavage site to improve spike stability and expression [41, 42]. The promoters, chimeric introns, and terminators are also optimized to further boost protein expression in mammalian cells. Combinations of N-terminal secretion tags and C-terminal linkers ensure high surface display density. Due to the variability of plasmid transfections in mammalian cell cultures, we included a triple FLAG epitope tag as a proxy for expression levels and as an internal control for signal normalization. Transfected cells expressing antigens are immunostained and analyzed by imaging or flow cytometry (Fig. 4). A 3C protease cleavage site and a Strep II tag are included in the C-terminal linker to enable the cleavage and rapid purification of surface-displayed antigens. Cleaving the antigen from cell surfaces for conventional biochemical and biophysical methods can save time and laborious recombinant purification. These features make mammalian antigen display a valuable tool for the current and future pandemic countermeasures [43, 44].

2 Materials

2.1 Media, Strains, Plasmids

1. DH5-alpha competent *E. coli*.
2. Mix & Go competent cells – strain Zymo 10B (Zymo Research T3019).
3. Superior Broth.
4. DMEM, high glucose, pyruvate.
5. Fetal Bovine Serum (FBS).
6. Opti-MEM I reduced serum medium, GlutaMAX Supplement.
7. pcDNA5/FRT/TO/intron/GFP [45] (Addgene #113547).

2.2 Generating Antigen Libraries

1. T7 DNA Ligase.
2. 10× CutSmart buffer (NEB B6004).
3. 10 mM ATP.
4. DNA Miniprep Kit.
5. Microcentrifuge.
6. Thermocycler.

2.2.1 Golden Gate Assemblies

1. Gene blocks for antigen mutagenesis.
2. AarI and activating oligo.
3. Drop-out plasmid (AddGene #172726).

2.2.2 Saturation Mutagenesis Library Generation

1. Template plasmid containing gene segment to be mutagenized, and BbvCI restriction enzyme cut site.
2. Mutagenic oligo mixture (IDT).
3. T4 Polynucleotide Kinase and buffer.
4. Universal secondary primer (IDT).
5. Exonuclease III.
6. Exonuclease I.
7. Nt.BbvCI.
8. Nb.BbvCI.
9. 100 mM DTT.
10. 50 mM NAD^+.
11. 10 mM dNTPs.
12. Phusion HiFi DNA polymerase (NEB M0530).
13. 5× Phusion HiFi buffer (NEB B0518).
14. Taq DNA Ligase.
15. DpnI.
16. Zymo clean and concentrate kit (Zymo D4005).

2.3 Expression of Surface-Displayed Antigens

2.3.1 Transient Transfection of HEK293Ts

1. HEK293T cell line (ATCC CRL-3216).
2. Dulbecco Modified Eagle Medium (DMEM).
3. Fetal Bovine Serum (FBS).
4. Penicillin-Streptomycin.
5. Trypsin-EDTA (0.25%), Phenol Red.
6. Opti-MEM Reduced Serum medium.
7. Lipofectamine 3000.
8. Mycoplasma detection kit.
9. Incubator at 37 °C and 5% CO_2.
10. 10 cm polystyrene-coated plates.
11. 6-well polystyrene-coated plates.

2.4 Immuno-fluorescence Microscopy

1. Glass-bottom imaging dishes.
2. BlockAid blocking solution.
3. Mouse anti-FLAG M2 antibody.
4. Antigen-specific antibody.
5. Goat anti-Mouse IgG(H + L), human ads-Alexa Fluor 488.
6. Goat anti-Human IgG Fc-Alexa Fluor 647.
7. PBS-BSA (1% BSA, 1× PBS, 2 mM EDTA pH 7.4).
8. Hoechst stain.
9. Widefield or confocal microscope equipped with visible and violet light sources and fluorescent filter cubes at 375/28, 480/30, 620/50.

2.5 Cleavage and Purification of Surface-Expressed Antigens

1. Expi293 cell line (Thermo Fisher A14527).
2. Expi293 Expression System Kit (Thermo Fisher A14635).
3. Incubator at 37 °C and 8% CO_2.
4. StrepTactin Superflow purification column (IBA 2-1206-025).
5. Superose 6 Increase 10/300 GL gel-filtration column (GE29-0915-96).
6. Protein storage buffer (2 mM Tris pH 8.0, 200 mM NaCl, 0.02%).
7. 3C Protease.
8. Wash buffer (100 mM Tris/HCl pH 8.0, 150 mM NaCl, 1 mM EDTA).
9. Elution buffer (100 mM Tris/HCl pH 8.0, 150 mM NaCl, 1 mM EDTA, 2.5 mM desthiobiotin).

2.6 Flow Cytometry

1. PBS.
2. Cell counter (e.g., Logos Biosystems L40002).
3. PBS-BSA buffer (1% BSA, 1× PBS, 2 mM EDTA pH 7.4).
4. Deep well grow blocks, 2 mL.
5. Microplate shaker.
6. Swinging bucket rotor.
7. Spectral cell analyzer (e.g., SONY SA3800).
8. Mouse anti-FLAG M2 antibody.
9. Goat anti-Mouse IgG(H + L), Human ads-Alexa Fluor 488.
10. Goat anti-Human IgG Fc-Alexa Fluor 647.

2.7 Fluorescence-Assisted Cell Sorting

1. PBS.
2. PBS-BSA buffer (1% BSA, 1× PBS, 2 mM EDTA pH 7.4).
3. Goat anti-Mouse IgG(H + L), Human ads-Alexa Fluor 488.
4. Goat anti-Human IgG Fc-Alexa Fluor 647.
5. Cell counter (e.g., Logos Biosystems L40002).
6. Deep well grow blocks, 2 mL.
7. Microplate shaker.
8. Swinging bucket rotor.
9. Cell sorter (e.g., Sony SH800S).

2.8 Data Analysis

1. Image analyzer software (e.g., FIJI [46]).
2. Flow cytometer analyzer packages (e.g., FlowCytometryTools [47]).

3 Methods

3.1 Generating Antigen Libraries

Below, we present two methods for generating antigen libraries (Fig. 3). The first method, termed Golden Gate (GG) assemblies, uses synthetic gene blocks to rapidly assemble antigens with defined mutations. This method is useful for antigen engineering, epitope mapping, and characterizing clinical variants with multiple mutations scattered throughout the protein. The second method generates saturating mutagenesis libraries for deep mutational scanning (DMS) [48]. DMS is useful for understanding antigen stability, molecular epistasis between mutations, and epitope mapping [38]. This method requires mutagenic primers and a pooled nickase-based primer extension [49].

3.1.1 Golden Gate Assemblies

Gene blocks that encode antigen fragments are ordered as double-stranded DNA gene blocks containing flanking AarI cut sites and unique overhangs matching the entry vector. For example, our entry vectors for SARS-CoV-2 spike include AarI cut sites and

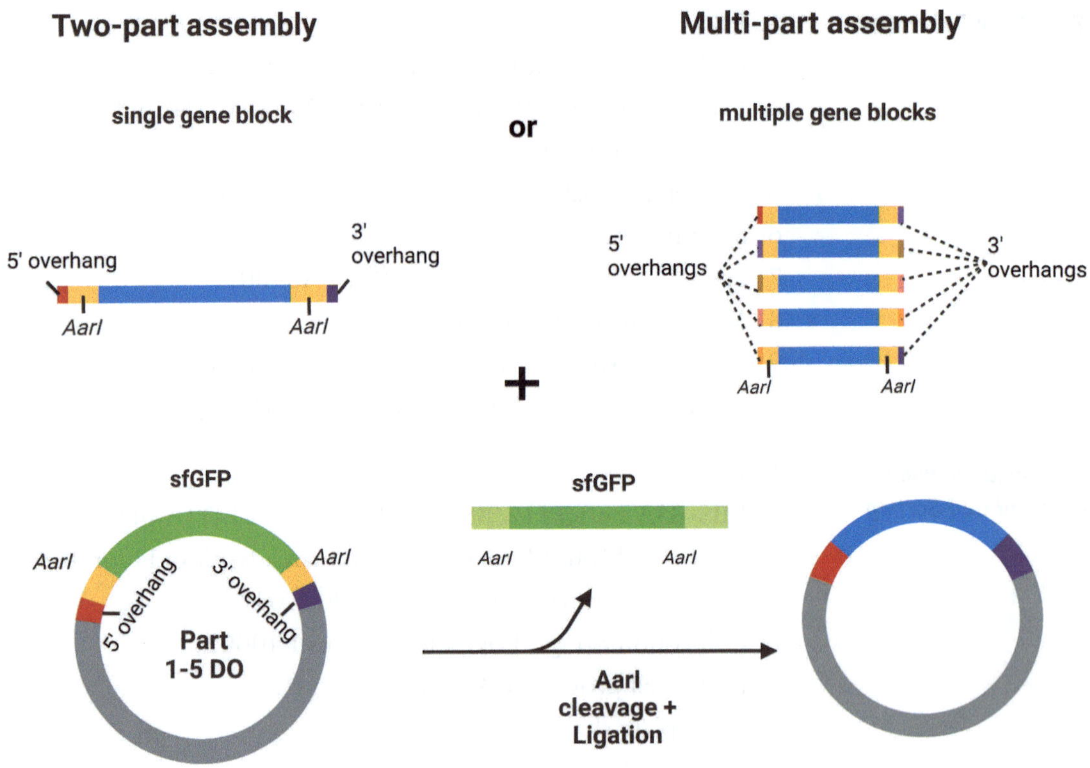

Fig. 3 Golden Gate assembly. An entry vector with a super-folder GFP (sfGFP in green) dropout cassette and AarI cut sites (yellow) is assembled with one or more gene blocks (blue) containing AarI cut sites and matching 5′ and 3′ overhangs. For a two-part assembly, the overhangs must match those of the entry vector. For a multi-part assembly, overhangs matching each part and the dropout cassette must be designed. The entire process can be automated using liquid handling robots for high-throughput experiments

overhangs matching the spike ectodomain sequence (*see* **Note 1**). Alternatively, the antigen sequence can be divided into sub-regions to reduce the cost of long gene block synthesis. In this case, unique Golden Gate overhangs will be necessary for each segment of the antigen. To ease cloning, this entry vector also includes a super-folder GFP (sfGFP) bacterial expression cassette that is removed when the spike DNA is ligated between the AarI sites. This allows rapid screening for colonies that have lost the sfGFP cassette (see below).

1. Design one or multiple gene blocks with specific overhangs matching the entry vector (*see* **Note 2**).
2. Assemble the GG reaction in PCR tubes with the gene blocks and dropout vector:
   (a) *Golden Gate assembly mix*
      (i) 0.25 μL T7 DNA Ligase.
      (ii) 0.25 μL AarI.

(iii) 0.2 μL AarI oligo [or 1 μL AarI oligo (1:5 dilution)].

(iv) 1 μL 10× CutSmart buffer.

(v) 1 μL 10 mM ATP.

(vi) 10 ng of a single gene block or 10 ng/gblock in case of a multi-part assembly.

(vii) 20–30 ng dropout plasmid.

(viii) Final volume: 10 μL.

(b) *Thermocycling (two-part assemblies)*

(i) 25 cycles: 37 °C for 1 min (digestion) and 16 °C for 2 min (ligation).

(ii) 37 °C for 30 min (final digestion).

(iii) 80 °C for 20 min (heat inactivation).

(iv) 4 °C hold.

(c) *Thermocycling (multi-part assemblies)*

(i) 35 cycles: 37 °C for 2 min (digestion) and 16 °C for 4 min (ligation).

(ii) 37 °C for 60 min (final digestion).

(iii) 80 °C for 20 min (heat inactivation).

(iv) 4 °C hold.

3. Transformations:

(a) Mix 4 μL of each GG reaction with 10 μL of NEB 5-alpha competent cells aliquoted into 1.5 mL Eppendorf tubes and incubate on ice for 30 min.

(b) Heat shock at 42 °C for 30 s and incubate on ice for 5 min.

(c) Add 200 μL of Superior Broth and incubate at 37 °C in the shaking incubator for 1 h.

(d) Spin down cells at 2000 rcf for 1 min.

(e) Remove the media and resuspend in 50 μL of fresh media.

(f) Plate on 10 cm LB-agar plates with carbenicillin (100 μg/mL). Let the plates dry and keep them at 37 °C overnight.

(g) Use a blue light to select colonies that have lost sfGFP. Colonies still containing the sfGFP drop-out gene will be fluorescent under blue light and should not be picked.

(h) Grow picked colonies in 5 mL of LB with carbenicillin (100 μg/mL) at 37 °C overnight with continuous shaking.

(i) Collect cells at 3000 rcf for 5 min and extract the plasmid DNA.

3.1.2 Saturation Mutagenesis Libraries

We generate saturating mutagenesis libraries via the protocol described in Wrenbeck et al. [49] (Fig. 7). That protocol outlines how to generate libraries with single amino acid substitutions. In our testing, increasing the mutagenic oligo to template ratio and increasing the number of PCR cycles generates libraries with more than one mutation per antigen. In addition to amino acid substitutions, deletions and insertions can be introduced into the template plasmid with similar efficiency by deleting or inserting the desired amino acid in the mutagenic oligos.

Mutagenic oligos are designed to contain the desired mutations and complement the wild-type template strand on either side of the programmed mutation. As part of this protocol, we provide a Python script for developing mutagenic oligo pools (https://github.com/finkelsteinlab/mutagenic-primer-design). Mutagenic oligos can be purchased as synthetic pools (IDT or Twist), or if a more limited library is required, purchased individually and pooled by hand.

1. Phosphorylate oligos:
   (a) Resuspend mutagenic oligo pool to a final concentration of 0.1 μM.
   (b) Combine in a PCR tube:
      (i) 20 μL 0.1 mM mutagenic oligo mixture.
      (ii) 2.4 μL T4 Polynucleotide Kinase buffer.
      (iii) 1 μL 10 mM ATP.
      (iv) 1 μL T4 Polynucleotide Kinase (10 U/μL).
   (c) Into a separate PCR tube add:
      (i) 18 μL.
      (ii) 3 μL T4 Polynucleotide Kinase buffer.
      (iii) 7 μL 100 μM secondary primer.
      (iv) 1 μL 10 mM ATP.
      (v) 1 μL T4 Polynucleotide Kinase (10 U/μL).
   (d) Incubate both tubes at 37 °C for 1 h.
   (e) Store phosphorylated oligos at −20 °C. The day of mutagenesis, dilute phosphorylated mutagenic oligo pool 1:10 and secondary primer 1:20 in H_2O.
2. Single-stranded DNA (ssDNA) template strand preparation:
   (a) Add the following into PCR tube(s):
      (i) 0.76 pmol template plasmid.
      (ii) 2 μL 10× CutSmart buffer.
      (iii) 1 μL 1:10 diluted Exonuclease III (10 U/μL).
      (iv) 1 μL Nt.BbvCI (10 U/μL).

- (v) 1 μL Exonuclease I (20 U/μL).
- (vi) to 20 μL.

(b) Thermocycling conditions:

- (i) 37 °C for 1 h.
- (ii) 80 °C for 20 min.
- (iii) 4 °C hold.

(c) Degraded template plasmid can be kept at 4 °C overnight if needed.

3. Comprehensive codon mutagenesis strand 1 (top strand):

(a) Add the following into each tube (100 μL final volume):

- (i) 36.7 μL H_2O.
- (ii) 20 μL 5× Phusion HiFi buffer.
- (iii) 4.3 μL 1:10 diluted phosphorylated mutagenic oligos.
- (iv) 10 μL 100 mM DTT.
- (v) 1 μL 50 mM NAD+.
- (vi) 2 μL 10 mM dNTPs.
- (vii) 1 μL Phusion HiFi polymerase (2 U/μL).
- (viii) 5 μL Taq DNA Ligase (40 U/μL).

(b) Thermocycling conditions (×15 cycles for steps ii-iv; add additional 4.3 μL of oligo mixture at the beginning of cycles 6 and 11):

- (i) 98 °C for 2 min.
- (ii) 98 °C for 30 s.
- (iii) 55 °C for 45 s.
- (iv) 72 °C for 1 min/kb of template plasmid.
- (v) 45 °C for 40 min.
- (vi) 4 °C hold.

4. Purify the DNA with a Zymo Clean and Concentrate kit. Follow the manufacturer's instructions. Elute with 15 μL of H_2O.

5. Degrade template strand:

(a) Transfer 14 μL of the purified DNA product to a PCR tube and add (20 μL final volume):

- (i) 2 μL 10× CutSmart buffer.
- (ii) 2 μL 1:50 diluted Exonuclease III (2 U/μL).

   (iii) 1 μL 1:10 diluted Nb.BbvCI (1 U/μL).
   (iv) 1 μL Exonuclease I (20 U/μL).
 (b) Thermocycling conditions:
   (i) 37 °C for 1 h.
   (ii) 80 °C for 20 min.
   (iii) 4 °C hold.

6. Synthesize complementary mutagenic strand:
   (a) To the PCR tube from step 5 add (100 μL final volume):
      (i) 37.7 μL H_2O.
      (ii) 20 μL 5× Phusion HiFi buffer.
      (iii) 3.3 μL 1:20 diluted phosphorylated secondary primer.
      (iv) 10 μL 100 mM DTT.
      (v) 1 μL 50 mM NAD+.
      (vi) 2 μL 10 mM dNTPs.
      (vii) 1 μL Phusion HiFi Polymerase (2 U/μL).
      (viii) 5 μL Taq DNA Ligase (40 U/μL).
   (b) Thermocycling conditions:
      (i) 98 °C for 30 s.
      (ii) 55 °C for 45 s.
      (iii) 72 °C for 5 min.
      (iv) 45 °C for 40 min.
      (v) 4 °C hold.
7. Column purification using Zymo Clean and Concentrate kit. Elute with 6 μL H_2O.
8. DpnI digest (20 μL final volume):
   (a) 5 μL cleaned plasmid from the previous step.
   (b) 2 μL DpnI.
   (c) 2 μL rCutSmart (10× buffer).
9. Thermocycling conditions:
   (a) 37 °C for 2 h.
   (b) 80 °C for 15 min.
   (c) 4 °C hold.
10. Column purification using Zymo Clean and Concentrate kit.

3.2 Antigen Expression

3.2.1 Transient Transfection of HEK293Ts

Handling cell lines requires aseptic conditions and a horizontal laminar flow cabinet. HEK293T cells should be cultured and maintained according to the manufacturer's protocol. Upon arrival, vials are stored at temperatures below −130 °C and only thawed when they are ready to use. The entire protocol can be completed in about a week (Fig. 2).

1. Culturing cells:
   - (a) Immediately after thawing, transfer cells into a 15 mL tube containing 9 mL Dulbecco's Modified Eagle's Medium supplemented with 10% FBS and 2% Penicillin-Streptomycin (complete medium). Spin down at 125 rfc for 5 min (*see* **Note 3**).
   - (b) Resuspend the pellet in complete medium. The amount of medium depends on the size of the culturing vessel (i.e., 10 cm plates, 25 or 75 cm^2 culture flasks). Maintain the vessel in an incubator with 5% CO_2 and 37 °C.
   - (c) Passage cells when they reach 80 to 90% confluency. Cells are ready to seed for transfection after two passages showing a consistent growing behavior.
   - (d) Cells grown up to 60–90% confluency into 10 cm plates can be frozen for future usage.
2. Seeding cells for transfection:
   - (a) 24 h before transfection, aspire the medium from a 10 cm plate containing 80% confluent cells.
   - (b) Slowly pour 6–10 mL of PBS through the plate wall, trying not to disturb the cell monolayer adhered to the bottom.
   - (c) Gently shake the plate and aspirate the PBS.
   - (d) Add 2 mL of Trypsin-EDTA and incubate at 37 °C for 5 min.
   - (e) Add 8 mL of complete medium and transfer cells into a centrifuge tube. Spin down at 125 rfc for 5 min.
   - (f) Resuspend the pellet with 10 mL complete medium and count the cells using an automated cell counter.
   - (g) Seed cells at a density of 0.2×10^6 cells into 6-well plates containing 2 mL per well of complete medium for flow cytometry. For microscopy, seed cells into 2 mL glass-bottom dishes.
   - (h) Incubate at 37 °C and 5% CO_2 for 24 h.
3. Transfection:
   - (a) Prepare a master mix containing 200 μL of OPTI-MEM per tube/sample and 3 μg of Lipofectamine 3000 per μg of DNA.

(b) For each sample/plasmid set up a sterile Eppendorf tube with 200 μL of Opti-MEM and add 1 μg of DNA per mL of culture (i.e., 2 μg of DNA for a 2 mL well in a 6-well plate).

(c) Add 200 mL of the master mix into each tube and gently mix by turning the tubes upside down or vortex at low speed. Vigorous vortexing is sometimes associated with low transfection efficiency and increased cytotoxicity.

(d) Incubate at room temperature for 15–20 min.

(e) Add each sample to the appropriate wells (6-well plates or glass-bottom dishes) and return plates to the 37 °C incubator. Wait 48 h before collecting cells.

3.3 Immuno-fluorescence Microscopy

3.3.1 Cell Preparation for Immunostaining

1. 48 h after transfection, aspirate the medium from the imaging dishes and gently wash with PBS without detaching the cell monolayer adhered to the bottom (Fig. 4).
2. Incubate the plates with 2 mL of freshly prepared 4% paraformaldehyde for 10 min at room temperature (*see* **Note 4**).

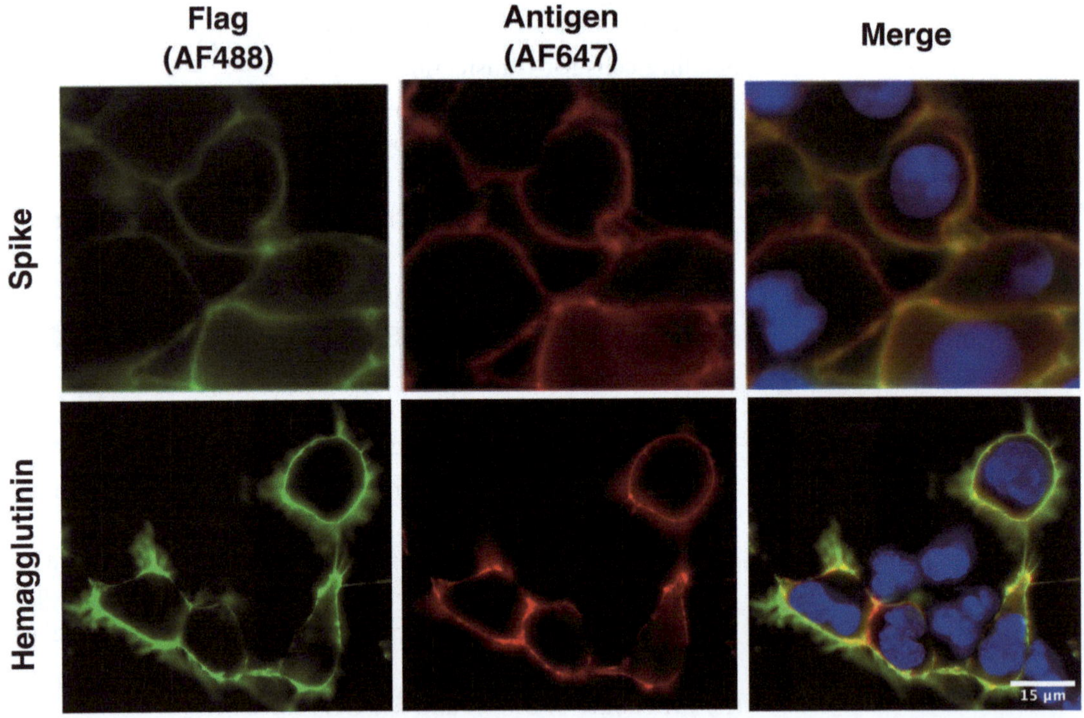

Fig. 4 Immunofluorescence imaging. Cells immunostained 48 h after transfection and imaged by fluorescence microscopy. REGN10933 [50] and CR9114 [51] were used for SARS-CoV-2 spike and hemagglutinin, respectively

3. To avoid non-specific background signal, the cells must be blocked for 30 min at room temperature with PBS-BSA buffer or with BlockAid blocking solution.
4. Wash cells with PBS before immunostaining.

3.3.2 Multi-color Immunostaining and Imaging

Since the primary antibodies are from different hosts, a simultaneous incubation with the unlabeled antibodies can be done in order to save time. Likewise, a simultaneous incubation with the secondary antibodies will speed-up the process.

1. Prepare a dilution of 1 μg/mL anti-FLAG M2 antibody and the same concentration of an anti-antigen antibody suitable for microscopy (i.e., REGN10933 [50] for SARS-CoV-2 spike or CR9114 [51] for HA) into PBS-BSA or BlockAid solution.
2. Aspirate the blocking solution from the imaging plates and add the antibody solution. For a 2 mL plate, add between 1 and 2 mL of the solution. Incubate at room temperature for 1 h.
3. Wash cells with PBS-BSA.
4. Dilute secondary antibodies (anti-mouse Alexa Fluor 488 and anti-human Alexa Fluor 647) at 10 μg/mL in PBS-BSA or BlockAid solution. Incubate in the dark at room temperature for 1 h.
5. Wash cells with PBS-BSA and add 1:10,0000 of Hoechst stain diluted in PBS-BSA. Incubate at room temperature for 15 min.
6. Wash cells with PBS-BSA once and add 1 mL of PBS-BSA or BlockAid solution before imaging.
7. Acquire images using fluorescent filter cubes at 375/28 (Hoechst), 480/30 (Alexa Fluor 488), and 620/50 (Alexa Fluor 647).

3.4 Cleavage and Purification of Surface-expressed Antigens

1. Transfect plasmids encoding antigens in HEK293Ts cells using Lipofectamine as suggested by the manufacturer's protocol and previously described [41].
2. Wash cells once 48 h after transfection with PBS. Resuspend to a density of 3–4 × 10^6 cells/mL in 3C cleavage buffer (150 mM NaCl, 50 mM Tris-HCl pH 8.0). Use five units of 3C protease/mL of resuspended cells.
3. Incubate mixture on a shaker at room temperature and 900 rpm for 1 h to cleave antigens from the cell surface.
4. Collect the antigen-containing supernatant by spinning tubes at 16,000 rcf for 1 min and transfer supernatant to a fresh tube. Supernatant can be kept on ice until analysis.
5. For electron microscopy imagining, further purify supernatant with 0.5 mL StrepTactin (IBA) column.
6. Wash column with base buffer (100 mM Tris/HCl pH 8.0, 150 mM NaCl, 1 mM EDTA).

7. Elute with 1 mL of elution buffer (100 mM Tris/HCl pH 8.0, 150 mM NaCl, 1 mM EDTA, 2.5 mM desthiobiotin).
8. Spikes can be further purified by size-exclusion chromatography using a Superose 6 Increase 10/300 column as previously described [42].

3.5 Flow Cytometry

3.5.1 Collect Cells

1. To collect cells 48 h after transfection, aspirate the medium from the 6-well plates and wash with 1 mL PBS without detaching the cell monolayer.
2. Resuspend the cells with 1 mL of PBS by gently pipetting until the cells are monodispersed. Transfer into 1.5 mL Eppendorf tubes.
3. Determine the cell density with an automated cell counter.
4. Spin down the cells for 1 min at 200 rcf. Aspire the supernatant and add 1 mL of chilled PBS-BSA to a density of 3.5×10^6 cells/mL. Keep tubes on ice.

3.5.2 Immunostaining

1. Prepare grow blocks with 450 mL of PBS-BSA and a predetermined concentration of the primary antibody per well. We recommend simultaneously incubating with anti-FLAG and anti-antigen antibodies to reduce incubation times and washing steps.
2. Add 50 mL of the resuspended cells (~1.5×10^5 cells) to each well.
3. Incubate at room temperature on a microplate shaker at 950 rpm for 1 h.
4. Spin down the blocks at 400 rcf for 5 min in a swinging bucket rotor.
5. Carefully decant the supernatant and wash cells by adding 450 mL of PBS-BSA to each well. Spin down again at 400 rcf for 5 min and repeat the washing step.
6. Incubate with 450 mL of the secondary antibody solution (5 μg/mL for Alexa Fluor 488 and 10 μg/mL of Alexa Fluor 647).
7. Incubate the plate at 950 rpm in a microplate shaker in a dark cold room at 4 °C for 25 min.
8. Spin down and wash with PBS-BSA twice.
9. Resuspend the cells with 300 mL of PBS-BSA and take them to the cell analyzer.

3.6 Flow Cytometry

1. HEK293T cells are used to establish forward scatter-area (FSC-A) and side scatter-area (SSC-A) gating. Singlet discrimination is established with forward scatter-height (FSC-H) vs forward scatter-area (FSC-A) and side scatter-height (SSC-H) vs side scatter-area (SSC-A) gates. Singlets reaching a minimum of 10 K counts are set as a stop condition (Fig. 5).

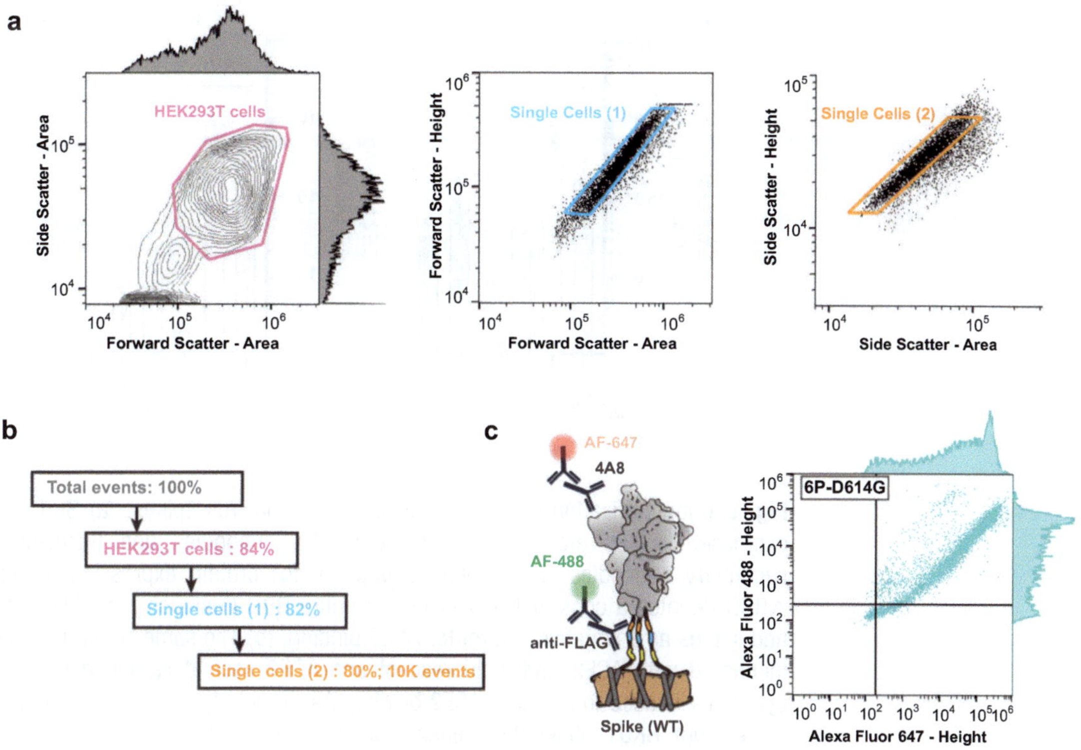

Fig. 5 Flow cytometry. (**a**) HEK293T cells are selected by gating the side scatter area vs. forward scatter area (left). Doublets are excluded with additional gating on the forward scatter height vs. forward scatter area (middle) and side scatter height vs. side scatter area (right). We recommend 10 K counts after gating. (**b**) Alexa Fluor 488 (AF-488) and Alexa Fluor 647 (AF-647) are used to measure antigen expression and antigenicity, respectively. (**c**) Schematic of a two color flow cytometry experiment (left). The anti-FLAG ant anti-spike 4A8 antibodies are fluorescently labeled with secondary antibodies. Right: a typical flow cytometry dataset from this experiment. The spike gene includes six stabilizing prolines, along with the globally prevalent D614G mutation to increase expression and stability

2. The singlet HEK293T cells are further analyzed in two fluorescent channels, Alexa Fluor 488 (AF488) and Alexa Fluor 647 (AF647), using manufacturer-recommended excitation and detection settings. The AF488 channel is used to measure antigen expression and AF647 is used to measure its antigenicity (Fig. 5).
3. Spectral unmixing should be applied to all data to reduce the effect of spectral spillover and autofluorescence on downstream calculations.

3.6.1 Flow Cytometry Analysis

Transient transfection efficiency varies among experiments. We normalize the data across multiple experiments to minimize day-to-day variation in transfection efficiency and cell quality. Our approach is to calculate the relative expression of all antigen variants against the wild-type antigen (Fig. 6). For example, the normalized expression of a spike variant is calculated by taking the median height of the AF-488 channel, M_x^{488}, and dividing it by the same value obtained for wild type (wt) spike, M_{wt}^{488}:

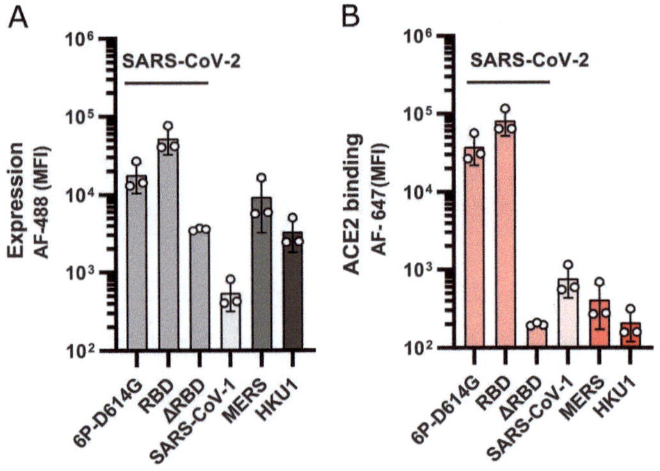

Fig. 6 Characterization of surface displayed coronavirus spikes. (**a**) Surface-expressed spikes are stained with anti-FLAG antibodies and fluorescent secondary antibodies. This signal is a proxy for protein expression levels. ΔRBD denotes a construct that lacks the entire RBD (residues 319–541 [63]) and serves as a negative control for ACE2 binding. (**b**) The same constructs are incubated with ACE2 and a fluorescent anti-ACE2 secondary antibody. Flow cytometry is used to measure ACE2 binding. As expected, spike-ΔRBD and the spike from HKU1 show the weakest signal. This serves as a measure of background fluorescence in these assays. All measurements are an average of three biological replicates. Error bars: S.D. of three replicates. (Data adapted from [37])

$$\text{Normalized expression} = \log_2\left(\frac{M_x^{488}}{M_{\text{wt}}^{488}}\right)$$

Some antigen variants will also have altered expression compared to the parental wild-type sequence. For example, pre-fusion stabilizing mutations will increase expression and stability, whereas destabilizing mutations may cause loss of expression. We apply a second round of normalization to correct for this by using the anti-FLAG signal to correct for changes in transfection efficiency and antigen expression when measuring antibody binding:

$$\text{Normalized binding} = \log_2\left(\frac{M_x^{647}/M_x^{488}}{M_{\text{wt}}^{647}/M_{\text{wt}}^{488}}\right)$$

In this case, the median signal measured by AF-647, M_x^{647}, is divided by the respective AF-488 signal for each sample and all samples are divided by this value obtained for the wild-type antigen.

3.6.2 Fluorescence-Assisted Cell Sorting (FACS)

Pooled antigen libraries can be analyzed via FACS, followed by next-generation sequencing (NGS). Transient transfection is not suitable for these experiments because individual cells can receive multiple plasmids, confounding the genotype-to-phenotype

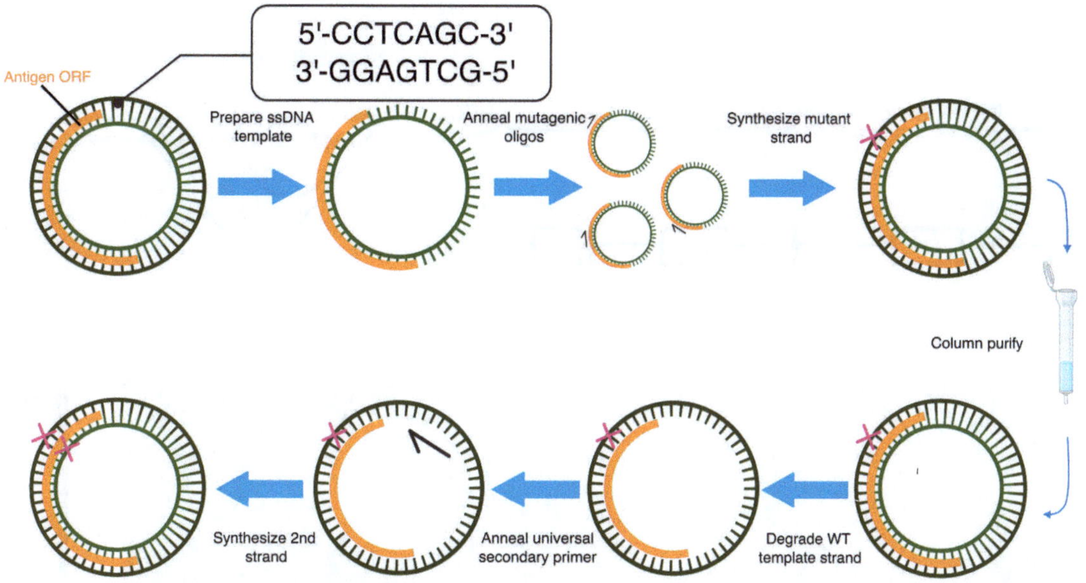

Fig. 7 Saturation mutagenesis library generation. Mutagenesis libraries are prepared via a one-pot protocol. The desired mutations are introduced through a pair of polymerase chain reactions using a mutagenic oligo pool. The parental plasmid encodes a seven basepair BbvCI restriction site that is sequentially nicked by Nt and Nb.BbvCI. First, the top strand is nicked by Nt.BbvCI and degraded. This strand is regenerated via primary PCR with the mutagenic oligo pool to introduce mutations into the top strand. Following a column purification, the bottom strand is nicked by Nb.BbvCI and degraded. Secondary PCR with a universal primer regenerates the bottom strand and fixes all mutation(s) that are now incorporated into the top strand. (This protocol is adapted from [49])

linkage. Instead, we recommend integrating antigen libraries into the genome, either via lentiviral integration, CRISPR knock-in, or via a serine integrase [52–55]. Lentiviral integration is a mature technology but suffers from slow viral amplification and selection. In addition, random integration into highly transcribed regions can lead to variable antigen expression levels. We prefer integration into an engineered landing pad in the AAVS1 locus (Fig. 8) [56]. This system uses BxbI integrase for efficient genomic insertion, followed by selection. Our lab routinely uses the engineered HEK-LLP cell lines described in Matreyek et al. to ensure single-copy integration and to reduce background noise. In order to integrate, antigen plasmids will need to contain the *attL* and *attR* attachment sites. The HEK-LLP cell line encodes iCasp9 for negative selection [57, 58]. iCasp9 is a fusion between Caspase 9 and the inducible dimerization domain FKBP1A. Cells that fail to integrate the donor plasmid will still contain iCasp9 in their active site, and will express the protein when induced. The small molecule AP1903 will trigger dimerization of expressed iCasp9 and will cause cell death through apoptosis [58]. We've observed integration efficiencies of 90% following negative selection. Integration, selection, and expansion of the HEK-LLP cells generally take 2 weeks.

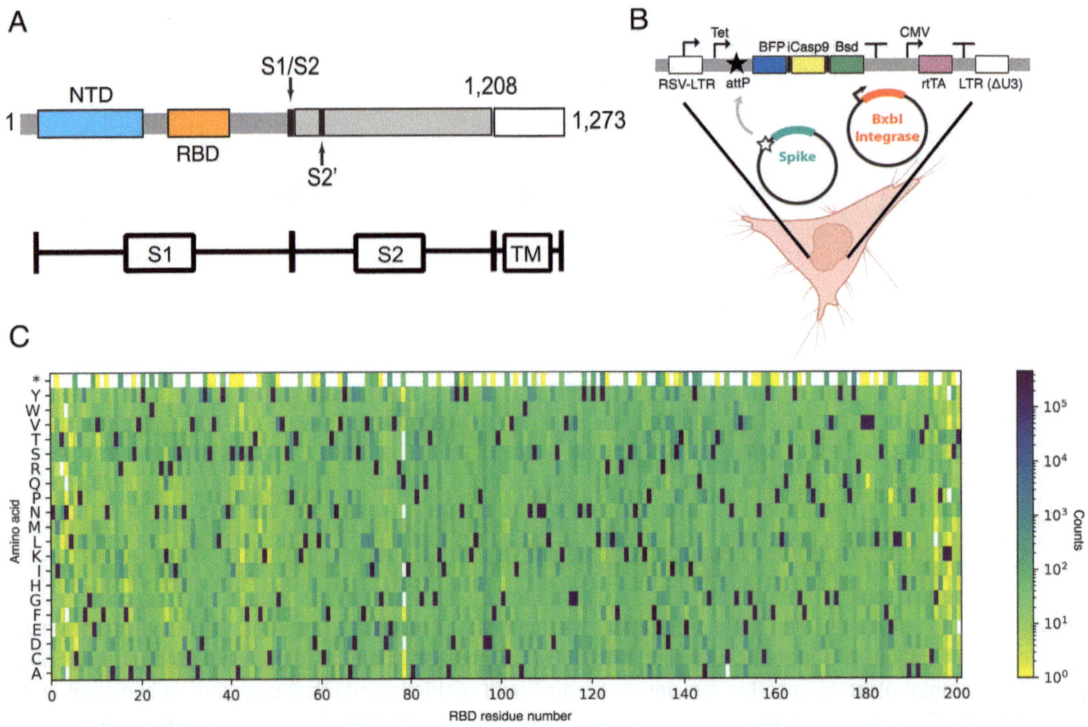

Fig. 8 Recombinase mediated integration and SARS-CoV-2 RBD mutagenesis. (**a**) SARS-CoV-2 spike domain map. SARS-CoV-2 spike protein is comprised of two domains S1 (residues 14–685) and S2 (residues 686–1273). S1 mediates receptor binding and contains the N-terminal domain (NTD) and the receptor binding domain (RBD) while S2 contains the fusion peptides required for membrane fusion. We only display the ectodomain, which lacks the transmembrane (TM) region (residues 1209–1273). (**b**) Schematic of Bxbl recombinase-mediated integration of antigen construct to the AAVS1 locus of HEK293T-LLP cells. (Adapted from [56]). The integration cassette at the AAVS1 locus contains a Tet inducible promoter, a blasticidin resistance (Bsd) gene, a blue fluorescent protein (BFP) for positive selection, and iCasp9 for negative selection. rtTA is the reverse Tet transactivator. LTR is a lentiviral long terminal repeat [56]. The AAVS1 locus contains a Bxbl1 attP recombination sequence. Similarly, the donor plasmid contains the corresponding attL and attR recombination sites that direct the Bxbl-mediated integration. (**c**) Heatmap of a saturating single amino-acid library created with nicking mutagenesis for 200 residues of the SARS-CoV-2 RBD. The wild-type sequence is dark blue

3.6.3 Collect Cells

1. Collect integrated cells expressing antigen libraries 72 h after selection, when they have become confluent. To collect cells, aspirate the media from the flask and resuspend in an equal volume of PBS.
2. Transfer resuspended cells to a 50 mL conical tube and spin down for 4 min at 400 rcf.
3. Aspirate PBS and resuspend cells again in fresh PBS. Pipette gently to monodisperse the cells.
4. Passage the cells at a 1:10 ratio of resuspended cells to fresh media into a new flask to maintain antigen libraries.

5. Spin down the remaining cells for 4 min at 400 rcf.
6. Aspirate the supernatant but leave a thin layer of PBS above the cells so they don't dry out. Keep cells on ice.

3.6.4 Immunostaining

1. Make primary staining master mix with PBS-BSA and 1 μg/mL primary antibody. We recommend simultaneously incubating with anti-FLAG and anti-antigen antibodies to save time. This can only be done if you've previously determined that your primary antibodies do not cross-react with each other.
2. Resuspend antigen libraries with primary staining master mix to a concentration of 8.0×10^6 cells/mL.
3. Incubate resuspended cells at room temperate on a microplate shaker at 950 rpm for 1 h. If needed, primary staining can be performed overnight at 4 °C.
4. While cells are staining chill a centrifuge with a swinging bucket rotor to 4 °C.
5. After primary staining is complete spin cells down in the pre-chilled centrifuge at 400 rcf for 5 min in a swinging bucket rotor.
6. Aspirate the supernatant.
7. Resuspend cells with 5 mL PBS-BSA and spin again at 400 rcf for 5 min.
8. Repeat the previous two steps and aspirate the supernatant.
9. Make secondary staining master mix with PBS-BSA and secondary antibodies (5 μg/mL of AF-488 and 10 μg/mL of AF-647 antibodies).
10. Resuspend cells with secondary staining master mix to a concentration of 1.6×10^7 cells/mL.
11. Incubate cells at 4 °C on an microplate shaker at 950 rpm for 20 min.
12. Spin cells down. Aspirate supernatant. Wash with 3 mL PBS-BSA.

3.6.5 Cell Sorting

1. Define the fluorescent channels and set gates to isolate single cells as described in Subheading 3.6.
2. The cell binning strategy is dependent on the type of experiment and the expected dynamic range, which must be calibrated ahead of time (see below). For example, when assessing antibody binding to a library of antigens, we use a four-bin sort. Each of the bins is defined as follows. Bin 1: cells with background fluorescence; bin 2: low affinity binding; bin 3: wild type-like binding; bin 4: antigens that bind the antibody with higher affinity than wt antigen. Bins are set using antigen mutants with lower and higher affinity than wild-type to the

antibody. To calibrate these values, cells expressing these mutants should be expressed and stained in parallel with antigen libraries, following the same protocol. To anticipate low-affinity antigens, we routinely perform a limited alanine scan of the expected epitope prior to performing bigger sorts.

3. Spin down sorted cells at max speed for 4 min. Aspirate supernatant.
4. Use Promega Wizard Genomic DNA purification kit to isolate genomic DNA (gDNA).
5. PCR amplifies the necessary region of gDNA for NGS. PCR amplifications should be limited to 17–20 cycles to minimize bias.

3.6.6 Data Analysis

Raw paired-end FASTQ files from NGS are processed and merged using fastp [59]. Illumina adapters are removed from all reads, then forward and reverse reads are paired and merged when they overlap by ≥30 bases and both reads contain fewer than 40% of bases with phred score less than 15. Extract barcodes (UMIs) by specifying their length at both the 5′ and 3′ ends of the merged reads.

Align nucleotide sequences of merged reads to the wild-type sequence using bowtie2 [60] and output to SAM files using samtools [61]. To identify amino acid mutations from wild-type sequence, translate the query sequence from the longest contiguous alignment to amino acids and compare to wild-type sequence. To measure changes in antigen expression or anti-antigen mAb binding first calculate each variant's mean fluorescent intensity based on its distribution across the bins, then normalize against the wild-type distribution [62].

4 Notes

1. The entry vector (Addgene # 172726) was designed for SARS-CoV-2 spike display. When adapting this workflow for different antigens, modify the overhangs to match the sequence of your antigen.
2. We recommend optimizing codon usage for expression in human cells. Take care to remove all AarI cut sites in all DNA blocks. Some antigens may include a native transmembrane domain. As the entry vector also contains a transmembrane (TM) domain and a linker with an anti-FLAG epitope, we recommend that the antigen's native TM domain is entirely removed. Removing the native TM domain will avoid artifacts due to unanticipated tethering of the antigen, leading to occluded anti-FLAG epitopes.

3. Cells should be tested for Mycoplasma contamination before use and regularly thereafter. Immortalized cell lines (such as HEK293T) are more prone to be genetically unstable. Therefore, discard the plates and flasks after 4–6 weeks of passaging. Performing experiments with cells that have been passaged more than 20 times is not recommended due to the genotypic and phenotypic drift that might arise from the selective pressure of culture conditions.
4. Fixation with PFA or methanol will partially permeabilize the cell membrane, resulting in antibodies entering the cell and staining intracellular antigens. Skip the fixation step to stain only the antigens displayed on the cell surface. Extra care must be taken while manipulating the cells, especially during the washing steps. These cells are prone to dissociating from the surface and may be decanted along with the aspirated liquids.

Author Contributions A.A., A.Q., K.J., and I.J.F. conceived the project. A.Q., A.A., and K.J. performed all experiments, analyzed the data, and prepared figs. J.C. wrote the bioinformatics processing pipeline and analyzed some data. I.J.F. secured funding and supervised the project. A.A., A.Q., and I.J.F. wrote the manuscript with input from all co-authors.

Inclusion and Diversity Statement The Finkelstein lab is committed to elevating people of underrepresented geographical locations, ethnicities, genders, abilities, and other forms of diversity in science. A.Q. self-identifies as part of underrepresented gender and ethnic groups in STEM.

Funding This work was supported by the Welch Foundation grant F-1808 to I.J.F., the Bill & Melinda Gates Foundation (INV-034714 to I.J.F.), and NIST (70NANB22H017 to I.J.F.).

References

1. Marani M, Katul GG, Pan WK, Parolari AJ (2021) Intensity and frequency of extreme novel epidemics. Proc Natl Acad Sci 118(35): e2105482118. https://doi.org/10.1073/pnas.2105482118
2. Bok K, Sitar S, Graham BS, Mascola JR (2021) Accelerated COVID-19 vaccine development: milestones, lessons, and prospects. Immunity 54(8):1636–1651
3. Weldon WC, Wang B-Z, Martin MP, Koutsonanos DG, Skountzou I, Compans RW (2010) Enhanced immunogenicity of stabilized trimeric soluble influenza hemagglutinin. PLoS One 5(9):e12466. https://doi.org/10.1371/journal.pone.0012466
4. Sanders RW et al (2013) A next-generation cleaved, soluble HIV-1 env trimer, BG505 SOSIP.664 gp140, expresses multiple epitopes for broadly neutralizing but not non-neutralizing antibodies. PLoS Pathog 9(9):e1003618. https://doi.org/10.1371/journal.ppat.1003618
5. Lu Y, Welsh JP, Swartz JR (2013) Production and stabilization of the trimeric influenza hemagglutinin stem domain for potentially broadly protective influenza vaccines. Proc Natl Acad

Sci 111(1):125–130. https://doi.org/10.1073/pnas.1308701110

6. Krarup A et al (2015) A highly stable prefusion RSV f vaccine derived from structural analysis of the fusion mechanism. Nat Commun 6(1). https://doi.org/10.1038/ncomms9143
7. Pallesen J et al (2017) Immunogenicity and structures of a rationally designed prefusion MERS-CoV spike antigen. Proc Natl Acad Sci 114(35). https://doi.org/10.1073/pnas.1707304114
8. Battles MB et al (2017) Structure and immunogenicity of pre-fusion-stabilized human metapneumovirus f glycoprotein. Nat Commun 8(1). https://doi.org/10.1038/s41467-017-01708-9
9. Crank MC et al (2019) A proof of concept for structure-based vaccine design targeting RSV in humans. Science 365(6452):505–509. https://doi.org/10.1126/science.aav9033
10. Hsieh C-L et al (2022) Structure-based design of prefusion-stabilized human metapneumovirus fusion proteins. Nat Commun 13(1). https://doi.org/10.1038/s41467-022-28931-3
11. Byrne PO, McLellan JS (2022) Principles and practical applications of structure-based vaccine design. Curr Opin Immunol 77:102209
12. Sanders RW, Moore JP (2021) Virus vaccines: proteins prefer prolines. Cell Host Microbe 29(3):327–333
13. Peleg Y et al (2021) Community-wide experimental evaluation of the PROSS stability-design method. J Mol Biol 433(13):166964. https://doi.org/10.1016/j.jmb.2021.166964
14. Campeotto I et al (2017) One-step design of a stable variant of the malaria invasion protein RH5 for use as a vaccine immunogen. Proc Natl Acad Sci 114(5):998–1002. https://doi.org/10.1073/pnas.1616903114
15. Goldenzweig A et al (2016) Automated structure- and sequence-based design of proteins for high bacterial expression and stability. Mol Cell 63(2):337–346. https://doi.org/10.1016/j.molcel.2016.06.012
16. Shroff R et al (2020) Discovery of novel gain-of-function mutations guided by structure-based deep learning. ACS Synth Biol 9(11):2927–2935. https://doi.org/10.1021/acssynbio.0c00345
17. Impagliazzo A et al (2015) A stable trimeric influenza hemagglutinin stem as a broadly protective immunogen. Science 349(6254):1301–1306. https://doi.org/10.1126/science.aac7263
18. Nachbagauer R et al (2020) A chimeric hemagglutinin-based universal influenza virus vaccine approach induces broad and long-lasting immunity in a randomized, placebo-controlled phase i trial. Nat Med 27(1):106–114. https://doi.org/10.1038/s41591-020-1118-7
19. Liao H-Y et al (2020) Chimeric hemagglutinin vaccine elicits broadly protective CD4 and CD8 t cell responses against multiple influenza strains and subtypes. Proc Natl Acad Sci 117(30):17757–17763. https://doi.org/10.1073/pnas.2004783117
20. Martinez DR et al (2021) Chimeric spike mRNA vaccines protect against sarbecovirus challenge in mice. Science 373(6558):991–998. https://doi.org/10.1126/science.abi4506
21. Cohen AA et al (2022) Mosaic RBD nanoparticles protect against challenge by diverse sarbecoviruses in animal models. Science 377(6606). https://doi.org/10.1126/science.abq0839
22. Boder ET, Wittrup KD (1997) Yeast surface display for screening combinatorial polypeptide libraries. Nat Biotechnol 15(6):553–557. https://doi.org/10.1038/nbt0697-553
23. Han T et al (2011) Fine epitope mapping of monoclonal antibodies against hemagglutinin of a highly pathogenic H5N1 influenza virus using yeast surface display. Biochem Biophys Res Commun 409(2):253–259. https://doi.org/10.1016/j.bbrc.2011.04.139
24. Gaiotto T, Hufton SE (2016) Cross-neutralising nanobodies bind to a conserved pocket in the hemagglutinin stem region identified using yeast display and deep mutational scanning. PLoS One 11(10):e0164296. https://doi.org/10.1371/journal.pone.0164296
25. Gaiotto T et al (2021) Nanobodies mapped to cross-reactive and divergent epitopes on a (H7N9) influenza hemagglutinin using yeast display. Sci Rep 11(1). https://doi.org/10.1038/s41598-021-82356-4
26. Greaney AJ et al (2021) Mapping mutations to the SARS-CoV-2 RBD that escape binding by different classes of antibodies. Nat Commun 12(1):4196. https://doi.org/10.1038/s41467-021-24435-8
27. Starr TN et al (2021) Prospective mapping of viral mutations that escape antibodies used to treat COVID-19. Science 371(6531):850–854. https://doi.org/10.1126/science.abf9302
28. Francino-Urdaniz IM et al (2021) One-shot identification of SARS-CoV-2 S RBD escape

mutants using yeast screening. Cell Rep 36(9). https://doi.org/10.1016/j.celrep.2021.109627

29. Brouwer PJM et al (2020) Potent neutralizing antibodies from COVID-19 patients define multiple targets of vulnerability. Science 369(6504):643–650. https://doi.org/10.1126/science.abc5902
30. Liu L et al (2020) Potent neutralizing antibodies against multiple epitopes on SARS-CoV-2 spike. Nature 584(7821):450–456. https://doi.org/10.1038/s41586-020-2571-7
31. Rogers TF et al (2020) Isolation of potent SARS-CoV-2 neutralizing antibodies and protection from disease in a small animal model. Science 369(6506):956–963. https://doi.org/10.1126/science.abc7520
32. Wang C et al (2021) A conserved immunogenic and vulnerable site on the coronavirus spike protein delineated by cross-reactive monoclonal antibodies. Nat Commun 12(1). https://doi.org/10.1038/s41467-021-21968-w
33. Voss WN et al (2021) Prevalent, protective, and convergent IgG recognition of SARS-CoV-2 non-RBD spike epitopes. Science 372(6546):1108–1112. https://doi.org/10.1126/science.abg5268
34. Silva RP et al (2023) Identification of a conserved S2 epitope present on spike proteins from all highly pathogenic coronaviruses. eLife 12. https://doi.org/10.7554/elife.83710
35. Hamilton SR et al (2003) Production of complex human glycoproteins in yeast. Science 301(5637):1244–1246. https://doi.org/10.1126/science.1088166
36. Grant OC, Montgomery D, Ito K, Woods RJ (2020) Analysis of the SARS-CoV-2 spike protein glycan shield reveals implications for immune recognition. Sci Rep 10(1):14991. https://doi.org/10.1038/s41598-020-71748-7
37. Javanmardi K et al (2021) Rapid characterization of spike variants via mammalian cell surface display. Mol Cell 81(24):5099–5111.e8. https://doi.org/10.1016/j.molcel.2021.11.024
38. Javanmardi K et al (2022) Antibody escape and cryptic cross-domain stabilization in the SARS-CoV-2 Omicron spike protein. Cell Host Microbe 30(9):1242–1254.e6. https://doi.org/10.1016/j.chom.2022.07.016
39. Allen JD et al (2021) Site-specific steric control of SARS-CoV-2 spike glycosylation. Biochemistry 60(27):2153–2169. https://doi.org/10.1021/acs.biochem.1c00279
40. Yao H et al (2020) Molecular architecture of the SARS-CoV-2 virus. Cell 183(3):730–738.e13. https://doi.org/10.1016/j.cell.2020.09.018
41. Hsieh C-L et al (2020) Structure-based design of prefusion-stabilized SARS-CoV-2 spikes. Science 369(6510):1501–1505. https://doi.org/10.1126/science.abd0826
42. Schaub JM et al (2021) Expression and characterization of SARS-CoV-2 spike proteins. Nat Protoc 16(11):5339–5356. https://doi.org/10.1038/s41596-021-00623-0
43. Tan TJ et al (2023) High-throughput identification of prefusion-stabilizing mutations in SARS-CoV-2 spike. Nat Commun 14(1):2003
44. Ouyang WO et al (2022) Probing the biophysical constraints of SARS-CoV-2 spike n-terminal domain using deep mutational scanning. Sci Adv 8(47):eadd7221
45. Shelton SB, Shah NM, Abell NS, Devanathan SK, Mercado M, Xhemalçe B (2018) Crosstalk between the RNA methylation and histone-binding activities of MePCE regulates p-TEFb activation on chromatin. Cell Rep 22(6):1374–1383. https://doi.org/10.1016/j.celrep.2018.01.028
46. Schindelin J et al (2012) Fiji: an open-source platform for biological-image analysis. Nat Methods 9(7):676–682. https://doi.org/10.1038/nmeth.2019
47. Yurtsev E, Friedman J, Gore J (2015) FlowCytometryTools: version 0.4.5. Zenodo. https://doi.org/10.5281/ZENODO.32991
48. Fowler DM, Fields S (2014) Deep mutational scanning: a new style of protein science. Nat Methods 11(8):801–807. https://doi.org/10.1038/nmeth.3027
49. Wrenbeck EE, Klesmith JR, Stapleton JA, Adeniran A, Tyo KEJ, Whitehead TA (2016) Plasmid-based one-pot saturation mutagenesis. Nat Methods 13(11):928–930. https://doi.org/10.1038/nmeth.4029
50. Hansen J et al (2020) Studies in humanized mice and convalescent humans yield a SARS-CoV-2 antibody cocktail. Science 369(6506):1010–1014. https://doi.org/10.1126/science.abd0827
51. Dreyfus C et al (2012) Highly conserved protective epitopes on influenza b viruses. Science 337(6100):1343–1348. https://doi.org/10.1126/science.1222908
52. Elegheert J et al (2018) Lentiviral transduction of mammalian cells for fast, scalable and high-level production of soluble and membrane proteins. Nat Protoc 13(12):2991–3017. https://doi.org/10.1038/s41596-018-0075-9

53. Yarnall MTN et al (2022) Drag-and-drop genome insertion of large sequences without double-strand DNA cleavage using CRISPR-directed integrases. Nat Biotechnol 41(4): 500–512. https://doi.org/10.1038/s41587-022-01527-4
54. Anzalone AV et al (2021) Programmable deletion, replacement, integration and inversion of large DNA sequences with twin prime editing. Nat Biotechnol 40(5):731–740. https://doi.org/10.1038/s41587-021-01133-w
55. Durrant MG et al (2022) Systematic discovery of recombinases for efficient integration of large DNA sequences into the human genome. Nat Biotechnol 41(4):488–499. https://doi.org/10.1038/s41587-022-01494-w
56. Matreyek KA, Stephany JJ, Chiasson MA, Hasle N, Fowler DM (2019) An improved platform for functional assessment of large protein libraries in mammalian cells. Nucleic Acids Res. https://doi.org/10.1093/nar/gkz910
57. Gargett T, Brown MP (2014) The inducible caspase-9 suicide gene system as a âœsafety switchâ to limit on-target, off-tumor toxicities of chimeric antigen receptor t cells. Front Pharmacol 5. https://doi.org/10.3389/fphar.2014.00235
58. Ramos CA et al (2010) An inducible caspase 9 suicide gene to improve the safety of mesenchymal stromal cell therapies. Stem Cells 28(6):1107–1115. https://doi.org/10.1002/stem.433
59. Chen S, Zhou Y, Chen Y, Gu J (2018) Fastp: an ultra-fast all-in-one FASTQ preprocessor. Bioinformatics 34(17):i884–i890. https://doi.org/10.1093/bioinformatics/bty560
60. Langmead B, Salzberg SL (2012) Fast gapped-read alignment with bowtie 2. Nat Methods 9(4):357–359. https://doi.org/10.1038/nmeth.1923
61. Danecek P et al (2021) Twelve years of SAMtools and BCFtools. GigaScience 10(2). https://doi.org/10.1093/gigascience/giab008
62. Starr TN et al (2020) Deep mutational scanning of SARS-CoV-2 receptor binding domain reveals constraints on folding and ACE2 binding. Cell 182(5):1295–1310.e20. https://doi.org/10.1016/j.cell.2020.08.012
63. Huang Y, Yang C, Xu X, Xu W, Liu S (2020) Structural and functional properties of SARS-CoV-2 spike protein: potential antivirus drug development for COVID-19. Acta Pharmacol Sin 41(9):1141–1149. https://doi.org/10.1038/s41401-020-0485-4

Part IV

Analysis of Glycoproteins

Chapter 13

Analysis of Native and Permethylated N-Glycan Isomers Using MGC-LC-MS Techniques

Andrew I. Bennett, Oluwatosin Daramola, Md Mostofa Al Amin Bhuiyan, Vishal Sandilya, and Yehia Mechref

Abstract

Glycosylation is an important post-translational modification that affects many critical cellular functions such as adhesion, signaling, protein stability, and function, among others. Abnormal glycosylation has been linked to many diseases. As such, the investigation of glycans and their roles in disease pathway and progression is important. Glycan analysis can be challenging, however, due to such factors as the heterogeneity of glycans and isomers as well as the poor ionization efficiency provided by mass spectrometry analyses. This chapter presents efficient methods that overcome these and other challenges for the analysis of native and permethylated *N*-glycan isomers in biological samples. Instructions regarding the packing of the MGC column, the *N*-glycan sample prep, and the LC-MS conditions are also provided.

Key words Glycomics, Liquid chromatography-mass spectrometry, Isomers, Mesoporous graphitized carbon (MGC)

1 Introduction

Glycans are an important post-translational modification of proteins that affect many different biological processes and interactions including cell-cell signaling [2, 18], protein stability [16], immune cell trafficking [17, 20], and cell adhesion [11, 15]. Approximately 50% of mammalian proteins are glycosylated [1]. Aberrant glycosylation has been linked to many different diseases including metabolic disorders [3, 19], Alzheimer's disease [8, 14], and cancers [5, 7, 9, 10, 12, 21]. The use of glycan isomers as potential biomarkers in various diseases has been investigated with promising results [6, 13, 20]. As such, sensitive and accurate methods for the analysis of glycan isomers from biological samples are needed.

Here, we describe the necessary steps for the analysis of native and permethylated *N*-glycans using mesoporous graphitized carbon (MGC) columns interfaced with liquid chromatography-mass

Steven B. Bradfute (ed.), *Recombinant Glycoproteins: Methods and Protocols*, Methods in Molecular Biology, vol. 2762,
https://doi.org/10.1007/978-1-0716-3666-4_13,

spectrometry (LC-MS). This MGC-LC-MS approach provides enhanced isomeric separation of glycans, thus enabling a more accurate representation of glycans and glycan isomers in a biological sample (Figs. 1 and 2).

In this chapter, the procedures outlining the preparation of the MGC column will first be presented. The use of mesoporous graphitized carbon provides increased separation, column stability, and reproducibility compared to porous graphitized carbon columns [4]. Following this, sample preparation steps for native and permethylated glycans are described, including denaturation, enzyme digestion, reduction, permethylation, and cleaning for LC-MS analysis (Fig. 3).

The enzymatic release of *N*-glycans is particularly significant. The enzyme peptide *N*-glycosidase F (PNGase F) is of foundational importance for glycomics analysis. PNGase F conveniently cleaves the entire *N*-glycan from the peptide backbone. Permethylation of glycans is also an established procedure in the analysis of glycans for several reasons. Permethylation occurs when the hydrogens of the hydroxide groups on sugars are replaced by methyl groups. This change has been shown to increase sensitivity 100× in MS analysis compared to native glycans. Permethylation also provides more compatible interaction for separation of glycans using C18 (and similar) column materials. Lastly, it helps to stabilize the glycan, preventing rearrangement of sialic acids and fucose monomers during the ionization process. One downside to performing permethylation is that it adds an additional step to the sample preparation process, which can contribute to unintended sample alteration, contamination, or loss. The analysis of native glycans is therefore still of value if it can provide acceptable results without the additional step of permethylation. To conclude this chapter, the LC-MS conditions for both permethylated and native *N*-glycans will be presented. The effects of varying mobile phase compositions on separation efficiency using an MGC column are shown in Fig. 4.

2 Materials

2.1 Biological Reagents

1. Bovine fetuin.
2. RNAse B.
3. Human blood serum (HBS).
4. Immunoglobulin G (IgG).
5. Trypsin-EDTA 1× (0.25% trypsin/2.21 mM EDTA).
6. PNGase F.
7. Cell lines used in these protocols include MDA-MB-231BR (231BR), MDA-MB-231 (231), and CRL-1620 (CRL).

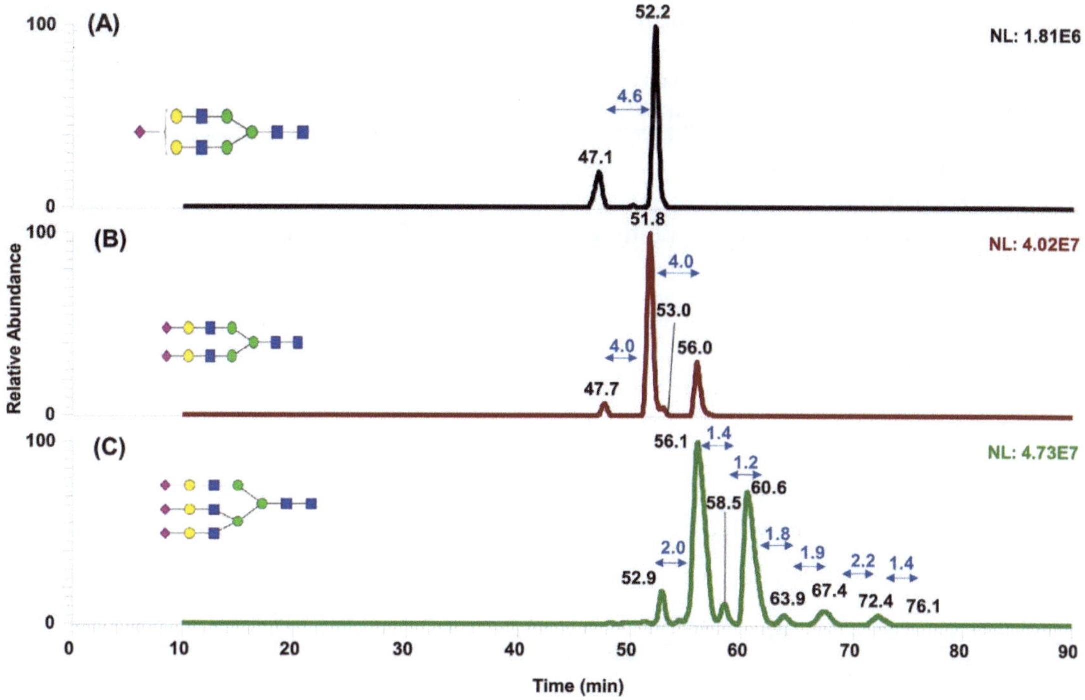

Fig. 1 Separation of isomeric glycans derived from bovine fetuin. MPA was 98% water, 2% ACN, 0.1% DFA. MPB was 50% ACN, 50% IPA, 0.1% DFA. (**a**) Separation of glycan HexNAc4Hex5NeuAc1. (**b**) Separation of glycan HexNAc4Hex5NeuAc2. (**c**) Separation of glycan HexNAc5Hex6NeuAc3. (Reprinted with permission from Ref. [4])

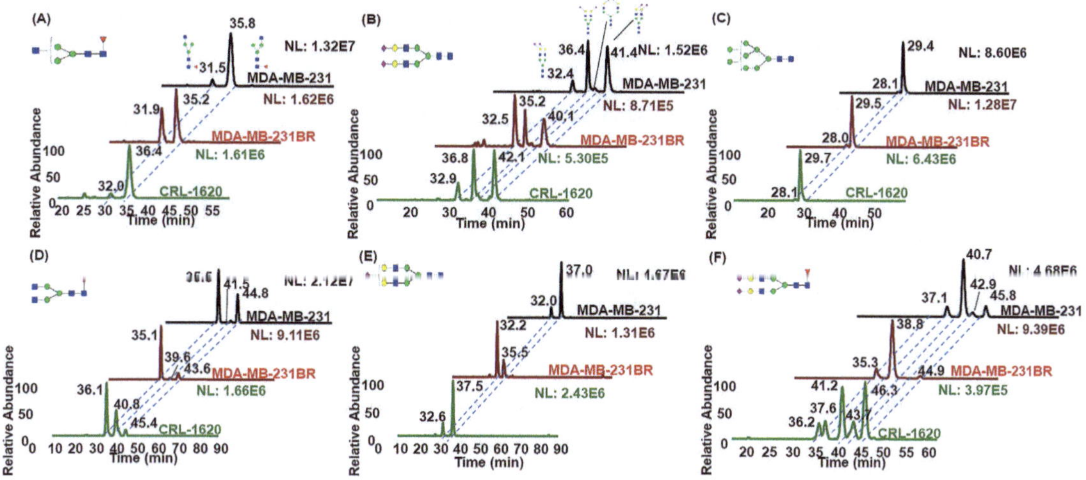

Fig. 2 Isomeric separation of permethylated glycans derived from MDA-MB-231 (black), MDA-MB-231BR (red), and CRL-1620 (green) cell lines. MPA was 98% water, 2% ACN, 0.1% DFA. MPB was 50% ACN, 50% IPA, 0.1% DFA. Traces are for extracted ion chromatograms of glycans (**a**) HexNAc3Hex3Fuc1, (**b**) HexNAc4Hex5NeuAc2, (**c**) HexNAc2Hex8, (**d**) HexNAc4Hex3Fuc1, (**e**) HexNAc4Hex5NeuAc1, and (**f**) HexNAc4Hex5-Fuc1NeuAc2. (Reprinted with permission from Ref. [4])

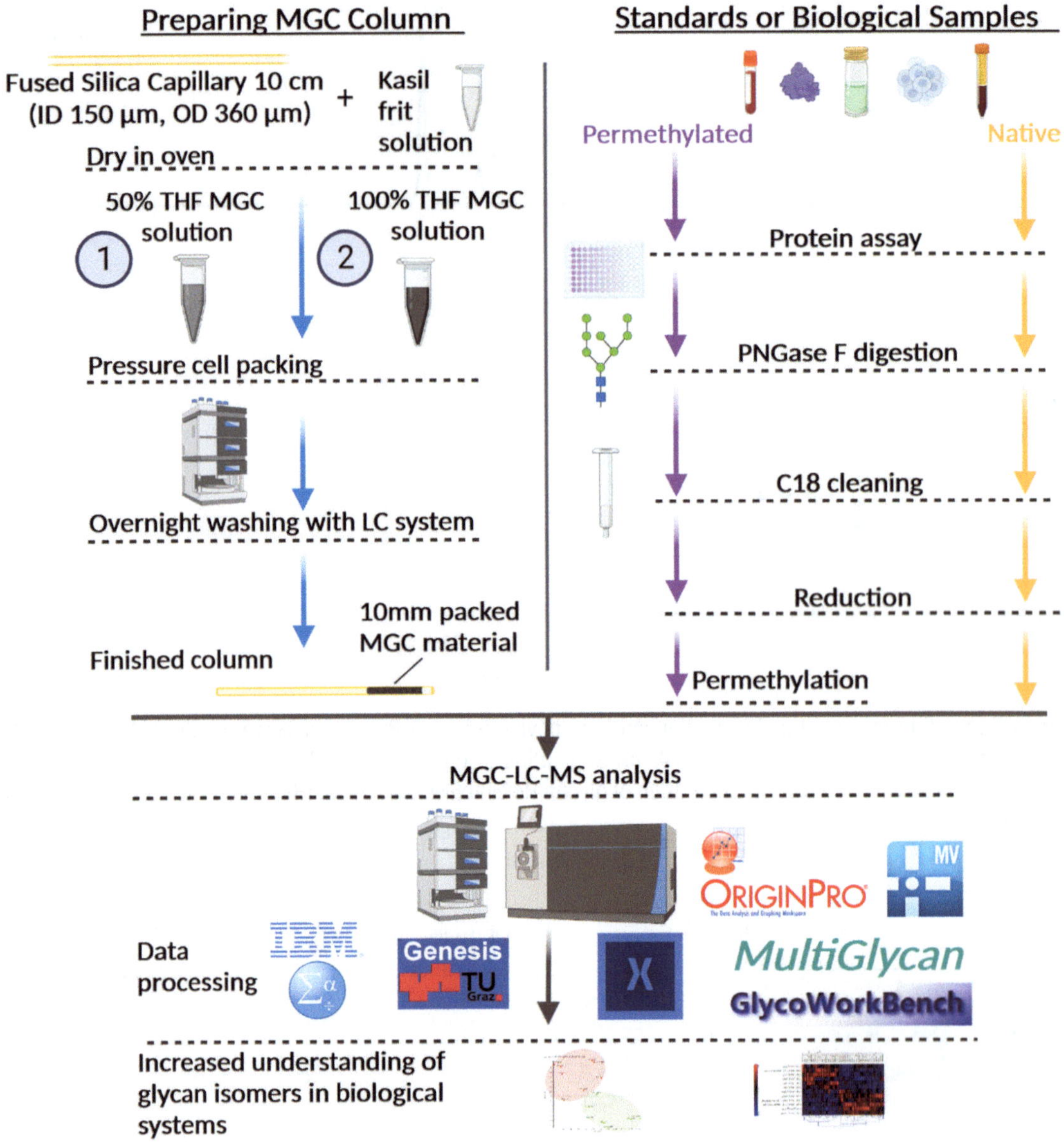

Fig. 3 Workflow for the preparation of the MGC column, permethylated *N*-glycans, and native *N*-glycans

2.2 MGC Column Materials

1. Mesoporous graphitized carbon (MGC).
2. Tetrahydrofuran (THF).
3. Fused silica capillary (inner diameter [ID] of 150 μm, outer diameter [OD] of 360 μm).
4. Kasil frit kit.
5. Pressure cell.

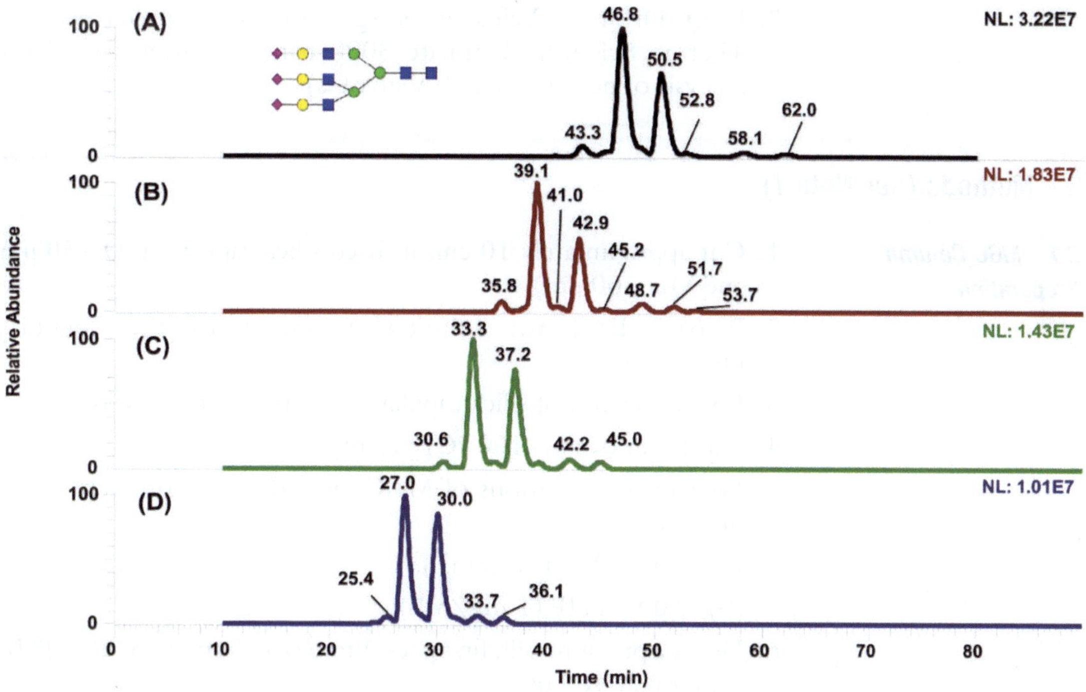

Fig. 4 Effect of different MPB compositions on isomeric separation of permethylated glycans for a MGC column. MPA was 95% water, 2% ACN with 0.1% formic acid and remained the same for all separations. The same glycan, HexNAc5Hex6NeuAc3, is shown for all conditions. (**a**) 25% isopropanol (IPA) and 75% ACN, 0.1% DFA. (**b**) 50% IPA and 50% ACN, 0.1% DFA. (**c**) 75% IPA and 25% ACN, 0.1% DFA. (**d**) 100% IPA, 0.1% DFA. (Reprinted with permission from Ref. [4])

2.3 Other Materials

1. HPLC-grade acetonitrile (ACN).
2. Methanol.
3. Ethyl alcohol.
4. MS-grade formic acid (FA).
5. Sodium deoxycholate (SDC).
6. Borane-ammonia complex (97% purity).
7. Dimethyl sulfoxide (DMSO >99.9%).
8. Iodomethane.
9. Sodium hydroxide beads (20–40 mesh).
10. HPLC-grade water.
11. Trifluoropropionic acid (TFP).
12. Difluoroacetic acid (DFA).

2.4 Instruments

1. Thermo Scientific Ultimate 3000 nanoLC system interfaced with a Q Exactive HF mass spectrometer (used for native *N*-glycan analysis).

2. LTQ Orbitrap Velos mass spectrometer coupled with a Thermo Scientific Ultimate 3000 nanoLC system (used for analysis of permethylated *N*-glycans).

3 Methods (*See* Note 1)

3.1 MGC Column Preparation

1. Cut approximately 10 cm of fused silica capillary (ID 150 μm and OD 360 μm).
2. Prepare Kasil frit solution according to the vendor's instructions.
3. Insert one end of silica capillary into frit solution for 3 s.
4. Let dry in oven (~100 °C) overnight.
5. Prepare two solutions of MGC in THF, approximately 1 mL of each:
   (a) 50%THF (1 mg/mL).
   (b) 100% THF (1 mg/mL).
6. Using a pressure cell, first pack 2 mm of column with 50% THF solution at 400 psi.
7. Pack an additional 10 mm (for about 12 mm total) of MGC material using 100% THC solution at 800 psi.
8. Connect MGC column to nanoLC system and wash with 95% ACN overnight.
9. The MGC column should be about 10 mm when properly packed.

3.2 PNGase F Digestion

1. Transfer samples (*see* **Note 2**) with 50–100 μg of proteins to a 1.5 mL sample tube and add 50 mM ammonium bicarbonate (ABC) buffer solution until the final volume is 100 μL.
2. Denature proteins by heating the sample in a water bath at 90 °C for 15 min.
3. Dilute the PNGase F enzyme (500 units/μL) five times to give 100 unit/μL. Add the volume needed based on the concentration of the protein, mix gently (*see* **Note 3**), and incubate in a water bath at 37 °C for 18 h.
4. Remove the sample from the 37 °C water bath, cool down to room temperature, and then spin down in a centrifuge.
5. Dry the sample using a benchtop vacuum concentrator.

3.3 SPE-C18 Cleanup

1. Resuspend the dried sample in 300 μL of 5% acetic acid.
2. Wash the SPE-C18 cartridges with 3 mL of methanol and then equilibrate with 3 mL of 5% acetic acid (*see* **Note 4**).

3. Transfer the resuspended samples into the SPE-C18 cartridges and wash with 300 μL of 5% acetic acid three times. Collect all flow-through and dry overnight in a vacuum concentrator.

3.4 Reduction

1. Add 10 μL of 10 μg/μL borane ammonia solution to the dried samples. Vortex and spin down, then incubate at 60 °C for 1 h.
2. After incubation, remove the residual borane from the reduced sample by the addition of methanol to generate methyl borate that will be evaporated while drying in the vacuum concentrator.
   (a) Add 1 mL methanol to the sample, then dry in speed vac. Repeat four more times.
3. For native glycans, resuspend the cleaned and dried sample in mobile phase A (80% water, 20% acetonitrile, 0.1% formic acid) used for LC separation to give a final concentration of 5 μg/μL, depending on the starting concentration of protein.
   (a) *For permethylated glycans, skip to Subheading 3.5.*
4. Centrifuge the sample at 14.8k rpm for 10 min and transfer to a properly labeled LC sample vial for LC-MS/MS analysis (*see* **Note 5**).

3.5 Permethylation (See Note 6)

If native glycans are to be analyzed, skip this permethylation step and proceed with LC-MS analysis. For permethylated glycans:

1. Add 30 μL DMSO and 1.2 μL H_2O to the sample. Vortex and spin down.
2. Place the column into a 2 mL tube with spacers to allow flow through.
3. Fill the column with DMSO almost entirely, leaving only 2–3 mm of space at the top.
4. Add NaOH beads to the column, filling halfway, then centrifuge at 1800 rpm for 2 min.
5. Wash the column with 200 μL DMSO, then centrifuge at 1800 rpm for 2 min.
6. While the column is centrifuging, add 20 μL iodomethane (CH3I) to the sample, vortex, and spin down.
7. Empty flow through of DMSO, and change to a clean, labeled tube to collect permethylated glycans.
8. Add the sample to the column, and incubate at room temperature in the dark for 25 min.
9. Add 20 μL of iodomethane to the column, then incubate for 15 min.
10. Centrifuge at 1800 rpm for 2 min. Then add 30 μL of ACN to the column. Centrifuge at 1800 rpm again for 2 min.

11. Use a syringe to expel the residual sample on the column into a collection tube.
12. Dry the sample.
13. Resuspend the sample in 80% water, 20% ACN (containing 0.1% formic acid).

3.6 nanoLC Conditions (Native)

1. Prepare mobile phases:
   (a) Mobile phase A (MPA) is 98% HPLC water, 2% ACN, and 0.1% formic acid.
   (b) Mobile phase B (MPB) is 100% ACN, and 0.1% formic acid.
2. The injection amount of native *N*-glycans is equal to 5 μg of digested proteins.
3. Load native glycans directly onto an MGC column without a trap (*see* **Note 7**).
4. Use a 2 or 5 μL sample loop for efficient loading of sample onto the column (*see* **Note 8**).
5. The separation is done on an MGC 64 Å column (10 mm × 150 μm ID, in-house packed column).[1] The column compartment temperature should be set at 75 °C and a flow rate between 0.300 and 0.400 μL/min can be used.
6. For native *N*-glycans apply a multistep mobile phase gradient: start at 1% MPB for 15 min, gradually increasing to 20% over 10 min, and 35% over 50 min. Then, ramp up to 80% of MPB in 2 min and keep constant for 9 min to wash the system. Finally, decrease to 1% B in 1 min and keep it constant for 4 min to equilibrate the column (Fig. 5).

3.7 nanoLC Conditions (Permethylated)

1. Prepare mobile phases:
   (a) Mobile phase A (MPA) is 98% HPLC water, 2% ACN, and 0.1% DFA.
   (b) Mobile phase B (MPB) is 50% IPA, 50% ACN, and 0.1% DFA.
2. An Acclaim PepMap 100 C18 trap column (75 μm × 2 cm, 3 μm particle size) should be used for permethylated glycans.
3. A multistep gradient will be used for the separation, where MPB is 20% for 10 min at the start.
4. Increase MPB to 50% after 20 min.
5. Increase to 95% at 60 min until 87 min.
6. Reduce MPB to 20% at 88–90 min, to equilibrate the column.

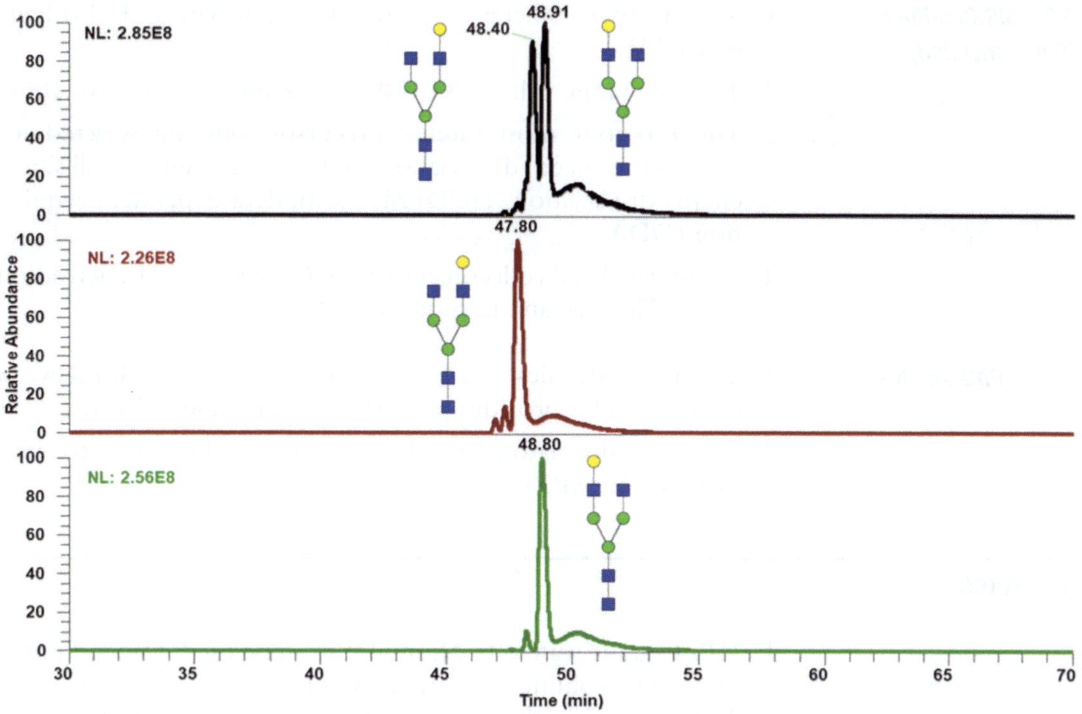

Fig. 5 Isomeric separation of left and right galactosylated arms of HexNac(4)Hex(4) native N-glycan standards on MGC

3.8 MS Conditions (Native)

1. After the LC separation, the native *N*-glycans are analyzed by mass spectrometer via a nano ESI source in positive ion mode.
2. Set the spray voltage at 2 kV, and the transfer tube temperature at 275 °C.
3. The full MS spectra is acquired by an Orbitrap mass analyzer with a mass range of 400–2000 m/z. The resolving power is 100,000 and the mass accuracy is 5 ppm.
4. The RF lens is set at 60% and the maximum injection time is 50 ms.
5. The dynamic exclusion parameters are as follows: repeat count 1; exclusion duration of 60 s; mass tolerance of 10 ppm; and an intensity threshold of 5.0e4.
6. The MS/MS Orbitrap scan will be generated in a data-dependent manner. For this purpose, the duty cycle is 3 s and 20 of the most intense ions from the full MS scan are selected for an HCD MS/MS scan with a normalized collision energy (NCE) of 23% and a 10 ms activation time.
7. The isolation mode is Quadrupole with an isolation window of 2 m/z. The resolution of the mass analyzer is 30,000, with a fixed scan range of 100–2000 m/z, and a maximum injection time of 60 ms and three dependent scans.

3.9 MS Conditions (Permethylated)

1. Set mass spectrometer to positive ion mode with an ESI voltage of 1.6 kV.
2. The scan range will be 400–2000 m/z and resolution of 100k.
3. The top four most intense precursor ions are selected for collision-induced dissociation (CID) and higher collisional energy dissociation (HCD) MS2 with data-dependent acquisition (DDA).
4. The normalized collision energy for CID is 35%. The activation Q is 0.25, with an injection time of 30 ms.

3.10 Data Analysis

1. The raw data files can be first processed using MultiGlycan (vendor). This provides peak areas for identified glycans.
2. The ion chromatograms (EICs) can then be extracted with Xcalibur 2.2 SP1 or Freestyle software (*see* **Note 9**).

4 Notes

1. HPLC grade water, ACN, and FA must be used. Depending on the total volume of either mobile phase A or B (e.g., 250–500 mL), the volume of each composition should be added accordingly. Also, both mobile phase A and B must be sonicated/degassed to remove gas after mixing. Store mobile phase at room temperature.
2. Samples can be in either solid phase or liquid phase. Solid samples, such as some commercial glycoprotein standards, should be completely dissolved in an aliquot of 100 μL of 50 mM ABC buffer solution. Liquid samples should be diluted using 50 mM ABC buffer solution until the final volume reaches 100 μL.
3. Each sample requires 5 units of PNGase F per μg of protein, so the amount of protein will determine the volume of the enzyme to be added for effective digestion.
4. Because the cartridge is 1 mL, add 1 mL of methanol and acetic acid three times to wash with 3 mL.
5. Centrifuging the sample before loading it onto the LC system is critical to remove any undissolved particles.
6. Permethylation must be done quickly as water interferes with the reaction. It is recommended to only do 4–8 samples at a time.
7. You can choose to use a trap instead of direct injection if you have a compatible HILIC trap.

8. If you are loading directly onto the column using the flow rate of 300 nL to 400 nL to shorten the loading time, it is necessary to use a shorter sample loop (2–5 μL) instead of the 20 μL sample loop.
9. Glycan compositions should be analyzed using full MS data. Mass tolerance for extracted ion chromatograms (EICs) can be set between 3–10 ppm and the peak smoothing from 7 to 15-point boxcar setting.

Acknowledgments

This work is supported by NIH (1R01GM130091-04), Robert A. Welch Foundation (No. D-0005), and The CH Foundation.

References

1. Apweiler R, Hermjakob H, Sharon N (1999) On the frequency of protein glycosylation, as deduced from analysis of the SWISS-PROT database. Biochim Biophys Acta 1473:4–8
2. De Vreede G, Morrison HA, Houser AM et al (2018) A drosophila tumor suppressor gene prevents tonic TNF signaling through receptor N-glycosylation. Dev Cell 45:595–605 e594
3. Freeze HH, Chong JX, Bamshad MJ et al (2014) Solving glycosylation disorders: fundamental approaches reveal complicated pathways. Am J Hum Genet 94:161–175
4. Gautam S, Banazadeh A, Cho BG et al (2021) Mesoporous graphitized carbon column for efficient isomeric separation of permethylated glycans. Anal Chem 93:5061–5070
5. Guo H, Abbott KL (2015) Functional impact of tumor-specific N-linked glycan changes in breast and ovarian cancers. Adv Cancer Res 126:281–303
6. Hedlund M, Ng E, Varki A et al (2008) alpha 2-6-Linked sialic acids on N-glycans modulate carcinoma differentiation in vivo. Cancer Res 68:388–394
7. Kailemia MJ, Park D, Lebrilla CB (2017) Glycans and glycoproteins as specific biomarkers for cancer. Anal Bioanal Chem 409:395–410
8. Kizuka Y, Kitazume S, Taniguchi N (2017) N-glycan and Alzheimer's disease. Biochim Biophys Acta 1861:2447–2454
9. Mechref Y, Hu Y, Garcia A et al (2012) Identifying cancer biomarkers by mass spectrometry-based glycomics. Electrophoresis 33:1755–1767
10. Mehta A, Herrera H, Block T (2015) Glycosylation and liver cancer. Adv Cancer Res 126: 257–279
11. Oyama M, Kariya Y, Kariya Y et al (2018) Biological role of site-specific O-glycosylation in cell adhesion activity and phosphorylation of osteopontin. Biochem J 475:1583–1595
12. Pan S, Brentnall TA, Chen R (2016) Glycoproteins and glycoproteomics in pancreatic cancer. World J Gastroenterol 22:9288–9299
13. Rudd PM, Endo T, Colominas C et al (1999) Glycosylation differences between the normal and pathogenic prion protein isoforms. Proc Natl Acad Sci U S A 96:13044–13049
14. Schedin-Weiss S, Winblad B, Tjernberg LO (2014) The role of protein glycosylation in Alzheimer disease. FEBS J 281:46–62
15. Singh C, Shyanti RK, Singh V et al (2018) Integrin expression and glycosylation patterns regulate cell-matrix adhesion and alter with breast cancer progression. Biochem Biophys Res Commun 499:374–380
16. Solá RJ, Griebenow K (2009) Effects of glycosylation on the stability of protein pharmaceuticals. J Pharm Sci 98:1223–1245
17. Sperandio M, Gleissner CA, Ley K (2009) Glycosylation in immune cell trafficking. Immunol Rev 230:97–113
18. Tzeng SF, Tsai CH, Chao TK et al (2018) O-Glycosylation-mediated signaling circuit drives metastatic castration-resistant prostate cancer. FASEB J 32:6869–6682
19. Van Scherpenzeel M, Willems E, Lefeber DJ (2016) Clinical diagnostics and therapy

monitoring in the congenital disorders of glycosylation. Glycoconj J 33:345–358

20. Veillon L, Huang Y, Peng W et al (2017) Characterization of isomeric glycan structures by LC-MS/MS. Electrophoresis 38:2100–2114

21. Wooding KM, Peng W, Mechref Y (2016) Characterization of pharmaceutical IgG and biosimilars using miniaturized platforms and LC-MS-MS. Curr Pharm Biotechnol 17:788–801

Chapter 14

Targeted Glycoproteomics Analysis Using MRM/PRM Approaches

Cristian D. Gutierrez Reyes, Akeem Sanni, Moyinoluwa Adeniyi, Damir Mogut, Hector R. Najera Gonzalez, Parisa Ahmadi, Mojgan Atashi, Sherifdeen Onigbinde, and Yehia Mechref

Abstract

MS-target analyses are frequently utilized to analyze and validate structural changes of biomolecules across diverse fields of study such as proteomics, glycoproteomics, glycomics, lipidomics, and metabolomics. Targeted studies are commonly conducted using multiple reaction monitoring (MRM) and parallel reaction monitoring (PRM) techniques. A reliable glycoproteomics analysis in intricate biological matrices is possible with these techniques, which streamline the analytical workflow, lower background interference, and enhance selectivity and specificity.

Key words MRM, PRM, Glycopeptides

1 Introduction

Protein glycosylation is among the most important and common post-translational modifications (PTMs). It has a major influence on various functional aspects of proteins, including structural and modulatory functions [1–3]. Changes in glycosylation have been correlated to neurodegenerative diseases [4, 5], cancer [6, 7], and viral infections [8, 9]. Currently, liquid chromatography-mass spectrometry (LC-MS) is routinely applied for the identification and quantification of protein site glycosylation. Although LC-MS analysis enables effective characterization of these PTMs, the site heterogeneity and the low glycan abundance make a full description of the glycoprotein site challenging. In this regard, MS-targeted approaches facilitate the analysis of low abundance analytes such as *N*- and *O*-glycopeptides by filtering the precursors of interest from the ionized sample. This action allows the separation of the analyte target from other sample components. Triple quadrupole

Steven B. Bradfute (ed.), *Recombinant Glycoproteins: Methods and Protocols*, Methods in Molecular Biology, vol. 2762, https://doi.org/10.1007/978-1-0716-3666-4_14,

mass spectrometers were the initial choice for targeted studies as they enable the performance of MRM [10–13]. In this case, the initial quadrupole isolates the targeted precursors by functioning as a mass filter, and fragmentation occurs in the second quadrupole. The fragments are then selected, isolated, and detected in the third quadrupole [14]. The development of PRM made the parallel detection of all the MS/MS transitions in a single experiment possible using Orbitrap instruments [15–18].

The targeted analysis of *N*- and *O*-glycopeptides using MRM and PRM represents an accurate, sensitive, and selective alternative to determine the abundance of these important biomolecules.

2 Materials

2.1 Model Glycoproteins

1. Fetuin from serum bovine.
2. Alpha-1-acid glycoprotein (AGP).
3. Standard human blood serum (HBS).

2.2 Reagents

1. DL-dithiothreitol (DTT).
2. Iodoacetamide (IAA).
3. Ammonium bicarbonate (ABC).
4. Trifluoroacetic acid (TFA).
5. Formic acid (FA).
6. HPLC-grade acetonitrile (MeCN).
7. HPLC-grade water.
8. Trypsin/Lys-C mix.
9. Mass spectrometry-grade Glu-C.
10. *O*-Glycoprotease (IMPa).
11. Acclaim PepMap 100 C18 trapping column (75 μm i.d. × 2 cm, 3 μm, 100 Å).
12. Acclaim PepMap 100 C18 column (75 μm i.d. × 15 cm, 2 μm, 100 Å).
13. C18 and HILIC Top Tips.

3 Methods

3.1 Enzymatic Digestion of N-glycopeptides

The analytical workflow for the sample preparation is depicted in Fig. 1a.

1. For HBS (*see* **Notes 1** and **2**) and model glycoproteins, the equivalent volume of 100 μg of protein was diluted to a total volume of 100 μL with 50 mM ABC buffer.

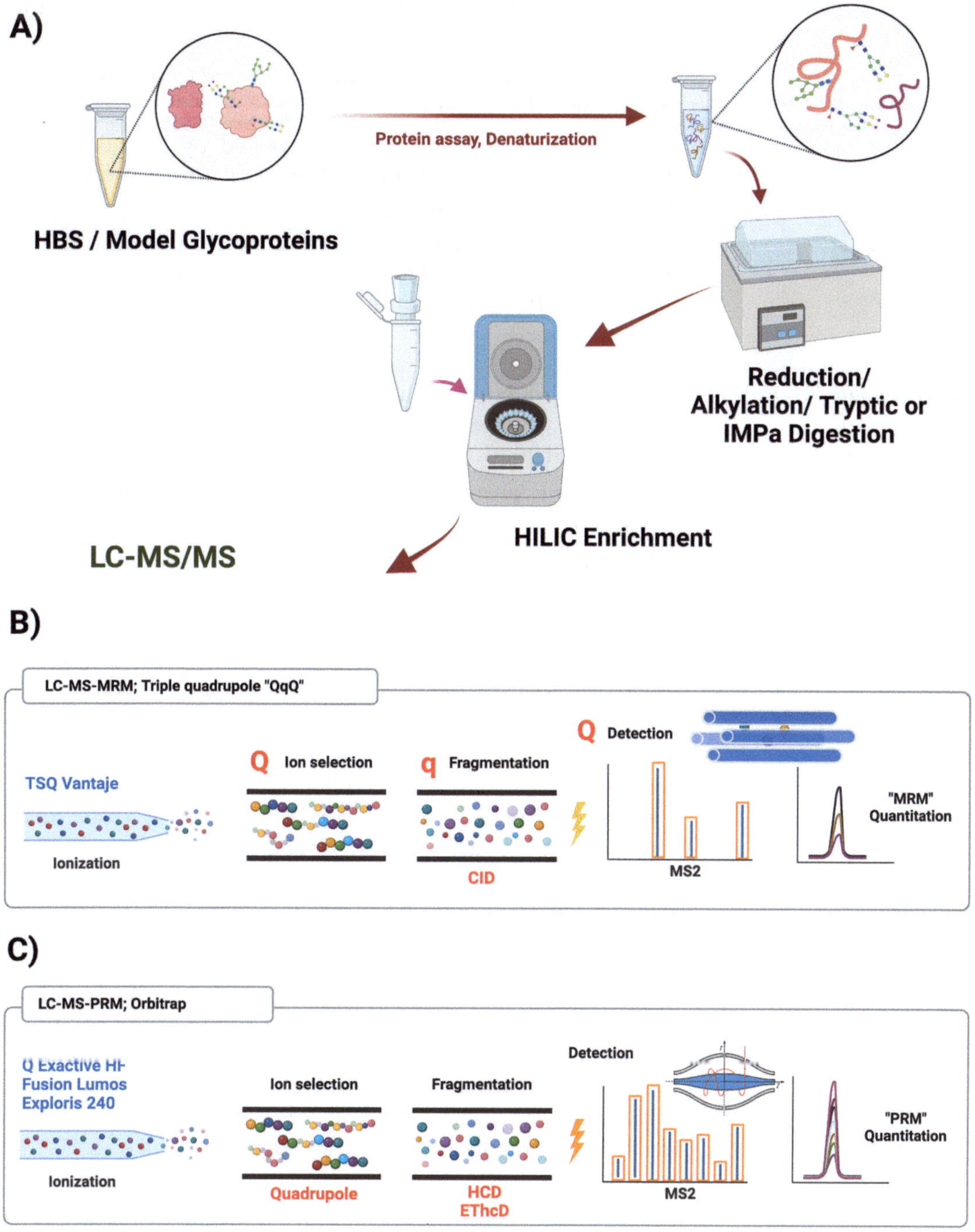

Fig. 1 Analytical workflow. (**a**) Sample preparation and (**b**) MRM quantitation of glycopeptides, and (**c**) PRM quantitation of glycopeptides. Glycan nomenclature: Fucose ▾; N-acetylglucosamine ■; mannose ●; galactose ○; sialic acid ◆

2. Denature the proteins by incubating in a 90 °C water bath for 15 min.
3. The denatured glycoproteins were subjected to alkylation using iodoacetamide (*see* **Note 3**).
4. The enzymatic digestion was completed by the addition of trypsin enzyme in a ratio of 1:25 w/w (trypsin: protein) and incubated at 37 °C overnight.
5. The glycopeptides were enriched from the tryptic digests using HILIC Top Tips (*see* **Note 4**) and reconstituted in a mobile phase A (MPA) solution containing 0.1% FA, 2% MeCN, and 98% of water.

3.2 Enzymatic Digestion of O-glycopeptides

1. For HBS (*see* **Notes 1** and **2**) and model glycoproteins, the equivalent volume of 100 μg of protein was diluted to a total volume of 100 μL with 50 mM ABC buffer.
2. Denature the proteins by incubating in a 90 °C water bath for 15 min.
3. The denatured glycoproteins were subjected to alkylation using iodoacetamide (*see* **Note 3**).
4. Trypsin enzyme was added to the sample in a ratio of 1:25 w/w (trypsin: protein), incubated at 37 °C overnight, and dried in a SpeedVac concentrator.
5. In the second enzymatic digestion, the tryptic digests are resuspended in 100 μL of 20 mM Tris-HCl solution (pH = 8). IMPa enzyme is added at a 1:10 enzyme to protein ratio and incubated at 37 °C for 20 h. The samples are finally dried in a SpeedVac concentrator and resuspended in MPA.

3.3 LC Instrumentation

1. Dionex UltiMate 3000 RSLCnano system with a reverse phase C18 Acclaim PepMap capillary column (75 μm i.d. × 150 mm, 3 μm, 100 Å) is used for the separation of the glycopeptides. The glycopeptides are purified online using an Acclaim PepMap C18 trapping column (75 μm i.d. × 20 mm, 3 μm, 100 Å) with a flow rate of 3 μL/min and MPA as a loading solution. MPA and MPB composed of MeCN with 0.1% FA are the aqueous and organic mobile phases used with three different analytical gradients across the investigated samples.
2. For model glycoproteins, the gradient is 60 min long with a temperature of 30 °C and a flow rate of 0.3 μL/min. Start at 2% of MPB for the initial 10 min, then increase to 45% in 30 min. Subsequently, ramp up to 80% in a period of 6 min and keep constant for 3 min. Finally, drop the percentage of MPB to 2% in 1 min and maintain at that condition for 10 min to pre-equilibrate the system.

3. For HBS *N*-glycopeptides, the gradient is 120 min long with a temperature of 30 °C and flow rate of 0.3 μL/min. Start at 2% of MPB for the initial 10 min, then increase to 38% in 11 min. During the next 80 min, gradually develop the MPB to 60%. Subsequently, ramp up to 90% in 3 min and keep constant for 5 min. Finally, drop the percentage of MPB to 2% in 1 min and maintain that condition for 10 min to pre-equilibrate the system.
4. For HBS *O*-glycopeptides, the gradient is 90 min long with a temperature of 40 °C and flow rate of 0.35 μL/min. Start at 2% of MPB for the initial 5 min, then increase to 30% in 35 min. During the next 32 min, gradually develop the MPB to 70%. Subsequently, ramp up to 90% in 1 min and keep constant for 7 min. Finally, drop the percentage of MPB to 2% in 1 min and maintain at that condition for 9 min to pre-equilibrate the system.

3.4 MS System and MRM Parameters

1. A TSQ Vantage triple quadrupole mass spectrometer was used to complete the MRM glycopeptide analysis. The transfer tube temperature is 275 °C, and the spray voltage is set at 1.6 kV. Positive ion mode was used for data-dependent acquisition mode (DDA) with two scan events. The first event is a Full MS scan in the range of 300 to 1500 *m/z* in Q1 with a peak width of 0.7 FWHM and a scan time of 0.5 s. In the second scan event (MS/MS), the five most intense ions were subjected to fragmentation using collision-induced dissociation (CID) with a normalized collision energy (NCE) of 35%. Figure 1b depicts the general MRM workflow.
2. The MRM experiments were performed at a scan time (dwell time) of 0.1 s and peak width of 0.7 FWHM in a mass range of 400–1500 *m/z*. The isolated precursor ions were subjected to fragmentation using CID with a NCE value of 35%. The workflow is depicted in Fig. 1b.

3.5 MS System and PRM Parameters

Three different instruments were used to perform the PRM analysis of glycopeptides. A Q Exactive HF, a Tribrid Fusion Lumos, and an Exploris 240. The spray voltage was set at 1.6 kV for the acquisition of *N*-glycopeptides and 2.0 kV for the acquisition of *O*-glycopeptides, with a transfer tube temperature of 275 °C.

1. Q Exactive HF. DDA was performed with two scan events in positive ion mode. The first event was a Full MS scan with a resolution of 60K, 1e6 of AGC target, 50 ms of maximum injection time, and a scan range of 400 to 1800 *m/z*. In the second event (MS/MS), the 20 most intense ions were subjected to HCD fragmentation with a NCE value of 25%.

The PRM mode is set with a resolution of 15K, 50 ms of maximum injection time, a loop count of 20, an isolation window of 1.6 *m/z*, and 1e5 of AGC target. The isolated precursor ions were subjected to fragmentation in the HCD cell with a NCE value of 25%. The workflow is depicted in Fig. 1c.

2. Orbitrap Fusion Lumos (*N*-glycopeptides derived from HBS and model glycoproteins). DDA was performed with two scan events in positive ion mode. The first event was a Full MS scan with a resolution of 120K, standard AGC target, injection time of 50 ms, and a scan range of 500 to 1800 *m/z*. In the second event (MS/MS), the 20 most intense ions were subjected to HCD fragmentation with an NCE value of 25%.

   The PRM mode was set with a resolution of 30K, automatic injection time, isolation window of 1.6 *m/z*, standard AGC target, and orbitrap detection. The isolated precursor ions were subjected to fragmentation in the HCD cell with a NCE value of 25%. The workflow is depicted in Fig. 1c.

3. Orbitrap Fusion Lumos (*O*-glycopeptides derived from HBS). DDA is performed with two scan events in positive ion mode. The first event was a Full MS scan with a resolution of 120K, standard AGC target, injection time of 50 ms, and a scan range of 500 to 1800 *m/z*. In the second event (MS/MS), the precursor ions selected with a duty cycle of 3 s were subjected to fragmentation using electron-transfer/higher-energy collision dissociation (EThcD) with an NCE of 25%.

   The PRM mode was set with a resolution of 30K, automatic injection time, isolation window of 1.6 *m/z*, standard AGC target, and orbitrap detection. The isolated precursor ions were subjected to fragmentation using EThcD with an NCE of 25%. The workflow is depicted in Fig. 1c.

4. Exploris 240. DDA is performed with two scan events in positive ion mode. The first event was a Full MS scan with a resolution of 120K, standard AGC target, automatic injection time, and a scan range of 400 to 1800 *m/z*. In the second event (MS/MS), the 20 most intense ions were subjected to HCD fragmentation with an NCE value of 25%.

   The PRM mode was set with a resolution of 30K, automatic injection time, isolation window of 1.6 *m/z*, and standard AGC target. The isolated precursor ions were subjected to fragmentation in the HCD cell with an NCE value of 25%. The workflow is depicted in Fig. 1c.

3.6 Full MS for MRM/PRM Analysis

1. Before MRM/PRM analysis, a DDA analysis was used to identify the target glycopeptides by determining their retention time and the *m/z* value of the dominant precursor in the investigated samples. The glycopeptide dissociation during

the MS/MS data acquisition is fundamental to achieving a reproducible and accurate fragmentation pattern across samples. Thus, the glycopeptide dissociation should be investigated using different levels of collision energy. The common parameters used to perform a DDA analysis for four different instruments are described in Subheadings 3.3, 3.4, and 3.5.

3.7 MRM/PRM Acquisition Parameters

The most common parameters used to perform an MRM and a PRM analysis of glycopeptides are:

1. **Name**. Glycopeptide identification; for example, Asn207 + 5-6-1-2 (*see* **Note 5**).
2. **Precursor (m/z)**. The *m/z* value of the most intense precursor ions or transition ions formed is information that is obtained during the ionization process during the DDA experiments. In the case of low abundant species where the precursor signals are not accurate enough, the targeted precursor ions and transition ions can be theoretically calculated using Glycoworkbench® [19] or other suitable software, **Note 6**. Exclusively for PRM analysis, it is recommended to increase the precursor *m/z* value by 0.5 units. Adjusting the precursor *m/z* facilitates the isolation of all monoisotopic and isotopic masses from the target.
3. **Charge (+)**. Positive charge value of the precursor.
4. **Retention time (rt)**. Set the rt 2 min before and 2 min after the peak signal. Retention times commonly change according to column life and other external parameters. Therefore, this parameter should be investigated in the preliminary DDA analysis.
5. **Collision energy (CE)**. The energy level applied for the glycopeptide dissociation is a key parameter in MRM and PRM analysis. The CE determines the formation of stable fragment ions across the analyzed samples.

Figure 2 shows a representative CID mass spectra of the vitronectin *N*-glycopeptide NGSLFAFR + 4-5-0-2 derived from standard HBS, with an NCE of 35%. The glycopeptide signals were acquired in a TSQ Vantage using the transition ions 137.78, 203.86, and 365.90 *m/z* to extract the MRM signals, the TIC can be observed in reference [10].

Figure 3 shows a representative HCD mass spectra of the serum fetuin *N*-glycopeptide LCPDCPLLAPLNDSR + 4-5-0-2 acquired in (a) Q Exactive HF, (b) Tribrid Fusion Lumos, and (c) Exploris 240, with an NCE of 25%. For the Q Exactive HF, the top three transitions observed in the mass spectra are 168.06, 316.13, and 1943.91 *m/z*. The top three transitions observed in the mass spectra of the Tribrid Fusion Lumos are 204.08, 274.09, and

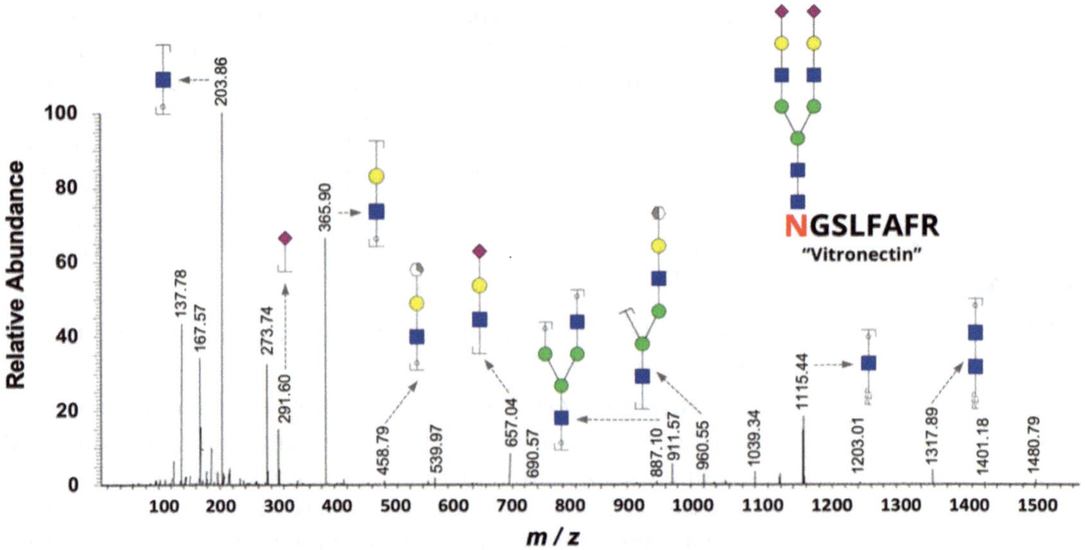

Fig. 2 Representative CID mass spectra of the *N*-glycopeptide NGSLFAFR + 4-5-0-2 derived from HBS vitronectin, with a NCE of 35%. Glycan nomenclature as in Fig. 1. (Figure reproduced from E. Song et al. [10])

366.13 *m/z*. For the Exploris 240, the top three transitions observed in the mass spectra are 168.08, 366.13, and 1943.91 *m/z*.

3.8 MRM Transitions

1. The fragmentation product ion *m/z* value is referred to as an MRM transition. Therefore, to monitor a glycopeptide of interest, the transition patterns must be known in advance. A comprehensive study describing the transitions used to extract the MRM signals of common *N*-glycopeptides derived from HBS, AGP, and fetuin can be found in our previously published data [10], as well as an evaluation of the CE effect in the fragmentation of the fetuin *N*-glycopeptides, and the changes in the signal by the incorporation of different transitions in the MRM quantitation.

3.9 PRM Transitions

1. The fragmentation product ion *m/z* value is referred to as a PRM transition. PRM provides a major advantage over MRM because of its ability to acquire full MS/MS scans. This facilitates the method setup, as the transitions for PRM do not need to be preselected during the data acquisition. Thus, many transitions will be available for the identification and quantification of the targeted glycopeptides. Here, we have investigated and described the precursor and transition m/z values of common *N*- and *O*-glycopeptides derived from different sample types. First, haptoglobin *N*-glycopeptides derived from HBS extracted from patients with cirrhosis and HCC, Table 1 [7]. Next, *N*-Glycopeptides derived from fetuin and subjected

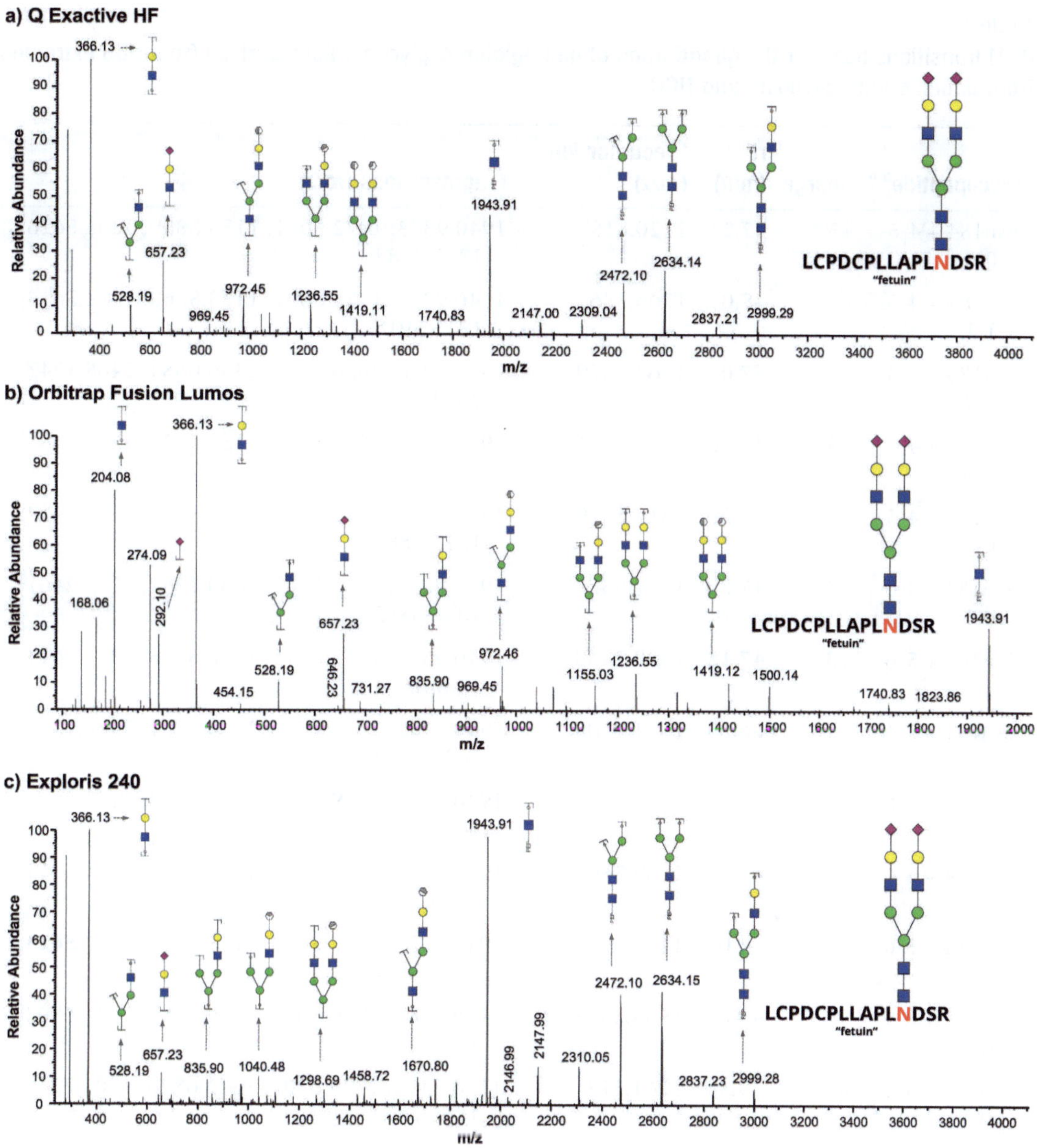

Fig. 3 Representative HCD mass spectra of the serum fetuin *N*-glycopeptide LCPDCPLLAPLNDSR + 4-5-0-2 acquired in (**a**) Q Exactive HF, (**b**) Tribrid Fusion Lumos, and (**c**) Exploris 240, with a NCE of 25%. Glycan nomenclature as in Fig. 1

to PRM analysis using three different instruments: the Q Exactive HF (Table 2), Tribrid Fusion Lumos (Table 3), and Exploris 240 (Table 4). Finally, *O*-Glycopeptides derived from standard HBS, Table 5.

Table 1
PRM transitions used for the quantitation of haptoglobin *N*-glycopeptides derived from HBS extracted from patients with cirrhosis and HCC

Glycopeptide[a,b]	Charge	rt (min)	Precursor ion (*m/z*)	Fragment ions (*m/z*)
Asn184 + 4-5-0-0	+3	42.24	1120.8183	1940.9333; 1072.5091; 1153.5359; 1234.5626; 1315.5888
Asn184 + 4-5-1-1	+3	48.01	1266.5361	1940.9345; 1072.5099; 1153.5363; 1234.5613; 1315.5918
Asn184 + 4-5-1-2	+3	57.07	1363.5679	1940.9336; 2086.9927; 2306.0684; 2468.1247; 2614.1797
Asn184 + 4-5-0-1	+3	48.16	1217.8501	1940.9348; 1072.5088; 1153.5328; 1234.5603; 1315.5876
Asn184 + 4-5-0-2	+3	54.73	1314.8819	1940.9337; 1072.5088; 1153.5358; 1234.5612; 1315.5887
Asn184 + 4-6-0-1	+3	47.70	1271.8677	1940.9342; 2468.1247; 2630.1767; 2792.2461; 1336.5882
Asn184 + 5-6-1-1	+3	47.18	1388.2468	1940.9331; 1072.5094; 1153.5355; 1234.5614; 1336.1014
Asn184 + 5-6-1-2	+4	56.36	1114.2108	1940.9355; 2086.9979; 2306.0781; 2468.1318; 2630.1824
Asn184 + 5-6-1-3	+3	57.30	1582.3104	1940.9361; 1459.6648; 1585.6160; 2101.8861; 2468.1117
Asn184 + 5-6-0-1	+3	47.48	1339.5609	1940.9336; 1072.5086; 1153.5362; 1234.5614; 1315.5884
Asn184 + 5-6-0-2	+3	54.08	1436.5927	1940.9337; 1072.5088; 1153.5349; 1234.5614; 1315.5827
Asn207 + 3-4-0-1	+3	49.81	1533.6345	1176.5484; 1056.5070; 1379.6285; 1541.6803; 1703.7228
Asn207 + 4-4-0-1	+2	49.44	1261.513	1176.5498; 1379.6307; 1541.6826; 1703.7349; 1865.7875
Asn207 + 4-5-0-0	+2	40.39	1363.0527	1176.5490; 1379.6296; 1703.7336; 1865.7869; 2068.8884
Asn207 + 4-5-1-1	+2	45.27	1298.5314	1176.5490; 1379.6298; 1541.6806; 1703.7309; 1865.7865
Asn207 + 4-5-1-2	+3	57.98	1011.7411	1176.5505; 1703.7195; 1865.7773; 2068.8518; 2011.8485
Asn207 + 4-5-0-1	+3	46.13	1108.7729	1176.5489; 1379.6279; 1541.6831; 1703.7328; 1865.7862
Asn207 + 4-5-0-2	+2	57.49	1444.0791	1176.5495; 1541.6829; 1703.7334; 1865.7890; 2068.8653

(continued)

Table 1
(continued)

Glycopeptide[a,b]	Charge	rt (min)	Precursor ion (*m/z*)	Fragment ions (*m/z*)
Asn207 + 5-6-1-1	+3	47.31	1060.0869	1176.5492; 1379.6293; 1541.6816; 1703.7332; 1865.7881
Asn207 + 5-6-1-2	+3	56.34	1133.4518	1176.5496; 1379.6278; 1541.6826; 1703.7333; 1865.7822
Asn207 + 5-6-1-3	+3	70.56	1230.4836	1176.5496; 1379.6244; 1541.6758; 1703.7313; 2068.8542
Asn207 + 5-6-0-1	+3	45.18	1327.5154	1176.5497; 1379.6282; 1541.6810; 1703.7330; 1865.7879
Asn207 + 5-6-0-2	+3	56.25	1084.7659	1176.5498; 1379.6283; 1541.6814; 1703.7339; 1865.7873
Asn207 + 5-6-0-3	+3	71.39	1181.7977	1176.5502; 1379.6268; 1541.6714; 1703.7332; 2068.8577
Asn207 + 6-7-1-1	+3	46.06	1278.8295	1176.5494; 1703.7345; 1932.9272; 2068.8702; 2230.9241
Asn207 + 6-7-1-2	+3	58.07	1255.1626	1176.5503; 1129.5302; 1210.5564; 2068.8720; 2230.9256
Asn207 + 6-7-1-3	+3	60.00	1352.1944	1176.5499; 1703.7356; 1875.9048; 1892.9331; 2068.8711
Asn207 + 6-7-0-1	+3	43.81	1449.2262	1176.5494; 1379.6287; 1541.6718; 1703.7314; 2068.8716
Asn207 + 6-7-0-2	+3	57.79	1206.4766	1176.5493; 1379.6289; 1703.7333; 2068.8687; 2230.9216
Asn207 + 6-7-0-4	+3	60.29	1303.5084	1176.5509; 1201.5585; 1642.6395; 1516.6875; 1220.5885
Asn241 + 4-4-0-1	+4	77.35	1123.4308	1998.0903; 2231.9111; 2069.8580; 1866.7745; 1177.5254
Asn241 + 4-5-0-0	+3	61.74	1182.8833	1998.0914; 1101.0855; 1182.1120; 1263.1389; 1344.1667
Asn241 + 4-5-1-2	+3	78.62	1139.8691	1998.0943; 2013.9092; 1290.7283; 1445.7031; 1486.7358
Asn241 + 4-5-0-1	+3	67.98	1382.6187	1998.0868; 1101.0865; 1182.1102; 1344.1648; 1445.7029
Asn241 + 4-5-0-2	+3	75.35	1236.9009	1999.0859; 1263.1281; 1344.1588; 1445.6924; 1526.7240
Asn241 + 5-6-1-2	+3	78.15	1333.9327	1998.0907; 1128.6021; 1347.7465; 1445.6995; 1526.7316

(continued)

Table 1
(continued)

Glycopeptide[a,b]	Charge	rt (min)	Precursor ion (*m/z*)	Fragment ions (*m/z*)
Asn241 + 5-6-0-1	+4	69.80	1128.4989	1998.0891; 1174.5344; 1417.6190; 1472.2158; 1580.6812
Asn241 + 5-6-0-2	+4	77.58	1019.2106	1998.0902; 1263.1382; 1709.2971; 1344.1647; 1182.1115
Asn241 + 5-6-0-3	+4	87.31	1091.9844	1998.0890; 1101.0842; 1263.1386; 1345.7034; 1526.7317
Asn241 + 6-7-0-1	+4	69.23	1164.7583	1998.0898; 1209.6461; 1445.7010; 1827.7471; 1451.6415
Asn241 + 6-7-0-2	+4	77.30	1110.4936	1998.0909; 1177.5336; 1445.7034; 1704.7222; 2231.9072
Asn241 + 6-7-0-3	+4	87.23	1183.2675	1998.0868; 1645.7706; 1445.7043; 1255.5942; 1150.5383

Information reproduced from C.D. Gutierrez Reyes et al. [7]
[a]Haptoglobin peptide + GlcNAc, Hex, Fuc, NeuAc (*N*-acetylglucosamine, Hexose, Fucose, N-acetylneuraminic acid)
[b]MVSHHN184LTTGATLINE = Asn184, NLFLN207HSE = Asn207, and VVLHPN241YSQVDIGLIK = Asn241

3.10 MRM/PRM Quantitation

1. Free Style software 1.8 (Thermo Scientific) is employed to extract the ion chromatograms from the Full MS scans and to verify the precursor m/z values and their retention times. The same software is used to extract the transition peak signals of the MRM quantitation along with the MS/MS data for each precursor subjected to PRM analysis. The five most intense fragment ions are assigned as transitions for the MRM and PRM quantitation. The theoretical validation of the glycopeptide fragment ions is performed using Glycoworkbench 2.0 software. Subsequently, a transition list is compiled using Microsoft Excel®, and uploaded to Skyline 20.2 software to extract the EICs and complete the peak area calculations.

4 Notes

1. The seven or fourteen most abundant proteins of the HBS samples are depleted using a multiple affinity removal spin cartridge [10] or a HPLC column [5], respectively (Agilent, Technologies, Santa Clara, CA, USA).
2. A 1–5 μL aliquot of the depleted or un-depleted HBS is subjected to protein quantification using the Pierce™ BCA Protein Assay Kit (Thermo Fisher Sci.) [10].

Table 2
PRM transitions used for the quantitation of fetuin *N*-glycopeptides, Q Exactive HF mass spectrometer

Glycopeptide[a,b]	Charge	rt (min)	Precursor ion (*m/z*)	Fragment ions (*m/z*)
Asn156 + 4-5-0-1	3	27.49	1219.0114	366.1381; 1943.9169; 2472.1028; 2634.1580; 2636.1345
Asn156 + 4-5-0-2	3	29.84	1316.0432	366.1379; 657.2321; 1943.9163; 2472.0994; 2634.1511
Asn156 + 4-5-0-2	4	29.84	987.2844	274.0921; 366.1381; 657.2324; 1943.9174; 2471.0938
Asn156 + 5-6-0-1	4	30.06	1005.6686	292.1024; 366.1380; 528.1906; 1006.5544; 1943.9159
Asn156 + 5-6-0-2	3	29.55	1437.7539	366.1379; 657.2322; 972.4589; 1419.1182; 1943.9163
Asn156 + 5-6-0-2	4	29.55	1078.5674	366.1378; 657.2320; 972.4603; 1236.5530; 1943.9163
Asn156 + 5-6-0-3	3	32.22	1151.3413	274.0911; 366.1381; 639.2226; 1099.5748; 1943.9119
Asn156 + 5-6-0-3	4	32.25	921.1746	292.1016; 366.1384; 867.9272; 921.4711; 1303.5681
Asn156 + 5-6-0-4	4	35.78	1224.1151	366.1378; 657.2319; 972.4601; 1943.9164; 2472.1038
Asn99 + 4-5-0-1	4	27.64	1397.2423	657.2329; 1292.6184; 1468.3472; 1590.3885; 1644.4052
Asn99 + 4-5-0-2	4	31.16	1470.0162	657.2321; 1292.6176; 1468.3463; 1590.3881; 1644.4045
Asn99 + 4-5-0-2	5	31.16	1176.2145	657.2321; 1292.9487; 1468.6772; 1522.6942; 1644.7340
Asn99 + 5-6-0-1	4	31.16	1488.5254	657.2329; 1252.6162; 1590.3904; 1644.4058; 1766.4470
Asn99 + 5-6-0-2	4	31.06	1561.2992	657.2318; 1292.9481; 1590.3876; 1712.0955; 1766.4454
Asn99 + 5-6-0-2	5	31.06	1249.2409	657.2324; 1292.9490; 1468.6774; 1590.3856; 1644.7340
Asn99 + 5-6-0-3	4	32.46	1634.0731	657.2323; 1292.6174; 1590.3884; 1711.7632; 1765.7793
Asn99 + 5-6-0-3	5	32.40	1307.4600	657.2323; 1292.6185; 1468.6820; 1590.3892; 1664.4054
Asn99 + 5-6-0-4	4	33.90	1706.8469	657.2317; 1292.9470; 1590.3824; 1712.0900; 1766.1063

(continued)

Table 2
(continued)

Glycopeptide[a,b]	Charge	rt (min)	Precursor ion (*m/z*)	Fragment ions (*m/z*)
Asn99 + 5-6-0-4	5	33.90	1365.6791	657.2322; 1292.6179; 1590.3887; 1644.4054; 1766.4460
Asn176 + 4-5-0-1	4	35.77	1233.4438	366.1380; 657.2325; 1438.0914; 1519.1179; 1620.6582
Asn176 + 4-5-0-2	4	32.43	1306.2176	366.1381; 1292.2845; 1468.0099; 1589.7181; 1643.7332
Asn176 + 4-5-0-2	5	31.07	1045.1757	366.1380; 564.7845; 1045.7833; 1046.4813; 1612.6240
Asn176 + 5-6-0-1	4	29.85	1324.7268	657.2319; 1252.6167; 1511.6360; 1943.9174; 2633.1528
Asn176 + 5-6-0-2	4	32.50	1397.5007	657.2327; 1292.9471; 1468.6776; 1590.3882; 1644.4047
Asn176 + 5-6-0-2	5	32.41	1118.2021	657.2327; 1280.0992; 1292.9496; 1468.3477; 1522.6973
Asn176 + 5-6-0-3	*4*	31.18	1470.2745	657.2320; 1292.6179; 1590.3840; 1644.7322; 1938.4089
Asn176 + 5-6-0-2	5	31.18	1176.4212	657.2322; 1292.9490; 1468.3434; 1522.6935; 1644.7335
Asn176 + 5-6-0-4	4	32.56	1543.0484	657.2323; 1430.1096; 1613.1742; 1644.4117; 1694.2001
Asn176 + 5-6-0-4	5	32.56	1234.6403	657.2323; 1292.9500; 1468.3474; 1522.6974; 1590.3885

[a]Fetuin peptide + GlcNAc, Hex, Fuc, NeuAc (N-acetylglucosamine, Hexose, Fucose, N-acetylneuraminic acid)
[b]LCPDCPLLAPLN156DSR = Asn156, RPTGEVYDIEIDTLETTCHVLDPTPLAN99CSVR = Asn99, VVHAVEVALATFNAESN176GSY LQLVEISR = Asn176

3. Alkylation protocol: Initially, the disulfide bonds from the denatured glycoproteins are reduced by the addition of 2.5 μL of 200 mM DTT solution and incubation of 45 min at 60 °C. Then, the alkylation of the produced thiol groups is completed by the addition of 10 μL of 200 mM IAA solution and incubation of 45 min at 37 °C. Finally, the non-reacted IAA is quenched with a second addition of 2.5 μL of 200 mM DTT solution and incubation of 30 min at 37 °C.

4. Glycopeptide enrichment: Equilibrate the HILIC Top Tips with 100 μL of water and repeated three times. Continue the sorbent equilibration by adding 100 μL of a loading buffer solution containing 80% MeCN, 20% water in 0.1% TFA. Reconstitute the dried samples in 50 μL of an elution solution

Table 3
PRM transitions used for the quantitation of fetuin *N*-glycopeptides, Tribrid Fusion Lumos mass spectrometer

Glycopeptide[a,b]	Charge	rt (min)	Precursor ion (*m/z*)	Fragment ions (*m/z*)
Asn156 + 4-5-0-1	3	27.49	1219.0114	138.0545; 168.0651; 204.0863; 274.0921; 366.1396
Asn156 + 4-5-0-2	3	29.84	1316.0432	138.0543; 168.0653; 204.0866; 274.0922; 366.1400
Asn156 + 4-5-0-2	4	29.84	987.2844	138.0545; 168.0652; 204.0864; 274.0920; 366.1397
Asn156 + 5-6-0-1	4	30.06	1005.6686	168.0653; 204.0863; 366.1398; 1006.0007; 1006.2545
Asn156 + 5-6-0-2	3	29.55	1437.7539	138.0546; 168.0652; 204.0864; 274.0920; 366.1398
Asn156 + 5-6-0-2	4	29.55	1078.5674	138.0553; 168.0661; 204.0875; 274.0934; 366.1416
Asn156 + 5-6-0-3	3	32.22	1151.3413	168.0652; 204.0864; 274.0920; 292.1027; 366.1397
Asn156 + 5-6-0-3	4	32.25	921.1746	756.4041; 903.4753; 921.4745; 1234.5905; 1235.5938
Asn156 + 5-6-0-4	4	35.78	1224.1151	204.0873; 274.0932; 292.1039; 366.1413; 657.2382
Asn99 + 4-5-0-1	4	27.64	1397.2423	138.0545; 168.0651; 204.0863; 274.0920; 366.1397
Asn99 + 4-5-0-2	4	31.16	1470.0162	204.0865; 366.1399; 1292.6212; 1590.4011; 1644.4216
Asn99 + 4-5-0-2	5	31.16	1176.2145	204.0866; 274.0922; 366.1400; 1292.6212; 1468.6877
Asn99 + 5-6-0-1	4	31.16	1488.5254	138.0545; 168.0651; 204.0863; 366.1397; 1292.6211
Asn99 + 5-6-0-2	4	31.06	1561.2992	204.0864; 366.1397; 657.2363; 1292.6208; 1590.3988
Asn99 + 5-6-0-2	5	31.06	1249.2409	204.0865; 274.0921; 366.1399; 1292.6215; 1468.3444
Asn99 + 5-6-0-3	4	32.46	1634.0731	204.0868; 274.0930; 366.1413; 657.2403; 1292.6296
Asn99 + 5-6-0-3	5	32.40	1307.4600	204.0873; 274.0937; 366.1426; 657.2440; 1292.6433
Asn99 + 5-6-0-4	4	33.90	1706.8469	204.0863; 274.0919; 366.1396; 657.2357; 1292.6208

(continued)

Table 3
(continued)

Glycopeptide[a,b]	Charge	rt (min)	Precursor ion (*m/z*)	Fragment ions (*m/z*)
Asn99 + 5-6-0-4	5	33.90	1365.6791	204.0873; 274.0936; 366.1423; 657.2426; 1292.6401
Asn176 + 4-5-0-1	4	35.77	1233.4438	204.0864; 274.0919; 292.1025; 366.1398; 1437.5962
Asn176 + 4-5-0-2	4	32.43	1306.2176	138.0545; 168.0654; 204.0865; 274.0918; 366.1397
Asn176 + 4-5-0-2	5	31.07	1045.1757	204.0871; 274.0930; 366.1417; 1045.6351; 1238.0748
Asn176 + 5-6-0-1	4	29.85	1324.7268	204.0864; 274.0920; 366.1397; 1511.6420; 1512.1403
Asn176 + 5-6-0-2	4	32.50	1397.5007	138.0545; 204.0863; 366.1397; 1292.9552; 1644.4115
Asn176 + 5-6-0-2	5	32.41	1118.2021	204.0862; 366.1395; 1292.6185; 1292.9562; 1293.2919
Asn176 + 5-6-0-3	4	31.18	1470.2745	204.0865; 274.0921; 366.1399; 1292.6212; 1644.4133
Asn176 + 5-6-0-2	5	31.18	1176.4212	204.0866; 274.0922; 366.1400; 1292.6217; 1292.9584
Asn176 + 5-6-0-4	4	32.56	1543.0484	204.0864; 274.0920; 366.1398; 1694.2069; 1694.7092
Asn176 + 5-6-0-4	5	32.56	1234.6403	204.0863; 274.0919; 292.1027; 366.1398; 1292.6202

[a]Fetuin peptide + GlcNAc, Hex, Fuc, NeuAc (N-acetylglucosamine, Hexose, Fucose, N-acetylneuraminic acid)
[b]LCPDCPLLAPLN156DSR = Asn156, RPTGEVYDIEIDTLETTCHVLDPTPLAN99CSVR = Asn99, VVHAVEVALATFNAESN176GSY LQLVEISR = Asn176

containing 0.1% of TFA, and place them into the Top Tip HILIC spin column. Wash the glycopeptides with 100 μL of loading buffer three times. Finally, the glycopeptides are eluted with 100 μL of the elution solution three times and dried in a SpeedVac concentrator.

5. The amino acid position of the glycosylation in the protein is followed by the attached glycan structure using a four-digit nomenclature, where "1-1-1-1" stands for GlcNAc,Hex,Fuc, NeuAc (GlcNAc = *N*-acetylglucosamine, Hex = mannose or galactose, Fuc = fucose, and NeuAc = sialic acid).

Table 4
PRM transitions used for the quantitation of fetuin *N*-glycopeptides, Exploris mass spectrometer

Glycopeptide[a,b]	Charge	rt (min)	Precursor ion (*m/z*)	Fragment ions (*m/z*)
Asn156 + 4-5-0-1	3	27.49	1219.0114	274.0923; 366.1395; 1943.9221; 1944.9249; 2634.1606
Asn156 + 4-5-0-2	3	29.84	1316.0432	274.0918; 366.1389; 1943.9199; 1945.9227; 2634.1575
Asn156 + 4-5-0-2	4	29.84	987.2844	274.0921; 366.1393; 1943.9222; 1944.9244; 1945.9253
Asn156 + 5-6-0-1	4	30.06	1005.6686	893.4733; 1006.5574; 1077.5945; 1148.6317; 1303.5630
Asn156 + 5-6-0-2	3	29.55	1437.7539	366.1391; 1943.9205; 1944.9224; 1945.9236; 2999.2917
Asn156 + 5-6-0-2	4	29.55	1078.5674	366.1395; 1943.9221; 1944.9247; 1945.9255; 2472.1086
Asn156 + 5-6-0-3	3	32.22	1151.3413	274.0911; 366.1378; 1943.9155; 1944.9170; 1945.9175
Asn156 + 5-6-0-3	4	32.25	921.1716	274.0924; 366.1398; 1943.9235; 1944.9271; 1945.9285
Asn156 + 5-6-0-4	4	35.78	1224.1151	366.1383; 1943.9174; 1944.9196; 1945.9202; 2472.1025
Asn99 + 4-5-0-1	4	27.64	1397.2423	1280.1123; 1292.9570; 1317.6483; 1938.4225; 1938.9246
Asn99 + 4-5-0-2	4	31.16	1470.0162	1280.1069; 1292.9490; 1938.9259; 1939.4259; 2202.5195
Asn99 + 4-5-0-2	5	31.16	1176.2145	1280.1080; 1292.9504; 1468.6825; 1522.7010; 1938.9301
Asn99 + 5-6-0-1	4	31.16	1488.5254	539.3196; 732.4521; 1542.8599; 2085.0344; 2315.2041
Asn99 + 5-6-0-2	4	31.06	1561.2992	1280.1096; 1292.9541; 1938.9288; 1939.4302; 2202.5210
Asn99 + 5-6-0-2	5	31.06	1249.2409	1280.1108; 1292.9564; 1468.6851; 1938.9299; 1939.4305
Asn99 + 5-6-0-3	4	32.46	1634.0731	1280.1073; 1292.9497; 1938.9272; 1939.4269; 2202.5208
Asn99 + 5-6-0-3	5	32.40	1307.4600	1280.1071; 1292.9495; 1468.6818; 1938.9287; 1939.4299
Asn99 + 5-6-0-4	4	33.90	1706.8469	1280.1085; 1292.9525; 1938.9293; 1939.4299; 2466.1184

(continued)

Table 4
(continued)

Glycopeptide[a,b]	Charge	rt (min)	Precursor ion (m/z)	Fragment ions (m/z)
Asn99 + 5-6-0-4	5	33.90	1365.6791	1280.1085; 1292.9519; 1468.6819; 1938.9265; 1939.4277
Asn176 + 4-5-0-1	4	35.77	1233.4438	366.1397; 1280.1121; 1292.6420; 1938.4346; 1939.4297
Asn176 + 4-5-0-2	4	32.43	1306.2176	366.1383; 1280.1084; 1292.6198; 1468.3482; 1938.4247
Asn176 + 4-5-0-2	5	31.07	1045.1757	366.1399; 732.4523; 976.5841; 994.5944; 1204.7323
Asn176 + 5-6-0-1	4	29.85	1324.7268	366.1393; 1430.1158; 1511.1432; 1694.2081; 1943.9213
Asn176 + 5-6-0-2	4	32.50	1397.5007	1280.1127; 1292.9584; 1317.6478; 1468.6863; 1938.9241
Asn176 + 5-6-0-2	5	32.41	1118.2021	1280.1127; 1292.9584; 1468.6865; 1523.0377; 1938.9305
Asn176 + 5-6-0-3	*4*	31.18	1470.2745	1280.1079; 1292.9518; 1938.9290; 1939.4299; 2202.5237
Asn176 + 5-6-0-2	5	31.18	1176.4212	1280.1101; 1292.9542; 1468.6787; 1522.6969; 1938.9208
Asn176 + 5-6-0-4	4	32.56	1543.0484	657.2343; 1430.6144; 1693.7075; 1944.9221; 2472.1077
Asn176 + 5-6-0-4	5	32.56	1234.6403	657.2355; 1255.5326; 1438.1001; 1519.1258; 1621.1672

[a]Fetuin peptide + GlcNAc, Hex, Fuc, NeuAc (N-acetylglucosamine, Hexose, Fucose, N-acetylneuraminic acid)
[b]LCPDCPLLAPLN156DSR = Asn156, RPTGEVYDIEIDTLETTCHVLDPTPLAN99CSVR = Asn99, and VVHA-VEVALATFNAESN176GSY LQLVEISR = Asn176

6. The incorporation of theoretical precursor values in the target list implies a subsequent revision of the MRM or PRM signals and proves their validity.

Acknowledgments

Access to Exploris 240 mass spectrometer was provided by Prof. Luis Herrera-Estrella, President's Distinguished Professor of Plant Genomics and Director of the Institute of Genomics for Crop Abiotic Stress Tolerance at Texas Tech University. This work is supported by grants from the National Institutes of Health, NIH, including 1R01GM112490-08, 1R01GM130091-05, and

Table 5
PRM transitions used for the quantitation of HBS *O*-glycopeptides, Tribrid Fusion Lumos mass spectrometer

Glycopeptide[a,b]	Charge	rt (min)	Precursor ion (*m/z*)	Fragment ions (*m/z*)
Thr534 + 2-2-0-2	3	36.13	1038.9106	366.1393; 657.2353; 1038.5321; 1537.105; 1537.6078
Thr90 + 3-2-0-2	3	33.75	1024.9046	366.1397; 657.2348; 1024.5734; 1516.0979; 1537.6101
Ser43 + 2-2-0-2	3	43.02	1366.7106	178.1649; 366.1393; 1495.1281; 1576.6415; 1904.2604
Thr365 + 1-1-0-1	3	47.36	1049.6246	1049.4968; 1246.0692; 1428.6361; 1574.6853; 1575.1869
Thr626 + 2-2-0-2	3	35.58	1293.0416	178.1519; 1293.5675; 1466.1498; 1793.7618; 1939.2966
Thr241 + 3-3-1-2	4	60.74	1190.8946	204.0864; 366.1395; 657.2354; 1587.9376; 1907.2747
Thr1882 + 2-2-2-1	3	34.43	1115.2616	366.1397; 657.2356; 1526.587; 1651.13; 1651.6339
Thr1882 + 2-2-2-1	3	39.35	1134.2686	366.1384; 657.2327; 1133.5791; 1134.5688; 1680.6406

[a]Protein_Peptide + GlcNAc, Hex, Fuc, NeuAc (N-acetylglucosamine, Hexose, Fucose, N-acetylneuraminic acid)
[b]Q9BYP7_QT534GAECEETEVDQHVRQ = Thr534, P01042_KT90WQDCEYKDAAKA = Thr90, Q8WVL7_QS43LWVGNSDEDEEQDDKNEEWYRL = Ser43, P00747_PTAPPELT365PVVQDCYHGDGQSYRG = Thr365, Q9JIH7_ST626QVEPEEPEADQHQQLQYQQPS = Thr626, P35555_RT241GACQDVDECQAIPGLCQGGNCINTVGS3 = Thr241, and P35555_KT1882NDDQTMCLDINECERD = Thr1882

1U01CA225753-05. The work is also supported by grants from Robert A. Welch Foundation (No. D-0005) and The CH Foundation.

References

1. Gutierrez Reyes CD, Jiang P, Donohoo K et al (2021) Glycomics and glycoproteomics: approaches to address isomeric separation of glycans and glycopeptides. J Sep Sci 44(1): 403–425
2. Gutierrez-Reyes CD, Jiang P, Atashi M et al (2022) Advances in mass spectrometry-based glycoproteomics: an update covering the period 2017–2021. Electrophoresis 43(1–2): 370–387
3. Varki A et al (eds) (2022) Essentials of glycobiology. Cold Spring Harbor Laboratory Press, Cold Spring Harbor. (Copyright © 2022 by the Consortium of Glycobiology Editors, La Jolla, California. Published by Cold Spring Harbor Laboratory Press, Cold Spring Harbor, New York
4. Akasaka-Manya K, Manya H, Sakurai Y et al (2008) Increased bisecting and core-fucosylated N-glycans on mutant human amyloid precursor proteins. Glycoconj J 25(8): 775–786
5. Reyes CDG, Hakim MA, Atashi M et al (2022) LC-MS/MS isomeric profiling of N-glycans derived from low-abundant serum

glycoproteins in mild cognitive impairment patients. Biomolecules 12(11):1657

6. Machado E, Kandzia S, Carilho R et al (2011) N-Glycosylation of total cellular glycoproteins from the human ovarian carcinoma SKOV3 cell line and of recombinantly expressed human erythropoietin. Glycobiology 21(3):376–386
7. Gutierrez Reyes CD, Huang Y, Atashi M et al (2021) PRM-MS quantitative analysis of isomeric N-glycopeptides derived from human serum haptoglobin of patients with cirrhosis and hepatocellular carcinoma. Metabolites 11(8):563
8. Gong Y, Qin S, Dai L et al (2021) The glycosylation in SARS-CoV-2 and its receptor ACE2. Signal Transduct Target Ther 6(1):396
9. Cho BG, Gautam S, Peng W et al (2021) Direct comparison of N-glycans and their isomers derived from spike glycoprotein 1 of MERS-CoV, SARS-CoV-1, and SARS-CoV-2. J Proteome Res 20(9):4357–4365
10. Song E, Pyreddy S, Mechref Y (2012) Quantification of glycopeptides by multiple reaction monitoring liquid chromatography/tandem mass spectrometry. Rapid Commun Mass Spectrom 26(17):1941–1954
11. Selevsek N, Matondo M, Carbayo MS et al (2011) Systematic quantification of peptides/proteins in urine using selected reaction monitoring. Proteomics 11(6):1135–1147
12. Stahl-Zeng J, Lange V, Ossola R et al (2007) High sensitivity detection of plasma proteins by multiple reaction monitoring of N-glycosites. Mol Cell Proteomics 6(10):1809–1817
13. Addona TA, Abbatiello SE, Schilling B et al (2009) Multi-site assessment of the precision and reproducibility of multiple reaction monitoring-based measurements of proteins in plasma. Nat Biotechnol 27(7):633–641
14. Fu Q, Chen Z, Zhang S et al (2016) Multiple and selective reaction monitoring using triple quadrupole mass spectrometer: preclinical large cohort analysis. Methods Mol Biol 1410:249–264
15. Peterson AC, Russell JD, Bailey DJ et al (2012) Parallel reaction monitoring for high resolution and high mass accuracy quantitative, targeted proteomics. Mol Cell Proteomics 11(11):1475–1488
16. Delafield DG, Li L (2021) Recent advances in analytical approaches for glycan and glycopeptide quantitation. Mol Cell Proteomics 20: 100054
17. Sanda M, Yang Q, Zong G et al (2023) LC-MS/MS-PRM quantification of IgG glycoforms using stable isotope labeled IgG1 Fc glycopeptide standard. J Proteome Res 22(4): 1138–1147
18. Lin Y, Zhang J, Arroyo A et al (2022) A fucosylated glycopeptide as a candidate biomarker for early diagnosis of NASH hepatocellular carcinoma using a stepped HCD method and PRM evaluation. Front Oncol 12:818001
19. Ceroni A, Maass K, Geyer H et al (2008) GlycoWorkbench: a tool for the computer-assisted annotation of mass spectra of glycans. J Proteome Res 7(4):1650–1659

Chapter 15

Targeted Analysis of Permethylated *N*-Glycans Using MRM/PRM Approaches

Cristian D. Gutierrez Reyes, Akeem Sanni, Damir Mogut, Moyinoluwa Adeniyi, Parisa Ahmadi, Mojgan Atashi, Sherifdeen Onigbinde, and Yehia Mechref

Abstract

Targeted mass spectrometric analysis is widely employed across various omics fields as a validation strategy due to its high sensitivity and accuracy. The approach has been successfully employed for the structural analysis of proteins, glycans, lipids, and metabolites. Multiple reaction monitoring (MRM) and parallel reaction monitoring (PRM) have been the methods of choice for targeted structural studies of biomolecules. These target analyses simplify the analytical workflow, reduce background interference, and increase selectivity/specificity, allowing for a reliable quantification of permethylated *N*-glycans in complex biological matrices.

Key words MRM, PRM, Permethylated *N*-glycans

1 Introduction

Targeted mass spectrometry (MS) is a powerful technique for quantitative biomolecular analysis. The sensitivity and specificity of the technique have resulted in its popularity across various omics disciplines such as proteomics [1–4], glycomics [5–8], glycoproteomics [9], lipidomics [10–12], and metabolomics [13, 14]. Targeted approaches in MS particularly facilitate the analysis of low abundance analytes by filtering the precursors of interest from the undesired sample components which, in most cases, overload the signal of interest. Triple quadrupole mass spectrometers were the initial choice for targeted studies as they enable the performance of multiple reaction monitoring (MRM) [15, 16]. The initial quadrupole acts as a mass filter and isolates

Cristian D. Gutierrez Reyes and Akeem Sanni contributed equally with all other contributors.

Steven B. Bradfute (ed.), *Recombinant Glycoproteins: Methods and Protocols*, Methods in Molecular Biology, vol. 2762, https://doi.org/10.1007/978-1-0716-3666-4_15,

the targeted precursors, followed by its fragmentation in the second quadrupole. In the third quadrupole, the selected fragments are isolated for their subsequent detection [17]. Further development of parallel reaction monitoring (PRM) was made possible due to the availability of hybrid quadrupole-orbitrap mass analyzers; these instruments allow the parallel detection of all the transitions in a single experiment using the orbitrap [18]. The targeted approaches have immensely benefitted the quantitative analysis of biomarkers such as glycans for various diseases [2, 19, 20]. Glycosylation is a vital post-translational modification that renders functionality and stability to proteins [21, 22]. Glycans mediate many essential biological processes, including cell signaling [23–25], cell adhesion [15, 23, 24], and immune response [26–28]. Aberrations in glycan expressions have also been reported in various diseases [29–32]. Thus, glycans are widely investigated as potential biomarkers to understand their role in disease progression [2, 33–38]. Although native glycans have been extensively analyzed using MS [39, 40], permethylated glycans provide the advantage of displaying enhanced ionization efficiency in positive ion mode. Moreover, permethylation prevents sialic acid loss and fucose rearrangement in the glycans, thus stabilizing them in the gas phase [41]. Therefore, the targeted analysis of permethylated *N*-glycans represents an accurate, sensitive, and selective alternative to determine the abundance of these important biomolecules.

2 Materials

2.1 Model Glycoproteins

1. Ribonuclease B (RNase B) from bovine pancreas.
2. Fetuin from bovine serum.
3. Porcine thyroglobulin (PTG).
4. Human blood serum (HBS).
5. Pooled cerebral spinal fluid (CSF).

2.2 Cell Lines

1. MDA-MB-231.
2. MDA-MB-231BR.

2.3 Reagents

1. Ammonium bicarbonate (ABC) (50 mM).
2. Sodium hydroxide beads.
3. Borane-ammonia complex (10 μg/μL).
4. Iodomethane.
5. HyperSep™ C18 cartridges.
6. Acclaim PepMap 100 C18 trapping column (75 μm i.d. × 2 cm, 3 μm, 100 Å).

7. Acclaim PepMap 100 C18 column (75 μm i.d. × 15 cm, 2 μm, 100 Å).
8. Formic acid.
9. Acetic acid (5%).
10. Dimethyl sulfoxide (DMSO).
11. HPLC grade water.
12. Methanol.
13. Acetonitrile.
14. Empty micro spin columns.
15. PNGase F.
16. Protein denaturization buffer.

3 Methods

3.1 N-Glycan Release, Purification, Reduction, and Permethylation

1. For CSF, HBS, cell lines (*see* **Note 1**), and model glycoproteins, the equivalent volume of 20 μg of protein was diluted to a total volume of 100 μL with 50 mM ABC buffer (*see* **Note 2**).
2. The proteins were denatured by incubating in a 90 °C water bath for 15 min.
3. A thousand units of PNGase F were added into the samples and incubated in a 37 °C water bath overnight.
4. The released *N*-glycans were dried in a SpeedVac concentrator. Then, the samples are reconstituted in 300 μL of acetic acid 5% and purified through SPE C18 cartridges (*see* **Note 3**).
5. The *N*-glycan reduction was accomplished by the addition of 10 μL of 10 μg/μL ammonium borane solution, which was followed by incubation at 60 °C for 1 h. After incubation, the residual borane was removed by the addition of methanol, generating methyl borate that was evaporated while drying in the vacuum concentrator (*see* **Note 4**).
6. The reduced *N*-glycans were subjected to permethylation as follows. Dried *N*-glycans were resuspended in 30 μL of dimethyl sulfoxide (DMSO), 1.2 μL of water, and 20 μL of iodomethane (*see* **Note 5**). The solution was applied into a microspin column packed with sodium hydroxide beads that were subsequently incubated in darkness at room temperature for 25 min. After the initial incubation period, 20 μL of iodomethane was applied to the spin column, and the reaction was allowed to proceed for an additional 15 min. For the permethylated *N*-glycan recovery, the micro spin columns were spun down and washed with 50 μL of acetonitrile. Finally, the permethylated *N*-glycans were dried and resuspended in aqueous mobile phase for LC-MS analysis (*see* **Note 6**).

3.2 LC Conditions

1. A Dionex UltiMate 3000 RSLCnano system with a reverse phase C18 Acclaim PepMap capillary column (75 μm i.d. × 150 mm, 3 μm, 100 Å) was used for the separation of the permethylated *N*-glycans. The flow rate was set at 0.35 μL/min with a column temperature of 55 °C. The glycans were purified online using an Acclaim PepMap C18 trapping column (75 μm i.d. × 20 mm, 3 μm, 100 Å) with a flow rate of 3 μL/min and a loading solution containing 2% acetonitrile and 0.1% formic acid.
2. The mobile phase A (MPA) was an aqueous solution with 2% acetonitrile in 0.1% formic acid. The mobile phase B (MPB) was composed of acetonitrile with 0.1% formic acid. The chromatographic gradient started with 20% of MPB for 10 min while the sample was loading in the trapping column. Then it was continued by increasing the percentage of MPB to 42% in 1 min and reaching 55% in the following 39 min. Later, it was increased to 90% of MPB and was held for 10 min. Finally, the percent of MPB is decreased to 20% and kept constant to equilibrate the system for a new injection.

3.3 MS-MRM Conditions

1. For this study, a Thermo Scientific TSQ Vantage triple quadrupole mass spectrometer was used to perform the MRM analysis of permethylated *N*-glycans. The spray voltage was set at 1.6 kV with a capillary temperature of 275 °C. Data-dependent acquisition mode (DDA) with two scan events was performed in positive ion mode. The first event was a Full MS scan in the range of 400–1500 m/z in Q3 with a scan time of 0.7 s and peak width of 0.7 FWHM. The five most intense ions were subjected to MS/MS for the data-dependent scan event with a collision energy of 35. The workflow is depicted in Fig. 1.
2. The MRM mode was set with Q1 and Q3 peak width of 0.7. The cycle time was set at 2 s. The collision energy values and the selection of the transitions for the analyzed permethylated *N*-glycans are described in Subheading 3.

3.4 MS-PRM Conditions

1. For this study, a Q Exactive HF was used to perform the PRM analysis of permethylated *N*-glycans. The spray voltage was set at 1.6 kV with a capillary temperature of 275 °C. Data-dependent acquisition mode (DDA) with two scan events was performed in positive ion mode. The first event was a Full MS scan with a resolution of 60K, 1e6 of AGC target, and 50 ms of maximum injection time. The scan range was 400 to 1800 m/z, and the 10 most intense ions were subjected to HCD with a normalized collision energy of 35. The workflow is depicted in Fig. 1.

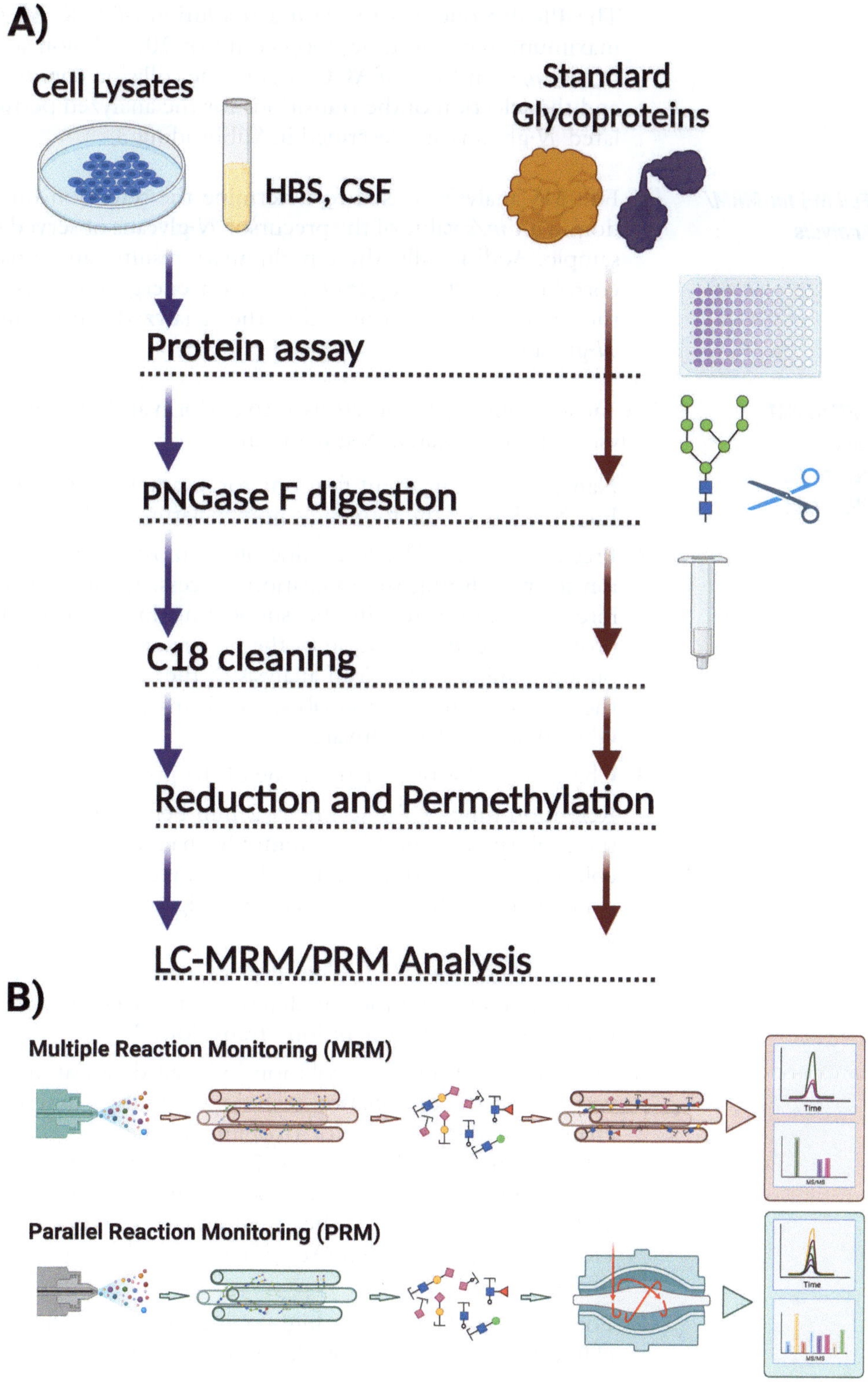

Fig. 1 Analytical workflows. (**a**) Sample analysis and (**b**) MRM/PRM quantitation

2. The PRM mode was set with a resolution of 15K, 50 ms of maximum injection time, loop count of 20, isolation window of 1.6 *m/z*, and 1e5 of AGC target. The collision energy values and the selection of the transitions for the analyzed permethylated *N*-glycans are described in Subheading 3.

3.5 Full MS for MRM/PRM Analysis

1. Full MS analysis is used to determine the retention time and dominant *m/z* value of the precursor *N*-glycans observed in the sample. Additionally, these preliminary results can be used to corroborate if the suggested collision energy values produce the expected transitions for the analyzed permethylated *N*-glycans.

3.6 MRM/PRM Acquisition Parameters for Transition List

The most common parameters used to perform an MRM and PRM analysis of permethylated *N*-glycans are:

1. Name. *N*-Glycan identification; for example, HexNAc,Hex, Fuc,Neu5Ac or the four-digit nomenclature "1-1-1-1".
2. Precursor (*m/z*). The *m/z* value of the most intense precursor ion formed during the ionization process. If the *N*-glycan of interest is not listed in the supporting information of this protocol, the identification of the precursor ions will be based on a preliminary Full Scan analysis of the sample evaluated or theoretical fragmentation obtained from applications such as GlycoWorkbench® software.
3. Charge (+). Positive charge value of the precursor.
4. Retention time (RT). Set the rt 2 min before and 2 min after the peak signal. The RTs commonly change according to the column life and other external parameters. Therefore, this parameter should be previously investigated in the Full Scan analysis.
5. Collision energy (CE). The level of energy applied for the *N*-glycan fragmentation will determine the accurate and stable formation of the fragment ion "transitions."
6. For the MRM analysis, collision-induced dissociation (CID) was used as the dissociation technique. Figure 2 describes the changes in the intensity of oxonium at the CEs of 30, 35, 40, and 45. According to Fig. 2, the optimum CE level for most of the tested *N*-glycan types (such as mannose, hybrid, and complex) was 35. The di-antennary di-sialofucosylated $HexNAc_4$,Hex_5,Fuc,$Neu5Ac_2$, however, had an optimum CE of 30.
7. For the PRM analysis, high-energy collision dissociation (HCD) was used as the dissociation technique. Figure 3 describes the changes in intensity of the transitions observed for the NCEs of 10, 15, 20, 25, 30, 35, and 40. According to Fig. 3, the optimum NCE levels for the tested *N*-glycan types

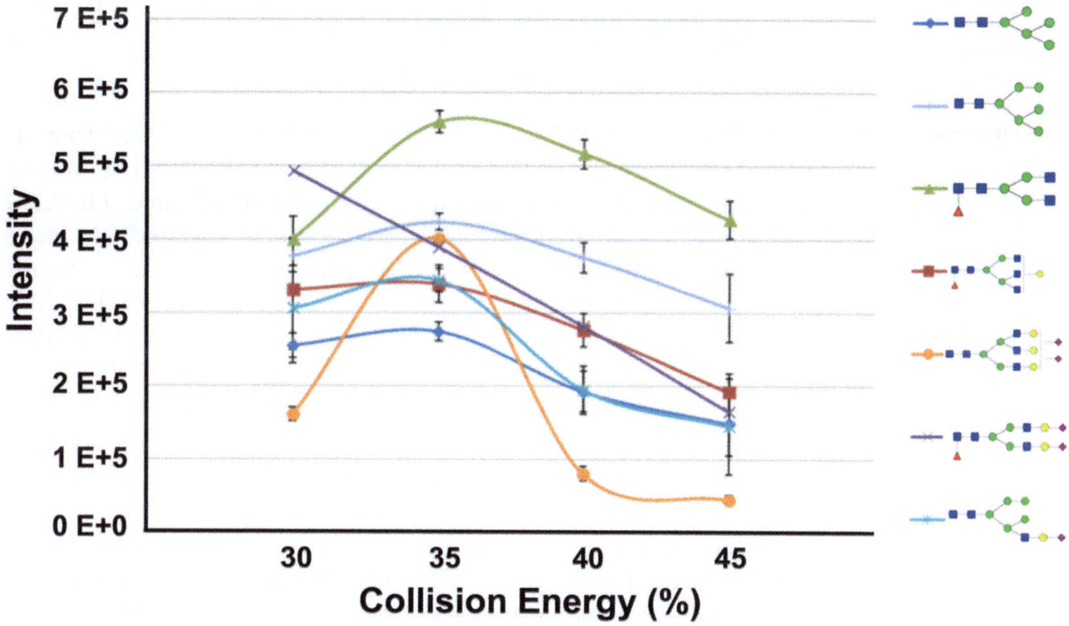

Fig. 2 Determination of collision energies for best intensity signal. Error bars represent the standard deviations of three measurements. Glycan nomenclature: Fucose ▾; N-acetyl glucosamine ■; mannose ●; galactose ○; sialic acid ◆. (Figure reproduced from Zhou et al. [8])

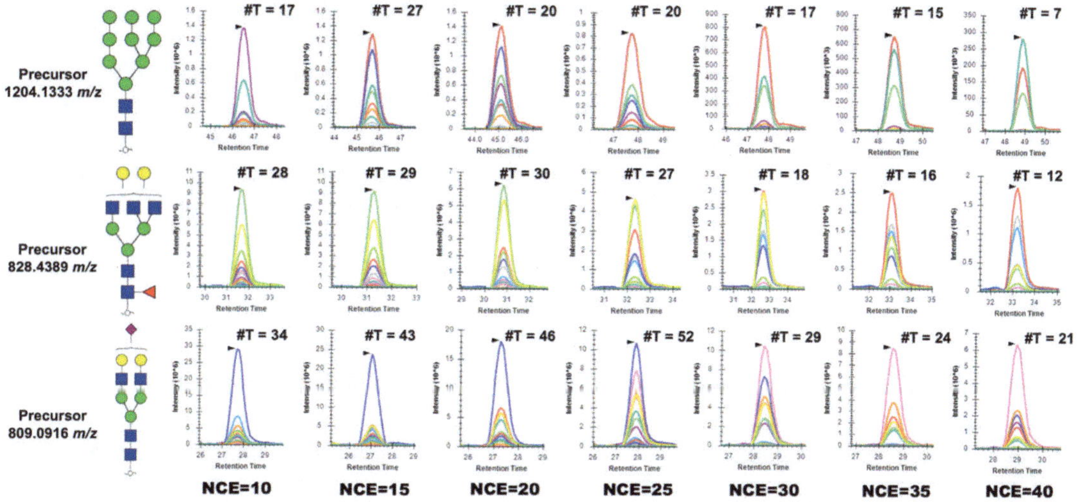

Fig. 3 Representative chromatograms showing collision energy optimization of permethylated glycans using PRM. Each color in the chromatogram represents different transitions. The NCE value increases from left (NCE = 10) to right (NCE = 40). #T = number of transitions, glycan nomenclature as in Fig. 2. (Figure reproduced from Byeong G. C. et al. [42])

(such as mannose, complex, and sialylated) were 20 or 25. A large number of transitions and large intensities were observed for the above-mentioned levels.

3.7 MRM Transitions

1. The fragmentation product ion *m/z* value is referred to as an MRM transition. Therefore, to monitor a *N*-glycan of interest, the transition patterns must be known in advance. Here, we have investigated and analyzed the permethylated *N*-glycans from some model standard glycoproteins: RNase B, fetuin, PTG, HBS, and the cell lines MDA-MB-231 and MDA-MB-231BR. The *N*-glycan transitions were divided into high mannose *N*-glycans (Table 1), complex *N*-glycans (Table 2), and hybrid *N*-glycans (Table 3).
2. Table 1 lists the precursor ion values of the most common high mannose *N*-glycans with their three most common transition ions. Depending on the amount of mannose residues, the transition fragments differ. *N*-glycans with three to five mannose residues fragment into 111, 230, and 262 *m/z* ions. If the number of mannose residues is above seven, the fragmentation patterns change to 187, 262, and 294 *m/z*.
3. Table 2 shows eight different transition patterns for complex *N*-glycans. The *N*-glycans with sialic acid had two distinct transition patterns. When fucose is present, the most abundant fragment ions are 312, 344, and 376 *m/z*. With no fucose residue in the *N*-glycan, instead of 312 *m/z* there, 825 *m/z* ion was predominant. Fragment ions 344 and 376 *m/z* are related to the cleavage of sialic acid residue. When sialic acid was not present in complex glycans, six different transition patterns were recognized. Ion 196 *m/z*, related to the fragmentation of HexNAc residue, occurred in five of the glycans. Cross-ring cleavage took place only if the *N*-glycan structure had three HexNAc and three-Man residues. When the number

Table 1
MRM transitions used for the quantitation of permethylated high Mannose *N*-glycans

Name	Precursor (*m/z*)	Charge	CE	Top three transitions (*m/z*)		
HexNAc2,Hex3	1164.625	+1	35	111	230	262
HexNAc2,Hex4	1368.725	+1	35			
HexNAc2,Hex5	795.933	+2	35			
HexNAc2,Hex6	897.983	+2	35			
HexNAc2,Hex7	1000.033	+2	35	230	262	294
HexNAc2,Hex8	1102.083	+2	35	187	262	294
HexNAc2,Hex9	1204.133	+2	35			
HexNAc2,Hex10	1306.182	+2	35			

Information reproduced from Zhou et al. [8]

Table 2
MRM transitions used for the quantitation of permethylated complex *N*-glycans

Name	Precursor (*m/z*)	Charge	CE	Top three transitions (*m/z*)		
HexNAc3,Hex3	705.883	+2	35	111	196	260
HexNAc3,Hex4,Fuc	894.978	+2	35	187	196	228
HexNAc5,Hex5,Fuc	1242.154	+2	35	196	432	464
HexNAc5,Hex4,Fuc	1140.104	+2	35	228	260	464
HexNAc4,Hex3	828.446	+2	35	196	228	260
HexNAc4,Hex3,Fuc	915.491	+2	35			
HexNAc4,Hex4	930.496	+2	35			
HexNAc5,Hex3	951.009	+2	35			
HexNAc4,Hex4,Fuc	1017.541	+2	35			
HexNAc5,Hex3,Fuc	1038.054	+2	35			
HexNAc5,Hex4	1053.059	+2	35			
HexNAc6,Hex3	1073.573	+2	35			
HexNAc5,Hex5	1155.109	+2	35			
HexNAc5,Hex6	1257.159	+2	35	196	228	464
HexNAc6,Hex7	988.184	+3	35			
HexNAc7,Hex5,Fuc	991.856	+3	35			
HexNAc4,Hex5,Fuc	1119.591	+2	35			
HexNAc6,Hex7,Fuc	1046.2142	+3	35			
HexNAc8,Hex5,Fuc	1073.565	+3	35			
HexNAc8,Hex6	1083.568	+3	35			
HexNAc8,Hex6,Fuc	1141.598	+3	35			
HexNAc3,Hex4,NeuAc	988.520	+2	35	344	376	825
HexNAc4,Hex4,NeuAc	1111.083	+2	35			
HexNAc4,Hex5,NeuAc	1213.133	+2	35			
HexNAc5,Hex4,NeuAc	1233.646	+2	35			
HexNAc5,Hex5,NeuAc	890.800	+3	35			
HexNAc4,Hex5,NeuAc2	929.482	+3	35			
HexNAc5,Hex6,NeuAc	958.833	+3	35			
HexNAc5,Hex6,NeuAc2	1079.224	+3	35			
HexNAc6,Hex7,NeuAc	1108.575	+3	35			
HexNAc5,Hex6,NeuAc3	1199.616	+3	35			
HexNAc7,Hex8,NeuAc	1258.317	+3	35			
HexNAc5,Hex6,NeuAc4	1320.007	+3	35			
HexNAc6,Hex7,NeuAc2	921.977	+4	35			
HexNAc8,Hex7,NeuAc	954.246	+4	35			
HexNAc6,Hex7,NeuAc3	1012.270	+4	35			
HexNAc7,Hex8,NeuAc2	1034.283	+4	35			
HexNAc8,Hex8,NeuAc	1095.565	+4	35			
HexNAc6,Hex7,NeuAc4	1102.563	+4	35			
HexNAc7,Hex8,NeuAc3	1124.577	+4	35			
HexNAc7,Hex8,NeuAc4	1214.87	+4	35			
HexNAc8,Hex9,NeuAc4	1327.176	+4	35			

(continued)

Table 2
(continued)

Name	Precursor (*m/z*)	Charge	CE	Top three transitions (*m/z*)		
HexNAc4,Hex4,Fuc,NeuAc	1198.128	+2	35	312	344	376
HexNAc4,Hex5,Fuc,NeuAc	1300.177	+2	35			
HexNAc5,Hex4,Fuc,NeuNc	1320.691	+2	35			
HexNAc5,Hex5,Fuc,NeuAc	948.830	+3	35			
HexNAc4,Hex5,Fuc,NeuAc2	987.512	+3	35			
HexNAc5,Hex6,Fuc,NeuAc	1016.863	+3	35			
HexNAc5,Hex5,Fuc,NeuAc2	1069.221	+3	35			
HexNAc5,Hex6,Fuc2,NeuAc	1074.893	+3	35			
HexNAc5,Hex6,Fuc,NeuAc2	1137.254	+3	35			
HexNAc6,Hex7,Fuc,NeuAc	1166.050	+3	35			
HexNAc6,Hex6,Fuc,NeuAc2	1218.963	+3	35			
HexNAc7,Hex5,Fuc,NeuAc2	1232.638	+3	35			
HexNAc8,Hex6,Fuc,NeuAc	1261.989	+3	35			
HexNAc5,Hex6,Fuc,NeuAc3	943.486	+4	35			
HexNAc6,Hex7,Fuc,NeuAc2	965.499	+4	35			
HexNAc5,Hex6,Fuc2,NeuAc3	987.008	+4	35			
HexNAc6,Hex6,Fuc,NeuAc3	1004.767	+4	35			
HexNAc6,Hex7,Fuc2,NeuAc2	1009.021	+4	35			
HexNAc6,Hex7,Fuc3,NeuAc2	1052.543	+4	35			
HexNAc6,Hex7,Fuc,NeuAc3	1055.792	+4	35			
HexNAc6,Hex7,Fuc2,NeuAc3	1099.315	+4	35			
HexNAc6,Hex7,Fuc3,NeuAc3	1142.837	+4	35			
HexNAc6,Hex7,Fuc,NeuAc4	1146.086	+4	35			
HexNAc7,Hex8,Fuc,NeuAc3	1168.099	+4	35			
HexNAc6,Hex7,Fuc2,NeuAc4	1189.608	+4	35			
HexNAc8,Hex8,Fuc,NeuAc3	1229.380	+4	35			
HexNAc6,Hex7,Fuc3,NeuAc4	1233.130	+4	35			

Information reproduced from Zhou et al. [8]

of HexNAc residues increased to six, then additionally to ions of 196 and 260 *m/z*, the intensity of 228 *m/z* ion was more prevalent. But if the number of galactose residues increased above two, the transition pattern changes from ion of 260 to 464 *m/z*, related to the cleavage of the HexNAc-Gal residue.

4. Table 3 illustrates that the hybrid *N*-glycan transition patterns can be divided into two different groups. If the *N*-glycans are sialylated, the transition pattern is fixed to 312, 344, and 376 *m/z* fragmentation ions. The non-sialylated *N*-glycans have three fragmentation patterns. The common fragmentation ions in all three patterns are 196 and 432 *m/z*. If there are five or more hexose or fucose residues, the additional ions that can occur are 260 or 228 *m/z*.

Table 3
MRM transitions used for the quantitation of permethylated hybrid N-glycans

Name	Precursor (*m/z*)	Charge	CE	Top three transitions (*m/z*)		
HexNAc3,Hex5	909.983	+2	35	196	260	432
HexNAc3,Hex5,Fuc	997.028	+2	35			
HexNAc4,Hex6,Fuc	814.763	+3	35			
HexNAc4,Hex9	940.826	+3	35	196	228	432
HexNAc3,Hex6	1012.033	+2	35	260	432	196
HexNAc3,Hex7	1201.128	+2	35			
HexNAc3,Hex5,NeuAc	1090.57	+2	35	312	344	376
HexNAc3,Hex4,Fuc,NeuAc	1177.615	+2	35			
HexNAc3,Hex8,NeuAc	931.482	+3	35			
HexNAc4,Hex6,NeuAc2	997.516	+3	35			
HexNAc4,Hex6,Fuc,NeuAc2	1055.546	+3	35			
HexNAc5,Hex7,Fuc,NeuAc	1084.896	+3	35			
HexNAc5,Hex8,Fuc3,NeuAc	1268.989	+3	35			
HexNAc5,Hex7,Fuc,NeuAc2	1205.287	+3	35			
HexNAc6,Hex8,Fuc,NeuAc	1234.638	+3	35			
HexNAc6,Hex8,Fuc,NeuAc2	1016.524	+4	35			
HexNAc6,Hex8,Fuc4,NeuAc	1237.384	+4	35			

Information reproduced from Zhou et al. [8]

Table 4
PRM transitions used for the quantitation of permethylated high mannose *N*-glycans

Name	Precursor (*m/z*)	Charge	CE	Top three transitions (*m/z*)		
HexNAc2,Hex8	1102.0834	+2	10	294.1910	1688.8477	1470.7322
HexNAc2,Hex7	1000.0335	+2	15			
HexNAc2,Hex9	1204.1333	+2	15	1878.9318	187.0965	1878.9318
HexNAc2,Hex5	795.9337	+2	10	1280.6482	187.0965	1280.6482
HexNAc2,Hex6	897.9782	+2	15	1484.7479	187.0965	1484.7479

Information reproduced from Byeong G. C. et al. [42]

3.8 PRM Transitions

1. The fragmentation product ion *m/z* value is referred to as a PRM transition. PRM provides a major advantage over MRM because of its ability to acquire full MS/MS scans. This facilitates the method setup, as the transitions for PRM do not need to be preselected during the data acquisition. Thus, a large number of transitions will be available for the identification and quantification of *N*-glycans. Here, we have investigated and analyzed the permethylated *N*-glycans from a pooled CSF sample. The *N*-glycan transitions were divided into high mannose *N*-glycans (Table 4) and complex *N*-glycans (Table 5).

Table 5
PRM transitions used for the quantitation of permethylated complex *N*-glycans

Name	Precursor (*m/z*)	Charge	CE	Top three transitions (*m/z*)		
HexNAc6,Hex7,Fuc2,NeuAc4	1189.6086	+4	15	344.1702	376.1966	825.4227
HexNAc6,Hex7,Fuc,NeuAc4	1146.0863	+4	20			
HexNAc6,Hex7,NeuAc4	1102.5640	+4	20			
HexNAc6,Hex7,Fuc,NeuAc3	1055.7929	+4	25			
HexNAc5,Hex6,Fuc,NeuAc3	1257.6458	+3	20			
HexNAc5,Hex6,NeuAc3	1199.6161	+3	10			
HexNAc5,Hex6,Fuc,NeuAc2	1137.2546	+3	10			
HexNAc5,Hex5,Fuc,NeuAc2	1069.2213	+3	10			
HexNAc6,Hex7,NeuAc3	1012.2706	+4	10	464.2490	825.4227	344.1702
HexNAc6,Hex7,NeuAc2	1228.9669	+3	20			
HexNAc6,Hex7,Fuc,NeuAc	1166.6054	+3	10			
HexNAc6,Hex7,Fuc,NeuAc2	965.4995	+4	10	464.2490	376.1966	344.1702
HexNAc5,Hex6,NeuAc	958.8336	+3	20			
HexNAc6,Hex7,Fuc,NeuAc	1079.2249	+3	15			
HexNAc3,Hex5,Fuc,NeuAc	785.4126	+3	10	187.0965	376.1966	344.1702
HexNAc3,Hex6,NeuAc	795.4161	+3	10			
HexNAc4,Hex5,NeuAc	809.0916	+3	15	376.1966	294.1911	344.172
HexNAc4,Hex5,NeuAc2	929.4828	+3	10			
HexNAc5,Hex5,NeuAc2	1011.1916	+3	10			
HexNAc6,Hex5,Fuc,NeuAc2	863.4496	+4	25	344.1702	294.1911	825.4227
HexNAc6,Hex4,Fuc,NeuAc	962.5056	+3	20			
HexNAc5,Hex4,Fuc,NeuAc	880.7968	+3	10	344.1702	260.1492	468.2803
HexNAc4,Hex4,Fuc,NeuAc	799.0881	+3	10			
HexNAc6,Hex7,Fuc2,NeuAc4	1189.6086	+4	15	344.1702	825.4227	376.1966
HexNAc6,Hex7,Fuc,NeuAc4	1146.0863	+4	20			
HexNAc6,Hex7,NeuAc4	1102.5640	+4	20			
HexNAc6,Hex7,Fuc,NeuAc3	1055.7929	+4	25			
HexNAc5,Hex6,Fuc,NeuAc3	1257.6458	+3	20			
HexNAc5,Hex6,NeuAc3	1199.6161	+3	10			
HexNAc5,Hex6,Fuc,NeuAc2	1137.2546	+3	10			
HexNAc5,Hex5,Fuc,NeuAc2	1069.2213	+3	10			
HexNAc5,Hex6,NeuAc	958.8336	+3	20	344.1702	376.1966	464.2490
HexNAc5,Hex5,Fuc2,NeuAc	1006.8598	+3	20	344.1702	294.1911	376.1966
HexNAc5,Hex4,Fuc2	818.4353	+3	10	260.1492	228.1492	189.1121
HexNAc6,Hex6,Fuc,NeuAc2	914.4745	+4	10	344.1702	376.1966	312.1440
HexNAc5,Hex5,Fuc	828.4389	+2	10	432.2228	464.2490	1552.7853
HexNAc5,Hex6	838.4424	+3	15	187.0965	858.4329	654.3331
HexNAc6,Hex4,Fuc	842.1144	+3	10	260.1492	432.2228	246.1336
HexNAc4,Hex5,Fuc3	862.7896	+3	10	468.2803	432.2228	189.1121
HexNAc4,Hex5,Fuc,NeuAc	867.1213	+3	10	344.1702	468.2803	464.2490

(continued)

Table 5
(continued)

Name	Precursor (*m/z*)	Charge	CE	Top three transitions (*m/z*)		
HexNAc5,Hex6,Fuc	1344.2013	+2	10	464.2490	432.2228	1756.8851
HexNAc5,Hex5,Fuc2	886.4686	+2	10	260.1492	468.2803	228.1230
HexNAc5,Hex5,NeuAc	890.8004	+3	20	344.1702	312.1440	246.1336
HexNAc6,Hex4,Fuc2	1349.7122	+2	20	228.1230	638.3382	432.2228
HexNAc6,Hex5,Fuc	910.1476	+2	10	260.1492	464.2490	432.2228
HexNAc6,Hex6,Fuc,NeuAc	824.1811	+4	25	344.1702	376.1966	246.1336
HexNAc5,Hex4,Fuc2,NeuAc	938.8266	+3	15	344.1702	468.2803	1726.8745
HexNAc5,Hex5,Fuc,NeuAc	948.8301	+3	25	344.1702	464.2490	432.2228
HexNAc5,Hex6,NeuAc	958.8336	+3	20	344.1702	464.2490	376.1966
HexNAc4,Hex5,Fuc2	804.7598	+3	10	187.0965	228.1230	648.2803
HexNAc6,Hex6,Fuc3	1214.6316	+3	15	432.2228	260.1492	638.3382
HexNAc6,Hex4,Fuc2	1349.7122	+2	20	228.1230	638.3382	432.2228
HexNAc5,Hex6,Fuc,NeuAc	1016.8634	+3	10	344.1702	376.1966	294.1911
HexNAc4,Hex5,Fuc2,NeuAc2	1045.5423	+3	10	344.1702	468.2803	376.1966
HexNAc6,Hex5,Fuc2	968.1774	+3	10	260.1492	464.2490	228.1230
HexNAc4,Hex5,Fuc,NeuAc2	987.5126	+3	10	344.1702	376.1966	468.2803
HexNAc6,Hex5,Fuc3	1026.2071	+3	10	260.1492	638.3382	432.2228
HexNAc6,Hex5,Fuc2	1088.5686	+3	20	260.1492	464.2490	228.1230
HexNAc6,Hex6,Fuc2,NeuAc	824.1811	+4	25	344.1702	376.1966	825.4227

Information reproduced from Byeong G. C. et al. [42]

2. Table 4 lists the precursor ion values of the most common high mannose *N*-glycans with their three most common transition ions. The most common fragments were 187.0965 *m/z* and 294.1910 *m/z*. These *N*-glycans are only composed of core GlcNAc and mannose, and therefore require lower NCE compared to more complex structures.
3. Table 5 shows the precursor ion values of the most common high mannose *N*-glycans with their three most common transition ions. The most common fragments were 187.0965 *m/z* and 294.1910 *m/z*. These *N*-glycans are only composed of core GlcNAc and mannose, and therefore require lower NCE compared to more complex structures.

3.9 MRM/PRM Quantitation

1. Xcalibur software 4.2 (Thermo Scientific) was employed to extract the ion chromatograms from a full MS scan, and to verify the precursor *m/z* values and their retention times. Using the same software, the MS/MS data was acquired for each precursor, and the three most intense fragment ions were assigned as transitions for the MRM and PRM quantitation. The theoretical validation of the permethylated *N*-glycan fragment ions was performed using Glycoworkbench 2.0 software. Subsequently, a transition list was compiled using Microsoft Excel®, and uploaded to Skyline 20.2 software to extract the EICs and complete the peak area calculations.

4 Notes

1. Cell lysis of cell lines: Breast cancer cell line MDA-MB-231 and its brain targeting sub-line MDA-MB-231BR were resuspended in 100 μL aliquots of phosphate buffered saline (PBS). The suspensions were then sonicated in iced water for 60 min. A 5 μL aliquot of the sonicated cells was subjected to BSA protein assay to determine protein concentrations. The extracted proteins were denatured in a 90 °C water bath for 15 min. Then, the samples were cooled to room temperature prior to the addition of a 2.4 mL aliquot of a 10 times diluted PNGase F solution. The enzymatic digestion was allowed to proceed at 37 °C in a water bath for 18 h. The *N*-glycans released from the glycoprotein samples were purified using a charcoal spin-column. The columns were washed with a 400 μL aliquot of 100% ACN and a 400 μL aliquot of 85% ACN with 0.1% TFA, three times. The spin-column was conditioned with a 400 μL aliquot of 5% ACN with 0.1% TFA, twice. Sample volumes were adjusted to 400 μL using 5% ACN with 0.1% TFA and washed four times with a 400 μL aliquot of 5% ACN with 0.1% TFA. The *N*-glycans were eluted using a 400 μL aliquot of 40% ACN with 0.1% TFA. Finally, the samples were dried in a SpeedVac concentrator.
2. For the analysis of permethylated *N*-glycans digested from the proteome extracted from cell lines, it is recommended to start with larger amounts of protein, between 100 to 500 μg according to the protein assay results.
3. SPE C18 cleaning: Dried digested *N*-glycan samples were resuspended with 300 μL of 5% acetic acid. The SPE C18 cartridges were washed with 3 mL of methanol and equilibrated with 3 mL of 5% acetic acid. Then, the resuspended samples were loaded to the SPE C18 cartridges and recovered with 300 μL of 5% acetic acid three times while all flow-through was collected and dried using the vacuum concentrator.

4. The addition of 1 mL of methanol should be repeated at least four times, or until the methyl borate is completely removed by evaporation.
5. The addition of iodomethane to the sample must be done before the preparation of the spin columns with sodium hydroxide beds and their washing with DMSO.
6. Immediately dry the final samples in the SpeedVac concentrator to avoid the formation of secondary products from the excess of iodine present in the sample.

Acknowledgments

This work was supported by grants from the National Institutes of Health, NIH (1R01GM112490-04, 1R01GM130091-01). The work is also supported by grants from Robert A. Welch Foundation (No. D-0005) and The CH Foundation.

References

1. Gillette MA, Carr SA (2013) Quantitative analysis of peptides and proteins in biomedicine by targeted mass spectrometry. Nat Methods 10(1):28–34
2. Hsiao Y-C, Chi L-M, Chien K-Y et al (2017) Development of a multiplexed assay for oral cancer candidate biomarkers using peptide immunoaffinity enrichment and targeted mass spectrometry. MCP 16(10):1829–1849
3. Deng L, Ibrahim YM, Hamid AM et al (2016) Ultra-high resolution ion mobility separations utilizing traveling waves in a 13 m serpentine path length structures for lossless ion manipulations module. Anal Chem 88(18): 8957–8964
4. Hüttenhain R, Malmström J, Picotti P et al (2009) Perspectives of targeted mass spectrometry for protein biomarker verification. Curr Opin Chem Biol 13(5–6):518–525
5. Tao S, Huang Y, Boyes BE et al (2014) Liquid chromatography-selected reaction monitoring (LC-SRM) approach for the separation and quantitation of sialylated N-glycans linkage isomers. Anal Chem 86(21):10584–10590
6. Wang J-R, Gao W-N, Grimm R et al (2017) A method to identify trace sulfated IgG N-glycans as biomarkers for rheumatoid arthritis. Nat Commun 8(1):1–14
7. Zhou S, Huang Y, Dong X et al (2017) Isomeric separation of permethylated glycans by porous graphitic carbon (PGC)-LC-MS/MS at high temperatures. Anal Chem 89(12): 6590–6597
8. Zhou S, Hu Y, DeSantos-Garcia JL et al (2015) Quantitation of permethylated N-glycans through multiple-reaction monitoring (MRM) LC-MS/MS. J Am Soc Mass Spectrom 26(4):596–603
9. Bermudez A, Pitteri SJ (2021) Enrichment of intact glycopeptides using strong anion exchange and electrostatic repulsion hydrophilic interaction chromatography. Methods Mol Biol (Clifton, NJ) 2271:107–120
10. Zhang C, Wang Y, Wang F et al (2017) Quantitative profiling of glycerophospholipids during mouse and human macrophage differentiation using targeted mass spectrometry. Sci Rep 7(1):1–13
11. Lee H-C, Yokomizo T (2018) Applications of mass spectrometry-based targeted and non-targeted lipidomics. BBRC 504(3): 576–581
12. Franco J, Ferreira C, Paschoal Sobreira TJ et al (2018) Profiling of epidermal lipids in a mouse model of dermatitis: identification of potential biomarkers. PLoS One 13(4):e0196595
13. Zhou J, Yin Y (2016) Strategies for large-scale targeted metabolomics quantification by liquid chromatography-mass spectrometry. Analyst 141(23):6362–6373
14. Tang Y, Zhu Y, Sang S (2020) A novel LC-MS based targeted metabolomic approach to study the biomarkers of food intake. Mol Nutr Food Res 64(22):2000615
15. An HJ, Gip P, Kim J et al (2012) Extensive determination of glycan heterogeneity reveals

an unusual abundance of high mannose glycans in enriched plasma membranes of human embryonic stem cells. MCP 11(4): M111.010660
16. Selevsek N, Matondo M, Carbayo MS et al (2011) Systematic quantification of peptides/ proteins in urine using selected reaction monitoring. Proteomics 11(6):1135–1147
17. Fu Q, Chen Z, Zhang S et al (2016) Multiple and selective reaction monitoring using triple quadrupole mass spectrometer: preclinical large cohort analysis. Methods Mol Biol 1410:249–264
18. Peterson AC, Russell JD, Bailey DJ et al (2012) Parallel reaction monitoring for high resolution and high mass accuracy quantitative, targeted proteomics. MCP 11(11):1475–1488
19. Keshishian H, Addona T, Burgess M et al (2009) Quantification of cardiovascular biomarkers in patient plasma by targeted mass spectrometry and stable isotope dilution. MCP 8(10):2339–2349
20. Malchow S, Loosse C, Sickmann A et al (2017) Quantification of cardiovascular disease biomarkers in human platelets by targeted mass spectrometry. Proteomes 5(4):31
21. Jayaprakash NG, Surolia A (2017) Role of glycosylation in nucleating protein folding and stability. Biochem J 474(14):2333–2347
22. Shental-Bechor D, Levy Y (2008) Effect of glycosylation on protein folding: a close look at thermodynamic stabilization. PNAS 105(24):8256–8261
23. Cummings RD (2019) Stuck on sugars–how carbohydrates regulate cell adhesion, recognition, and signaling. Glycoconj J 36(4): 241–257
24. Zhao YY, Takahashi M, Gu JG et al (2008) Functional roles of N-glycans in cell signaling and cell adhesion in cancer. Cancer Sci 99(7): 1304–1310
25. Schultz MJ, Swindall AF, Bellis SL (2012) Regulation of the metastatic cell phenotype by sialylated glycans. Cancer Metastasis Rev 31(3): 501–518
26. Amon R, Reuven EM, Ben-Arye SL et al (2014) Glycans in immune recognition and response. Carbohydr Res 389:115–122
27. Crispin M, Ward AB, Wilson IA (2018) Structure and immune recognition of the HIV glycan shield. Annu Rev Biophys 47:499–523
28. Baum LG, Cobb BA (2017) The direct and indirect effects of glycans on immune function. Glycobiology 27(7):619–624
29. Freeze HH, Eklund EA, Ng BG et al (2015) Neurological aspects of human glycosylation disorders. Annu Rev Neurosci 38:105–125
30. Arigoni-Affolter I, Scibona E, Lin CW et al (2019) Mechanistic reconstruction of glycoprotein secretion through monitoring of intracellular N-glycan processing. Sci Adv 5(11): eaax8930
31. Adamczyk B, Tharmalingam T, Rudd PM (2012) Glycans as cancer biomarkers. Biochim Biophys Acta – Gen 1820(9):1347–1353
32. Wittenbecher C, Štambuk T, Kuxhaus O et al (2020) Plasma N-glycans as emerging biomarkers of cardiometabolic risk: a prospective investigation in the EPIC-potsdam cohort study. Diabetes Care 43(3):661–668
33. Watanabe Y, Allen JD, Wrapp D et al (2020) Site-specific glycan analysis of the SARS-CoV-2 spike. Science 369(6501):330–333
34. Walls AC, Tortorici MA, Frenz B et al (2016) Glycan shield and epitope masking of a coronavirus spike protein observed by cryo-electron microscopy. Nat Struct Mol 23(10):899–905
35. Pejchal R, Doores KJ, Walker LM et al (2011) A potent and broad neutralizing antibody recognizes and penetrates the HIV glycan shield. Science 334(6059):1097–1103
36. Helle F, Duverlie G, Dubuisson J (2011) The hepatitis C virus glycan shield and evasion of the humoral immune response. Viruses 3(10): 1909–1932
37. Peng W, Goli M, Mirzaei P et al (2019) Revealing the biological attributes of N-glycan isomers in breast cancer brain metastasis using porous graphitic carbon (PGC) liquid chromatography-tandem mass spectrometry (LC-MS/MS). J Proteome Res 18(10): 3731–3740
38. Bai H, Pan Y, Qi L et al (2018) Development a hydrazide-functionalized thermosensitive polymer based homogeneous system for highly efficient N-glycoprotein/glycopeptide enrichment from human plasma exosome. Talanta 186:513–520
39. Jensen PH, Karlsson NG, Kolarich D et al (2012) Structural analysis of N-and O-glycans released from glycoproteins. Nat Protoc 7(7): 1299–1310
40. Ashwood C, Pratt B, MacLean BX et al (2019) Standardization of PGC-LC-MS-based glycomics for sample specific glycotyping. Analyst 144(11):3601–3612
41. Zhou S, Veillon L, Dong X et al (2017) Direct comparison of derivatization strategies for LC-MS/MS analysis of N-glycans. Analyst 142(23):4446–4455
42. Cho BG, Gutierrez Reyes CD, Goli M et al (2022) Targeted N-glycan analysis with parallel reaction monitoring using a quadrupole-orbitrap hybrid mass spectrometer. Anal Chem 94(44):15215–15222

Chapter 16

Hydrophilic Interaction Liquid Chromatography (HILIC) Enrichment of Glycopeptides Using PolyHYDROXYETHYL A

Mona Goli, Peilin Jiang, Mojibola Fowowe, Md Abdul Hakim, and Yehia Mechref

Abstract

Glycosylation of proteins is an important post-translational modification that plays a role in a wide range of biological processes, including immune response, intercellular signaling, inflammation, and host-pathogen interaction. Abnormal protein glycosylation has been correlated with various diseases. However, the study of protein glycosylation remains challenging due to its low abundance, microheterogeneity of glycosylation sites, and low ionization efficiency. During the past decade, several methods for enrichment and for isolation of glycopeptides from biological samples have been developed and successfully employed in glycoproteomics research. In this chapter, we discuss the sample preparation protocol and the strategies for effectively isolating and enriching glycopeptides from biological samples, using PolyHYDROXYETHYL A as a hydrophilic interaction liquid chromatography (HILIC) enrichment technique.

Key words Glycopeptides, HILIC enrichment, TopTip C18 desalting, LC-MS/MS

1 Introduction

One of the proteins' most common post-translational modifications (PTMs) is glycosylation [1]. Glycosylation is crucial for cell signaling [2, 3], cell adhesion [4, 5], protein stability [6], localization [7], and immune response [8]. Glycosylated proteins, known as glycoproteins, account for over 50% of all mammalian proteins [9]. Given the significant role of glycoproteins in vital biological processes [10] and the correlation of their altered expressions with various diseases [11–14] and cancers [15–17], reliable quantitative and qualitative glycoproteomics methodologies have become increasingly critical for understanding these protein modification sites [18, 19].

Different methodologies and approaches have been developed over the last several decades to facilitate the characterization of glycopeptides [20, 21]. Because mass spectrometry (MS) provides

Steven B. Bradfute (ed.), *Recombinant Glycoproteins: Methods and Protocols*, Methods in Molecular Biology, vol. 2762, https://doi.org/10.1007/978-1-0716-3666-4_16,

high sensitivity and rich structural information, it can be used effectively in glycoproteomics analysis [22]. Despite advances in MS technologies and methods, however, glycoproteomics still faces technical challenges when analyzing glycoproteins derived from biological samples [18, 20, 23]. Glycoproteins are inherently low in abundance in biological systems [24]. Furthermore, co-eluting peptides impede the analysis of glycopeptides due to the microheterogeneity of glycosylation sites, glycan structure complexity, and low ionization efficiencies [18].

To address these concerns, selective enrichment and isolation techniques have been developed and employed in the glycoproteomics research [18, 21, 22, 25–27]. The purpose of this chapter is to summarize strategies required to characterize glycopeptides by the hydrophilic interaction liquid chromatography (HILIC) enrichment method and reverse-phase liquid chromatography-mass spectrometry (LC-MS/MS). There is also a detailed description of how samples are prepared from complex sample matrices such as cell lines, tissues, and human blood serum or plasma.

2 Materials

2.1 Cells Protein Extraction

1. Ammonium bicarbonate (ABC).
2. Sodium deoxycholate (SDC).
3. 2.0 mL microcentrifuge tubes, conical, with screw cap.
4. BeadBug microtube homogenizer.
5. 400 μm molecular biology grade zirconium beads.
6. Micro bicinchoninic acid (BCA) protein assay kit.

2.2 Tissue Protein Extraction

1. Phosphate-buffered saline (PBS).
2. Ammonium bicarbonate (ABC).
3. Sodium deoxycholate (SDC).
4. 2.0 mL microcentrifuge tubes, conical, with screw cap.
5. BeadBug microtube homogenizer.
6. 400 μm molecular biology grade zirconium beads.
7. Micro BCA protein assay kit.

2.3 Depletion of Abundant Proteins from Blood Serum

1. 0.22 μm spin filter.
2. Human 14 Multiple Affinity Removal Column.
3. Buffer A and Buffer B.

2.4 Buffer Exchange

1. Ammonium bicarbonate (ABC).
2. Spin 5 K MWCO.
3. Sorvall Legend X1R Centrifuge.

2.5 Protein Assay

1. Micro BCA protein assay kit.
2. 96 well plate.
3. Multiskan FC.

2.6 Tryptic Digestion

1. Ammonium bicarbonate (ABC).
2. Dithiothreitol (DTT).
3. Iodoacetamide (IAA).
4. HPLC-grade acetonitrile (ACN).
5. HPLC-grade water.
6. Formic acid (FA).
7. Mass spectrometry grade trypsin/Lys-C.

2.7 C18 Desalting

1. C18 TopTips.
2. HPLC-grade acetonitrile (ACN).
3. HPLC-grade water.
4. Formic acid (FA).
5. Three buffer solutions: buffer A includes 100% H_2O, 0.1% FA; buffer B includes 60% ACN, 40% H_2O, 0.1% FA; buffer C contains 100% ACN, 0.1% FA.

2.8 Glycopeptides Enrichment Using PolyHYDROXYETHYL A (HILIC)

1. PolyHydroxyethyl A (HILIC) TopTips.
2. HPLC-grade acetonitrile (ACN).
3. HPLC-grade water.
4. Trifluoroacetic acid (TFA) 99%, spectrometric grade.
5. Two buffer solutions: loading buffer includes 80% ACN, 20% H_2O, 1% TFA; elution buffer contains 100% H_2O, 0.1% TFA.

2.9 LC-MS/MS Analysis

1. Loading solution (98% HPLC-grade water, 2% acetonitrile, 0.1% formic acid).
2. Mobile Phase A (98% HPLC-grade water, 2% acetonitrile, 0.1% formic acid)
3. Mobile Phase B (100% acetonitrile, 0.1% formic acid) (*see* **Note 1**).
4. C18 Acclaim PepMap 100 column (Particle size 3 μm, 100 Å pore size, 75 μm id, 20 mM length).
5. C18 Acclaim PepMap RSLC column (Particle size 2 μm, 100 Å pore size, 75 μm id, 150 mM length).
6. Dionex 3000 UltiMate Nano LC system.
7. Orbitrap Fusion Lumos tribrid mass spectrometer, or Q-Exactive HF Hybrid Quadrupole Orbitrap mass spectrometer.

3 Methods

The procedure below describes detailed steps for isolating and enriching glycopeptides from biological samples. The workflows of the glycopeptides-enriched methods for cells/tissues and blood serum/plasma are depicted in Figs. 1 and 2, respectively.

3.1 Protein Extraction

The choice of protein extraction technique must be carefully made depending on the experimental goals. The protein extraction method described here is useful for the extraction of total protein from both cell and tissue samples.

3.2 Cells Protein Extraction

1. Add enough zirconium beads (400 μm, molecular biology grade) to cover 0.5 cm of a 2.0 mL microcentrifuge tube (conical, screw cap).
2. Add 100 μL of 50 mM ABC buffer (pH 8.0) to a cell pellet to resuspend it. Mix the resuspended cells by repetitive pipetting.

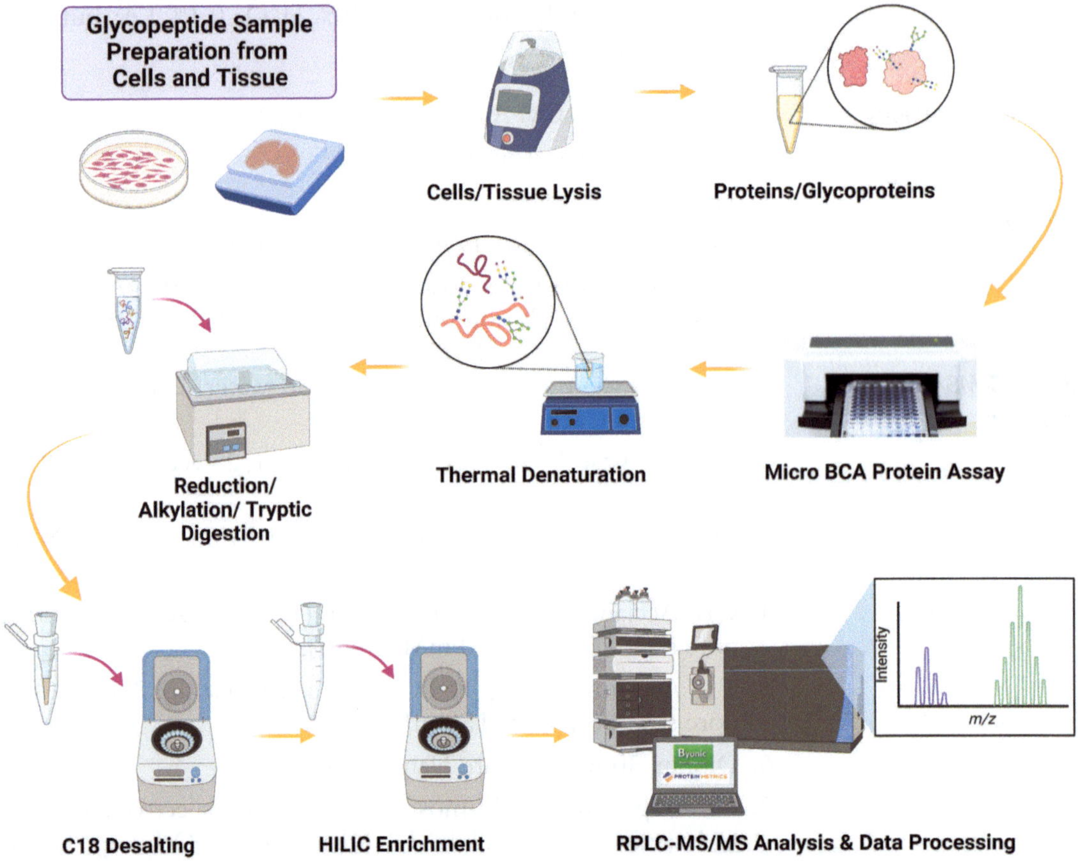

Fig. 1 Glycopeptide sample preparation workflow from cells and tissue

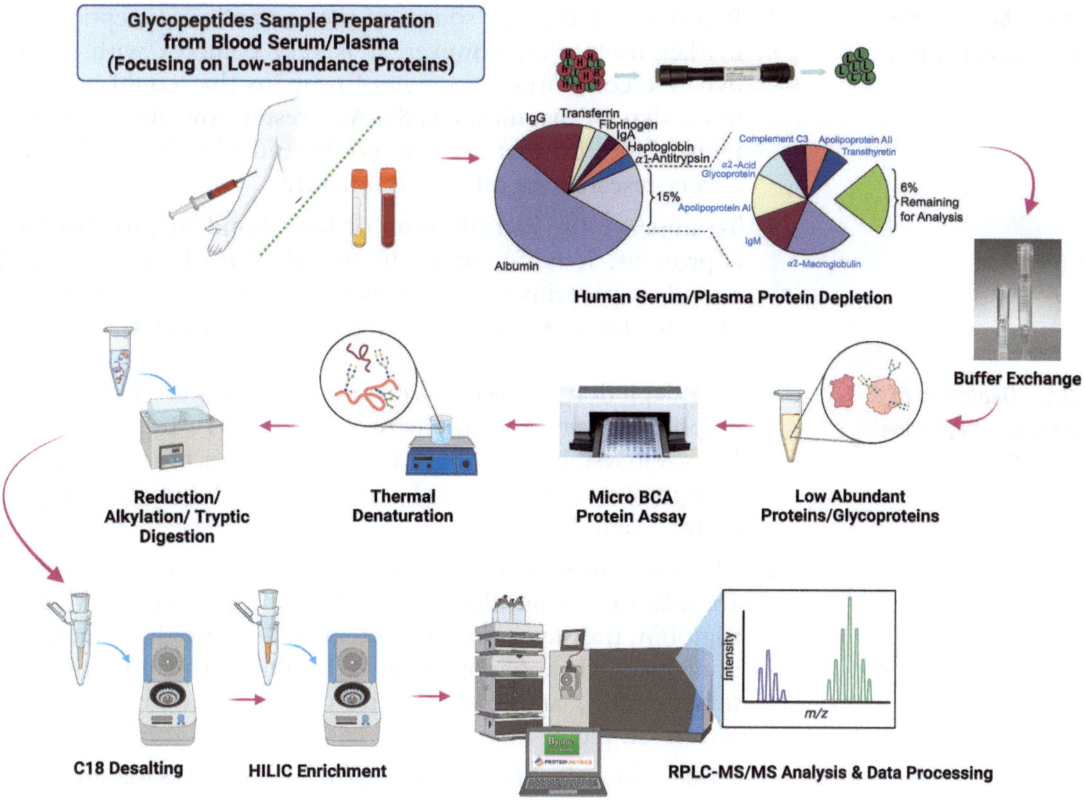

Fig. 2 Glycopeptide sample preparation workflow from blood serum/plasma, focusing on low-abundance proteins

3. Gently move the resuspended cells to the 2.0 mL microtube from step 1 that contains the zirconium beads.
4. Transfer 100 μL of a 5% SDC (aq.) solution to the microtube from the previous step to make 2.5% SDC and vortex the mixture (*see* **Note 2**).
5. Homogenize cells using a BeadBug microtube homogenizer at 4000 rpm (1541 × g) for 30 s (*see* **Notes 3** and **4**).
6. Sonicate the microtube in ice for 1 h (*see* **Note 5**).
7. Centrifuge at 21,000 g for 10 min.
8. Collect the supernatant as cell lysate.
9. Using a BCA protein assay kit, determine the total protein concentration in the prepared cell lysate.

3.3 Tissue Protein Extraction

1. Protein can be extracted from tissue samples using the same protocol described above for cell samples.
2. However, prior to protein extraction, salts and other contaminants must be removed from tissue samples with a few washes with ice-cold 10 mM PBS buffer solution.

3.4 Blood Serum Protein Extraction

1. Blood serum is a rich source of protein and glycopeptide biomarker molecules; however, it is very complex with a wide dynamic concentration range of proteins that could be up to ten orders of magnitude [28]. As a result, low-abundant proteins of interest that are commonly studied in biomarker discovery research are difficult to identify.
2. To improve the identification of low-abundant proteins/glycoproteins, it is important to first deplete the profusion of abundant proteins that form more than 90% of the proteome prior to the proteolytic digestion of serum samples.

3.5 Depletion of Abundant Proteins

1. The depletion of blood sera samples is achieved chromatographically using the commercial Human 14 Multiple Affinity Removal System Column (4.6 × 100 mM) for the fractionation of high-abundant proteins from human proteomics/glycoproteomics samples.
2. This column successfully removes fourteen interfering high-abundant proteins (albumin, antitrypsin, IgG, IgA, IgM, haptoglobin, transferrin, fibrinogen, alpha2-macroglobulin, apolipoprotein AI, apolipoprotein AII, alpha1-acid glycoprotein, transthyretin, and complement C3) from a human blood serum sample.
3. Upon removal of these proteins, the LC-MS/MS identification and quantification of low-abundant proteins/glycoproteins in the serum sample are greatly enhanced (*see* **Notes 6** and 7).

3.6 Buffer Exchange

After depletion, the depletion buffer in the low-abundant protein fraction is exchanged for 50 mM ABC buffer (pH 8.0) using a 5 KDa MWCO 4 mL spin concentrator following the steps below:

1. Transfer low-abundant proteins to 5 K filter and centrifuge for 30 min.
2. Add 3 mL water and centrifuge for 30 min. Discard the flow-through. Repeat this step twice.
3. Add 3 mL 50 mM ABC buffer and centrifuge for 30 min (*see* **Note 8**).
4. Transfer the solution from the filter cartridge to a clean tube. Wash the cartridge with 50 μL ABC buffer and combine.
5. Adjust the volume to 200 μL using 50 mM ABC buffer.
6. Take 10–15 μL to perform protein assay using the bicinchoninic acid (BCA) protein assay kit.

3.7 Tryptic Digestion

The proteolytic digestion of the proteins extracted from the cell and tissue samples and serum proteins will be carried out using the enzyme trypsin. The digestion protocol is described below:

1. Extract the desired proteins according to the protein assay and transfer them to a new Eppendorf tube (*see* **Note 9**).
2. Dilute the sample using 50 mM ABC buffer (*see* **Notes 10** and **11**).
3. Thermally denature the sample at 90 °C for 15 min.
4. Check pH (*see* **Note 12**).
5. Reduction: add 200 mM DTT to the sample at a DTT to sample volume ratio of 1:40.
6. Incubate samples at 60 °C for 45 min.
7. Alkylation: add 200 mM IAA to the sample. The volume of IAA to be added should be 4 times the volume of DTT that was added in the reduction step (*see* **Note 13**).
8. Incubate at 37.5 °C for 45 min.
9. Add 200 mM DTT a second time to quench the alkylation.
10. Incubate at 37.5 °C for 30 min.
11. Tryptic digestion: check the pH of the samples and ensure it is around 8.0 (*see* **Note 12**).
12. Dissolve trypsin in commercial suspension buffer or 0.1 M HCl.
13. Add trypsin solution to the sample at a trypsin mass to sample protein mass ratio of 1:25.
14. Incubate at 37.5 °C for 18 h (overnight).
15. Quench the tryptic digestion by adding neat formic acid to a final concentration of 0.5% (v/v) or by heating the reaction mixture in a water bath at 90 °C for about 10 min (*see* **Note 14**).
16. Centrifuge the samples at 1000 × g for 2 min, then centrifuge at high speed for 10 min.
17. Transfer the supernatant to a new Eppendorf tube, then dry the sample using a SpeedVac vacuum concentrator.

3.8 C18 Desalting

Prior to HILIC enrichment, sample clean-up steps are essential because salts will affect the enrichment efficiency of glycopeptides. Some methods, such as dialysis or solid phase extraction, can remove salts and purify samples. Here, a TopTip C18 desalting procedure is described:

1. Wash a TopTip C18 column with 50 μL of buffer B, spin down at 1000 × g for 1 min, then repeat this step two more times.
2. Wash the TopTip C18 column with 50 μL of buffer A, spin down at 1000 × g for 1 min, then repeat this step two more times.

3. Load the sample into the TopTip C18 column, spin down at 0.5 Kg for 2 min, then repeat this step one more time.
4. Wash the TopTip C18 column with 50 μL of buffer A, spin down at 1000 × g for 1 min, then repeat this step two more times.
5. Elute the sample with 50 μL of buffer B, spin down at 1000 × g for 1 min, then repeat this step two more times.
6. Continuously elute the sample with 50 μL of buffer C, spin down at 1000 × g for 1 min, then repeat this step one more time. Collect and dry the eluents.

3.9 Glycopeptides Enrichment Using PolyHYDROXYETHYL A (HILIC)

Due to the low abundance of glycopeptides and interference of non-glycopeptides, it is critical to enrich glycopeptides efficiently and selectively prior to LC-MS/MS analysis. HILIC enrichment protocol is presented below:

1. Dissolve dried samples in 50 μL of loading buffer.
2. Wash a TopTip PolyHYDROXYETHYL A (HILIC) spin column with 100 μL of H_2O, spin down at 1000 × g for 1 min, then repeat this step two more times.
3. Wash the TopTip PolyHYDROXYETHYL A (HILIC) spin column with 100 μL of loading buffer, spin down at 1000 × g for 1 min, then repeat this step two more times.
4. Load the sample into the TopTip PolyHYDROXYETHYL A (HILIC) spin column, spin down at 1000 × g for 1 min, then repeat this step two more times.
5. Wash the TopTip PolyHYDROXYETHYL A (HILIC) spin column with 100 μL of loading buffer, spin down at 1000 × g for 1 min, then repeat this step two more times.
6. Elute the sample with 50 μL of elution buffer, spin down at 1000 × g for 1 min, then repeat this step two more times. Collect and dry the eluents.

3.10 LC-MS/MS Analysis

After HILIC enrichment, the glycopeptide sample is ready for LC-MS/MS analysis. The separation is performed on a nano Pep-Map C18 column with online purification using a C18 trap column.

3.11 LC Condition

1. Resuspend the sample in 2% ACN, 98% H_2O, and 0.1% FA.
2. The injection amount depends on the sample type. Generally, for cell line and tissue samples, 50 μg of proteins are injected; for human blood serum, 2 μg of proteins are injected.
3. Set the column oven temperature to 29.5 °C and stabilize the temperature before starting the run.
4. The flow rate of the nano pump is set to 0.35 μL/min.

5. The flow rate of the loading pump is set to 3 μL/min.
6. The multistage gradient for glycoproteomic analysis is as follows:

   0–10 min, 3% B; the position of the 10-port valve is set to the 1–2 position at 0 min, where the trap column is connected to the loading pump.

   10–65 min, 3–20% B; the position of the 10-port valve is switched to the 10-1 position at 10 min, where the trap column is connected to the PepMap C18 column and nano pump.

   65–90 min, 20–30% B.

   90–110 min, 30–50% B.

   110–111 min, 50–80% B.

   111–115 min, keep at 80% B.

   115–116 min, 80-3% B; the position of the 10-port valve is switched to the 1–2 position at 115 min.

   116–120 min, 3% B.
7. Wash the column between two runs using a gradient at 90% B for 15 min and then condition for 10 min using 3% B.

3.12 MS Condition

1. Orbitrap system MS instruments such as Q Exactive HF Hybrid or Fusion Lumos tribrid are suitable for parameters setup.
2. The nanoESI source voltage is set to 1.6 kV in positive mode.
3. The temperature of the transfer tube is set to 275 °C.
4. The data is acquired in data-dependent acquisition (DDA) mode.
5. Full MS spectra are set to a resolution of 120 K.
6. Scan range is set to 400–2000 *m/z*.
7. The AGC target value is $1e^6$.
8. Maximum injection time is 50 ms, and detector type is Orbitrap.
9. For the tandem MS, the top 20 most intense precursor ions are selected for higher energy collision dissociation (HCD).
10. Resolution is set to 15 K.
11. The AGC target value is $1e^5$.
12. Maximum injection time is 50 ms.
13. Detector type is Orbitrap.
14. Charge states 2–8 are included.
15. Isolation window is 2.0 *m/z*.

16. Dynamic exclusion is 15 s.
17. Collision energy is set to stepped NCEs 15, 30, 45.

3.13 Data Processing

Glycoproteomics data acquired from Orbitrap mass spectrometers are initially processed by Byonic™ software (Protein Metrics by Dotmatics). The parameters are set as follows:

1. Cleavage sites, RK; cleavage side, C-terminal; digestion specificity, fully specific; missed cleavages, 2; precursor mass tolerance ±10 ppm; fragmentation type, QTOF/HCD; fragment mass tolerance ±20 ppm; recalibration (lock mass), none.
2. Protein modifications are set to Carbamidomethyl (C) as a fixed modification; oxidation (M) and acetyl (protein N-term) as variable modifications.
3. Protein database and glycan modifications are chosen based on the sample type.
4. After running the Byonic™, the result is analyzed by Byonic Viewer.
5. Glycopeptide identifications are filtered at a two-dimensional false discovery rate (2D FDR) $< 1\%$, PEP 2D < 0.01, $|\text{LogProb}| > 1$, score > 300, glycan is not empty.
6. The relative quantification of glycopeptides is based on the ion intensities acquired in the full scan.
7. For the manual data processing, Xcalibur Qual Browser is utilized to generate the extracted ion chromatogram (EIC) for the first monoisotopic peak of glycopeptides with a mass tolerance of ±10 ppm and Boxcar smoothing of 7 points.
8. The peak area of EIC is recorded to represent the abundance of the corresponding glycopeptide (*see* **Note 15**).

Figure 3 displays representative chromatograms of identified glycopeptides in Narcolepsy Type I depleted blood serum samples, which were generated using HILIC-enriched and non-enriched protocols. Meanwhile, Fig. 4 illustrates a Venn diagram that compares the number of identified glycopeptides in Narcolepsy Type I depleted blood serum samples using HILIC-enriched and non-enriched protocols. The comparison shows that the total glycopeptide intensity of the HILIC-enriched sample was observed to be higher than that of the non-enriched sample. Hence, the HILIC-enriched protocol is more efficient for identifying glycopeptides in biological samples.

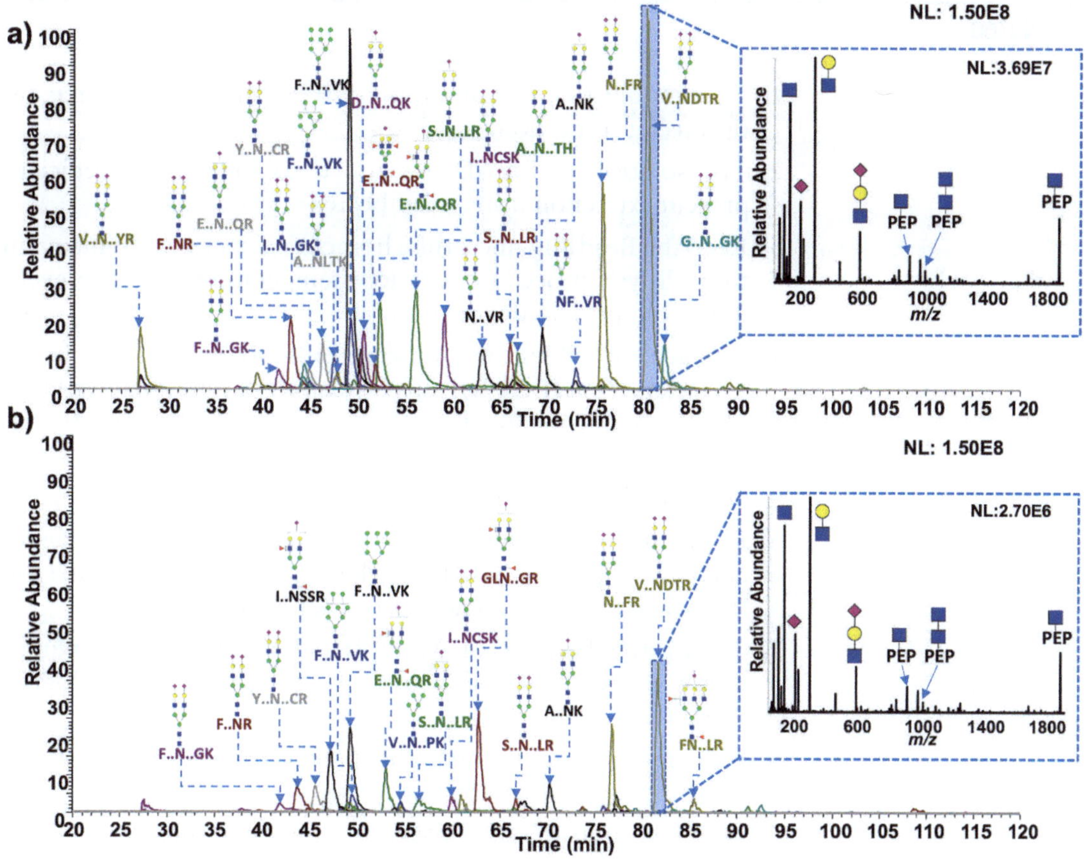

Fig. 3 Representative chromatograms of identified glycopeptides in narcolepsy type I blood serum samples using (**a**) HILIC-enriched and (**b**) non-enriched protocols generated from C18 (50 cm)-LC-MS/MS. Representative MS^2 identification is presented in inserted figures

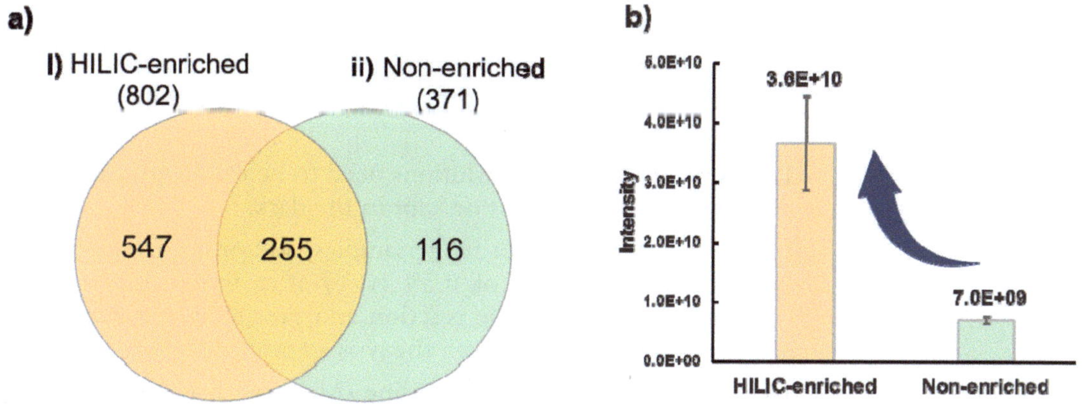

Fig. 4 Venn diagram comparing the number of identified glycopeptides in narcolepsy type I blood serum samples using (**a**) HILIC-enriched and non-enriched protocols. (**b**) The total glycopeptide intensity of the HILIC-enriched sample increased more than the non-enriched sample

4 Notes

1. Both mobile phases A and B need to be degassed after preparation using sonication.
2. The solution of 5% SDC is added to the cells for efficient protein extraction using bead beater.
3. Keep the BeadBug microtube homogenizer at 4 °C to maintain the cell lysis procedure at a low temperature and to prevent unwanted protein degradation.
4. The goal of cell lysis is to disrupt cells rapidly and completely. Therefore, repeat cell lysis steps at least five times followed by a 30 s pause to cool down, and be sure to have the clear supernatant at the end.
5. Sonication helps to improve protein dissolution.
6. Depleting the highly abundant proteins in glycoproteomics analysis is optional based on whether the analysis focuses on the whole glycoproteome or one that is low-abundant.
7. To use the Human 14 Multiple Affinity Removal Column, follow the manufacturer's instructions (Agilent Technologies, Santa Clara, CA, USA).
8. If the remaining solution in the filter cartridge exceeds 100 μL, centrifuge again for 10 min.
9. Normalization of the biological samples at this step is done based on the protein assay.
10. Normalization of the biological samples at this step is done based on the volume by adding the ABC buffer.
11. For cells and tissue samples, ten times diluting the samples with 50 mM ABC buffer is necessary before tryptic digestion to ameliorate the interference of the surfactant used in the lysis step.
12. The desired pH for tryptic digestion is around 8. If the pH is lower than 8, adjust the pH by adding a small amount of 100 mM ABC buffer.
13. Both DTT and IAA solutions need to be freshly prepared, and IAA solutions should be kept in the dark.
14. In the case of cells or tissue samples, add neat formic acid to a final concentration of 0.5% (v/v) after tryptic digestion to quench the enzymatic reaction and precipitate the remaining SDC in the samples from the lysis step.
15. Other software, such as Skyline (MacCoss Lab Software), can also be used to acquire the abundance of glycopeptides.

References

1. Hart GW, Copeland RJ (2010) Glycomics hits the big time. Cell 143(5):672–676. https://doi.org/10.1016/j.cell.2010.11.008
2. Tzeng SF, Tsai CH, Chao TK et al (2018) O-Glycosylation-mediated signaling circuit drives metastatic castration-resistant prostate cancer. FASEB J:fj201800687. https://doi.org/10.1096/fj.201800687
3. de Vreede G, Morrison HA, Houser AM et al (2018) A drosophila tumor suppressor gene prevents tonic TNF signaling through receptor N-glycosylation. Dev Cell 45(5):595–605.e594. https://doi.org/10.1016/j.devcel.2018.05.012
4. Singh C, Shyanti RK, Singh V et al (2018) Integrin expression and glycosylation patterns regulate cell-matrix adhesion and alter with breast cancer progression. Biochem Biophys Res Commun 499(2):374–380. https://doi.org/10.1016/j.bbrc.2018.03.169
5. Oyama M, Kariya Y, Kariya Y et al (2018) Biological role of site-specific O-glycosylation in cell adhesion activity and phosphorylation of osteopontin. Biochem J 475(9):1583–1595. https://doi.org/10.1042/bcj20170205
6. Solá RJ, Griebenow K (2009) Effects of glycosylation on the stability of protein pharmaceuticals. J Pharm Sci 98(4):1223–1245. https://doi.org/10.1002/jps.21504
7. Desko MM, Gross DA, Kohler JJ (2009) Effects of N-glycosylation on the activity and localization of GlcNAc-6-sulfotransferase 1. Glycobiology 19(10):1068–1077. https://doi.org/10.1093/glycob/cwp092
8. Sperandio M, Gleissner CA, Ley K (2009) Glycosylation in immune cell trafficking. Immunol Rev 230(1):97–113. https://doi.org/10.1111/j.1600-065X.2009.00795.x
9. Apweiler R, Hermjakob H, Sharon N (1999) On the frequency of protein glycosylation, as deduced from analysis of the SWISS-PROT database. Biochim Biophys Acta 1473(1):4–8. https://doi.org/10.1016/s0304-4165(99)00165-8
10. Shental-Bechor D, Levy Y (2008) Effect of glycosylation on protein folding: a close look at thermodynamic stabilization. Proc Natl Acad Sci U S A 105(24):8256–8261. https://doi.org/10.1073/pnas.0801340105
11. Kizuka Y, Kitazume S, Taniguchi N (2017) N-glycan and alzheimer's disease. Biochim Biophys Acta Gen Subj 1861(10):2447–2454. https://doi.org/10.1016/j.bbagen.2017.04.012
12. Van Scherpenzeel M, Willems E, Lefeber DJ (2016) Clinical diagnostics and therapy monitoring in the congenital disorders of glycosylation. Glycoconj J 33(3):345–358. https://doi.org/10.1007/s10719-015-9639-x
13. Magalhães A, Duarte HO, Reis CA (2021) The role of O-glycosylation in human disease. Mol Asp Med 79:100964. https://doi.org/10.1016/j.mam.2021.100964
14. Mondello S, Sandner V, Goli M et al (2022) Exploring serum glycome patterns after moderate to severe traumatic brain injury: a prospective pilot study. eClinicalMedicine 50:101494. https://doi.org/10.1016/j.eclinm.2022.101494
15. Mehta A, Herrera H, Block T (2015) Glycosylation and liver cancer. Adv Cancer Res 126:257–279. https://doi.org/10.1016/bs.acr.2014.11.005
16. Peng W, Goli M, Mirzaei P et al (2019) Revealing the biological attributes of N-glycan isomers in breast cancer brain metastasis using porous graphitic carbon (PGC) liquid chromatography-tandem mass spectrometry (LC-MS/MS). J Proteome Res 18(10):3731–3740. https://doi.org/10.1021/acs.jproteome.9b00429
17. Oliveira-Ferrer L, Legler K, Milde-Langosch K (2017) Role of protein glycosylation in cancer metastasis. Semin Cancer Biol 44:141–152. https://doi.org/10.1016/j.semcancer.2017.03.002
18. Yu A, Zhao J, Peng W et al (2018) Advances in mass spectrometry-based glycoproteomics. Electrophoresis 39(24):3104–3122. https://doi.org/10.1002/elps.201800272
19. Peng W, Gutierrez Reyes CD, Gautam S et al (2023) MS-based glycomics and glycoproteomics methods enabling isomeric characterization. Mass Spectrom Rev 42(2):577–616. https://doi.org/10.1002/mas.21713
20. Goli M, Yu A, Cho BG et al (2021) Chapter 8 – LC-MS/MS in glycomics and glycoproteomics analyses. In: El Rassi Z (ed) Carbohydrate analysis by modern liquid phase separation techniques, 2nd edn. Elsevier, Amsterdam, pp 391–441. https://doi.org/10.1016/B978-0-12-821447-3.00005-6
21. Banazadeh A, Veillon L, Wooding KM et al (2017) Recent advances in mass spectrometric analysis of glycoproteins. Electrophoresis 38(1):162–189. https://doi.org/10.1002/elps.201600357
22. Gutierrez-Reyes CD, Jiang P, Atashi M et al (2022) Advances in mass spectrometry-based

glycoproteomics: an update covering the period 2017-2021. Electrophoresis 43(1–2): 370–387. https://doi.org/10.1002/elps.202100188

23. Cummings RD, Pierce JM (2014) The challenge and promise of glycomics. Chem Biol 21(1):1–15. https://doi.org/10.1016/j.chembiol.2013.12.010
24. Xiao H, Sun F, Suttapitugsakul S et al (2019) Global and site-specific analysis of protein glycosylation in complex biological systems with mass spectrometry. Mass Spectrom Rev 38(4–5):356–379. https://doi.org/10.1002/mas.21586
25. Gutierrez Reyes CD, Jiang P, Donohoo K et al (2021) Glycomics and glycoproteomics: approaches to address isomeric separation of glycans and glycopeptides. J Sep Sci 44(1): 403–425. https://doi.org/10.1002/jssc.202000878
26. Liu L, Qin H, Ye M (2021) [Recent advances in glycopeptide enrichment and mass spectrometry data interpretation approaches for glycoproteomics analyses]. Se Pu 39(10): 1045–1054. https://doi.org/10.3724/sp.J.1123.2021.06011
27. Sun N, Wu H, Chen H et al (2019) Advances in hydrophilic nanomaterials for glycoproteomics. Chem Commun (Camb) 55(70): 10359–10375. https://doi.org/10.1039/c9cc04124a
28. Zhang H, Yi EC, Li XJ et al (2005) High throughput quantitative analysis of serum proteins using glycopeptide capture and liquid chromatography mass spectrometry. Mol Cell Proteomics 4(2):144–155. https://doi.org/10.1074/mcp.M400090-MCP200

Chapter 17

O-Glycoproteomics Sample Preparation and Analysis Using NanoHPLC and Tandem MS

Junyao Wang, Sherifdeen Onigbinde, Waziha Purba, Judith Nwaiwu, and Yehia Mechref

Abstract

Glycosylation refers to the biological processes that covalently attach carbohydrates to the peptide backbone after the synthesis of proteins. As one of the most common post-translational modifications (PTMs), glycosylation can greatly affect proteins' features and functions. Moreover, aberrant glycosylation has been linked to various diseases. There are two major types of glycosylation, known as *N*-linked and *O*-linked glycosylation. Here, we focus on *O*-linked glycosylation and thoroughly describe a bottom-up strategy to perform *O*-linked glycoproteomics studies. The experimental section involves enzymatic digestions using trypsin and *O*-glycoprotease at 37 °C. The prepared samples containing *O*-glycopeptides are analyzed using nanoHPLC coupled with tandem mass spectrometry (MS) for accurate identification and quantification.

Key words Glycosylation, *O*-Linked Glycopeptide, *O*-Glycoprotease, RPLC, Tandem MS

1 Introduction

There are two types of glycosylation. *N*-linked glycosylation attaches carbohydrate molecules to asparagine residues that are in a fixed peptide sequence [1]. Moreover, the carbohydrate attachments, namely *N*-glycans, share one consistent core structure [2]. On the other hand, *O*-linked glycosylation attaches *O*-glycans to serine or threonine residues in proteins [3]. Unlike *N*-glycosylation, the serine or threonine residues for *O*-glycosylation are not required to be in a specific peptide sequence. Furthermore, *O*-glycans do not share the same core structure [4]. All these factors make it more challenging to study *O*-glycosylation compared to *N*-glycosylation. Recently, *O*-glycoproteases have been introduced to tackle the above challenges [5]. One of these enzymes, known as immunomodulating metalloprotease (IMPa), immediately cleaves

Junyao Wang and Sherifdeen Onigbinde have equally contributed to this chapter.

Steven B. Bradfute (ed.), *Recombinant Glycoproteins: Methods and Protocols*, Methods in Molecular Biology, vol. 2762,
https://doi.org/10.1007/978-1-0716-3666-4_17,

the N-terminal to a serine or threonine residue attached by a mucin-type *O*-glycan [6]. The digestion efficiency of IMPa is not affected by the presence of sialic acid on the glycan moiety [5]. Therefore, this type of enzyme can be very helpful for locating *O*-glycosylation sites, as well as confirming the structure of the *O*-glycan attachment.

High-performance liquid chromatography coupled with mass spectrometry (HPLC-MS) has been recognized as a powerful tool for glycoproteomics studies [7, 8]. With online purification and various options of stationary phase, HPLC can achieve efficient separations of analytes in complex samples [9, 10], which greatly enhances selectivity and sensitivity during the subsequent MS detection. Advances in technology now allow mass spectrometry to provide a full MS scan with high resolution and mass accuracy [11]. In addition, more detailed information of analytes can be obtained by using suitable fragmentation techniques during the MS^n scans [12, 13].

Here, we demonstrate that *O*-glycosylation can be investigated via qualitative and quantitative analyses of *O*-glycopeptides. The entire process from sample preparation to data collection and processing is provided. Bovine fetuin and human serum were selected to represent standard glycoproteins and complex biological samples, respectively. Tryptic digestion was performed first to cleave all proteins, followed by IMPa digestion that specifically yielded *O*-glycopeptides. Then, samples were separated and analyzed using nanoHPLC-MS/MS with optimal instrument settings. The identification and quantification of *O*-glycopeptides were initially achieved using Byonic software [14] before manual validation.

2 Materials

All solutions should be prepared with HPLC-grade water and stored at room temperature unless otherwise stated.

2.1 Solutions for Enzymatic Digestion

1. Prepare 50 mM ammonium bicarbonate (ABC) buffer solution: Dissolve 197.64 mg ABC (molecular weight: 79.06) in 50 mL of water. Vortex mix and store at room temperature.
2. Prepare 200 mM Dithiothreitol (DTT) solution: Dissolve 3.085 mg DTT (molecular weight: 154.25) in 100 μL of 50 mL ABC buffer prepared in the first step. Vortex mix and store at room temperature in dark.
3. Prepare 200 mM Iodoacetamide (IAA) solution: Dissolve 7.400 mg IAA (molecular weight: 184.96) in 200 μL of 50 mL ABC buffer prepared in the first step. Vortex mix and store at room temperature in dark.

4. Obtain Mass Spec grade trypsin/Lys-C mix and resuspension buffer.
5. Prepare 20 mM Tris-HCl buffer solution: Weigh and transfer 53.28 mg Tris-HCl and 31.8 mg Tris-Base into a 50 mL centrifuge tube. Add 30 mL of water and vortex mix. Adjust the pH to 8 (*see* **Note 1**).
6. Obtain *O*-glycoprotease.

2.2 Mobile Phase for RPLC Separation

1. Mobile Phase A (MPA): 98% water, 2% acetonitrile (ACN), and 0.1% formic acid (FA).
2. Mobile Phase B (MPB): 100% ACN and 0.1% FA (*see* **Note 2**).

3 Methods

Carry out all steps at room temperature unless otherwise specified. See Fig. 1 for the flowchart.

3.1 Tryptic Digestion

1. Transfer samples (*see* **Note 3**) with 10 μg of proteins to a 1.5 mL sample tube and add 50 mM ammonium bicarbonate (ABC) buffer solution until the final volume is 50 μL.
2. Denature proteins by heating the sample in a water bath at 90 ° C for 15 min.
3. Add 1.25 μL (*see* **Note 4**) of 200 mM Dithiothreitol (DTT) to the 50 μL of the denatured protein sample. Vortex mix, spin down, and incubate in a water bath at 60 °C for 45 min to reduce the disulfide bonds.
4. Remove the sample from the 60 °C water bath, cool to room temperature, and spin down in a centrifuge.
5. Add 5 μL (*see* **Note 5**) of 200 mM Iodoacetamide (IAA) to the sample. Vortex mix, spin down, and incubate in a water bath at 37 °C for 45 min in the dark, for alkylation and to prevent the reformation of disulfide bonds.
6. Following the alkylation, add another 1.25 μL of 200 mM DTT solution and incubate in the 37 °C water bath for 30 min in the dark.
7. Remove the sample from the 37 °C water bath, cool to room temperature, and spin down in a centrifuge.
8. Take a trace amount of sample and test its pH value; ensure the pH is around 8 before adding trypsin (*see* **Note 6**).
9. Add 2 μg of trypsin (*see* **Note 7**) to the sample and incubate in the 37 °C water bath overnight (or 20 h).
10. After tryptic digestion, heat the sample in a 90 °C water bath for 10 min to quench the trypsin activity.

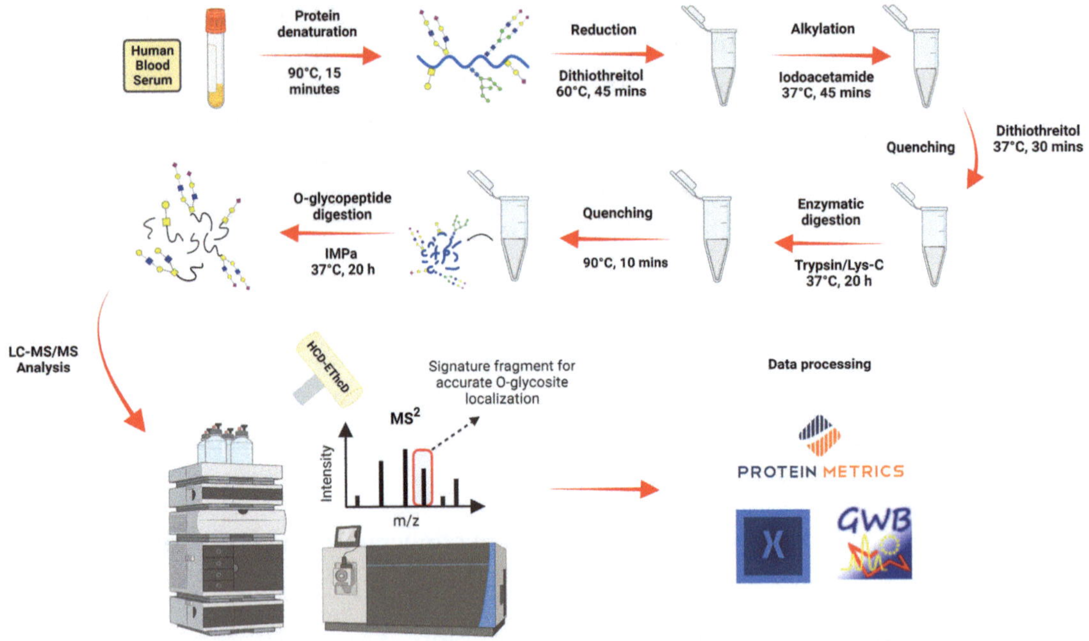

Fig. 1 Workflow of sample preparation and analysis for *O*-glycoproteomics

11. Remove the sample from the 90 °C water bath, cool to room temperature, and spin down.
12. Dry the sample in a spin vacuum dryer.

3.2 O-Glycoprotease Digestion

1. Add 20 mM Tris-HCl solution (pH = 8.0) to the dried sample and make the total reaction volume 50 μL.
2. Add an aliquot of 1 μL of NEB O-glycoprotease, gently mix the enzyme with the sample, and incubate in the 37 °C water bath overnight (or 20 h).
3. Remove the sample from the 37 °C water bath, cool to room temperature, and spin down in a centrifuge.
4. Dry the sample in the spin vacuum dryer, then reconstitute the dried sample in 20 μL of mobile phase A (98% water, 2% acetonitrile, 0.1% formic acid) used for LC separation. The final concentration will be 0.5 μg/μL, as the starting amount of protein was 10 μg.
5. Centrifuge the sample at 14.8 krpm for 10 min and transfer it to a properly labeled LC sample vial for LC-MS/MS analysis (*see* **Note 8**).

3.3 nanoLC Conditions

1. For each analysis, *O*-glycopeptides derived from 0.5 μg of digested proteins will be analyzed.

Table 1
Multistep gradient of nanoHPLC mobile phase for standard glycoprotein sample

	Retention [min]	Flow [μL/min]	%B	Curve
1	0.000	Run		
2	0.000	0.350	2.0	5
3	5.000	0.350	2.0	5
4	40.000	0.350	30.0	5
5	72.000	0.350	70.0	5
6	73.000	0.350	90.0	5
7	80.000	0.350	90.0	5
8	81.000	0.350	2.0	5
9	90.000	0.350	2.0	5
10	90.000	Stop run		

2. During the loading process, the sample is first passed through an Acclaim PepMap 100 C18 trap (75 μm × 2 cm, 3 μm particle size) for online purification.
3. *O*-glycopeptides are separated on a reversed-phase C18 Acclaim PepMap 100 Å capillary column (150 mM × 75 μm id). The column compartment temperature is set at 40 °C and the flow rate at 0.350 μL/min.
4. A multistep gradient with total elution time of 90 min (Table 1) is applied for the standard glycoprotein sample: starting at 2% MPB for 5 min, gradually increasing to 30% over 35 min, and to 70% over 32 min. Then, ramp up to 90% of MPB in one min and keep constant for 7 min to wash the system. Finally, decrease to 2% B in one minute and keep it constant for 9 min to equilibrate the column.
5. A multistep gradient with total elution time of 127 min (Table 2) is applied for human serum sample: starting at 3% MPB for 10 min, gradually increasing to 11% over 28 min, and to 60% over 70 min. Then, ramp up to 90% of MPB in 4 min and keep constant for another 4 min to wash the system. Finally, decrease to 3% B in 1 min and keep it constant for 10 min to equilibrate the column (*see* **Note 9**).

3.4 MS Conditions

1. After the nanoLC separation, the *O*-glycopeptides are analyzed by mass spectrometer via a nano ESI source in positive ion mode.
2. The spray voltage is 2 kV, and the transfer tube temperature is 275 °C.

Table 2
Multistep gradient of nanoHPLC mobile phase for complex sample (human serum)

	Retention [min]	Flow [μL/min]	%B	Curve
1	0.000	Run		
2	0.000	0.350	3.0	5
3	10.000	0.350	3.0	5
4	38.000	0.350	11.0	5
5	108.000	0.350	60.0	5
6	112.000	0.350	90.0	5
7	116.000	0.350	90.0	5
8	117.000	0.350	3.0	5
9	127.000	0.350	3.0	5
10	127.000	Stop run		

3. The full MS spectra is acquired by an orbitrap mass analyzer with a mass range of 500–1800 *m/z*. The resolving power is 120,000 and the mass accuracy is 5 ppm.
4. The RF lens is set at 60% and the maximum injection time is 50 ms.
5. The dynamic exclusion parameters are as follows: repeat count 1; exclusion duration of 60 s; mass tolerance of 10 ppm; and an intensity threshold of 5.0e4.
6. The MS/MS orbitrap scan is generated in a data-dependent manner. For this purpose, the duty cycle is 3 s, and 20 of the most intense ions from the full MS scan are selected for an HCD MS/MS scan with a normalized collision energy (NCE) of 35% and a 10 ms activation time.
7. The isolation mode is Quadrupole with an isolation window of 2 *m/z*. The resolution of the mass analyzer is 30,000, with a fixed scan range of 120–4000 *m/z*, and a maximum injection time of 60 ms and 3 dependent scans.
8. A second MS/MS scan is generated using an EThcD dissociation. The Quadrupole is utilized for ion isolation with a window of 1.6 *m/z*. The activation type is ETD with a 50 ms reaction time, ETD reagent target of 2.0e5, and a maximum ETD reagent injection time of 200 ms.
9. The ETD activation type is coupled with HCD as supplemental activation. The HCD collision energy is 25%. The generated fragments will be analyzed in the Orbitrap with a resolution of 30,000, a fixed scan range of 120–4000 *m/z*, and a maximum injection time of 200 ms.

3.5 Data Processing

1. The raw data files are first processed in Byonic software (version 4.1.10, Protein Metrics, Inc.) against a correlated proteome and glycome database for *O*-glycopeptide identification.
2. Mass tolerance is set at 10 ppm for precursors and 20 ppm for fragment ions. Carbamidomethyl (C) is a fixed modification, while oxidation (M), acetyl (protein N-terminal), and deamidation (N) are variable modifications.
3. In the digestion parameters, cleavage sites RK and ST are specified with cleavage sides at the C-terminal and N-terminal for trypsin and *O*-glycoprotease, respectively.
4. Software identification is followed by a manual check of the monoisotopic mass of precursor ions using Xcalibur software (Thermo Scientific). The tandem mass spectra are also validated using Glycoworkbench [15] (see examples in Figs. 2 and 3).

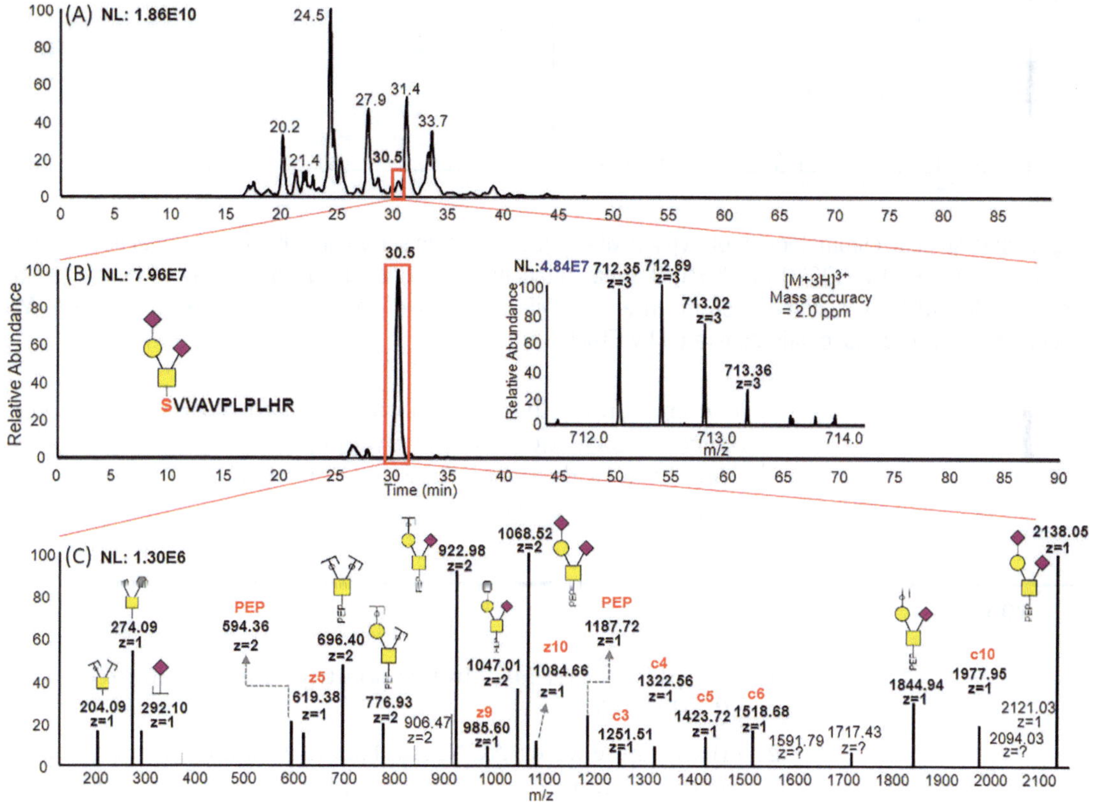

Fig. 2 (**a**) Base peak chromatogram of trypsin/IMPa digested bovine fetuin (FETUA_BOVIN), (**b**) Extracted ion chromatogram (EIC) of *O*-glycopeptide (Peptide backbone: SVVAVPLPLHR; *O*-glycan attachment: $GalNAc_1Gal_1Neu5Ac_2$) derived from bovine fetuin, inset: full MS spectrum of the *O*-glycopeptide, (**c**) Tandem MS spectrum of the *O*-glycopeptide generated by EThcD

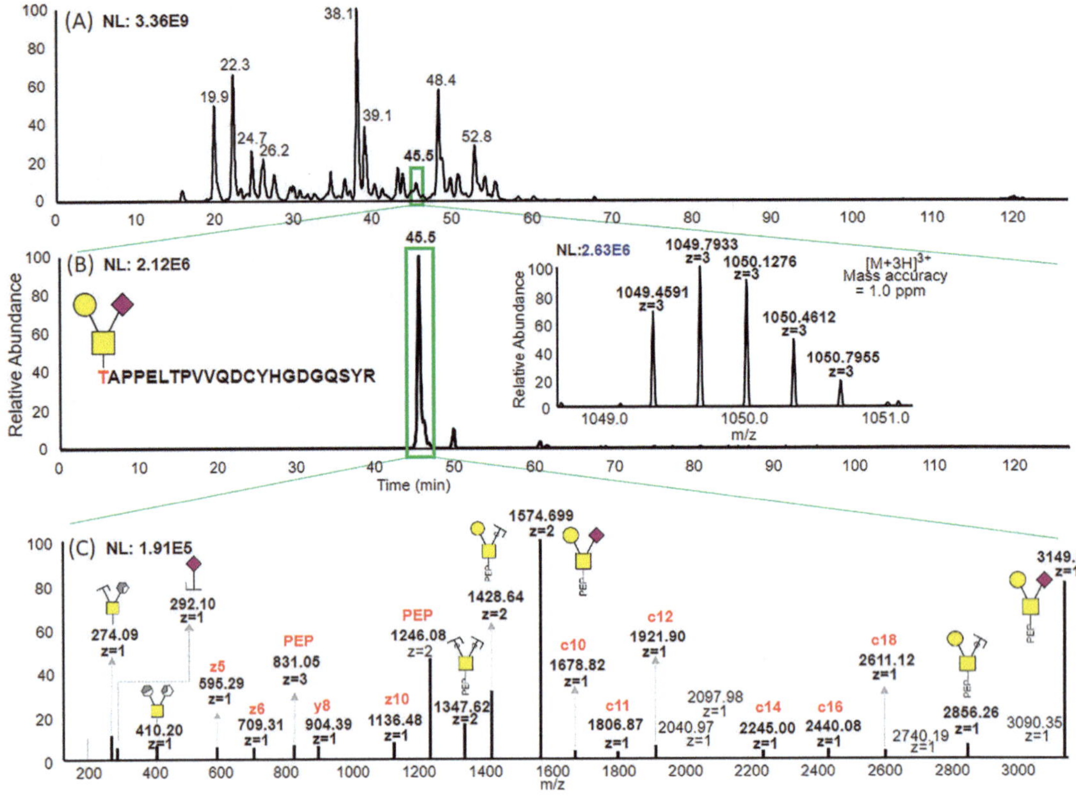

Fig. 3 (**a**) Base peak chromatogram of trypsin/IMPa digested human serum, (**b**) EIC of *O*-glycopeptide (Peptide backbone: TAPPELTPVVQDCYHGDGQSYR; *O*-glycan attachment: $GalNAc_1Gal_1Neu5Ac_1$) derived from plasminogen (PLMN_HUMAN) in human serum, inset: full MS spectrum of the *O*-glycopeptide, (**c**) Tandem MS spectrum of the *O*-glycopeptide generated by EThcD

5. The area under the peaks is calculated to represent the abundance of each identified *O*-glycopeptide. Relative quantification of the *O*-glycopeptides is accomplished by normalizing the area by total abundance.

4 Notes

1. The pH of Tris-HCl buffer solution can be adjusted using 50 mM HCl.
2. HPLC grade water, ACN, and FA must be used. Depending on the total volume of either mobile phase A or B (e.g., 500 mL or 1000 mL), the volume of each composition should be adjusted accordingly. Also, both mobile phase A and B must be sonicated to remove gas after mixing. Store mobile phase at room temperature.

3. Samples can be in either solid phase or liquid phase. Solid samples, such as some commercial glycoprotein standards, should be completely dissolved in an aliquot of 50 mL of 50 mM ABC buffer solution. Liquid samples should be diluted using 50 mM ABC buffer solution until the final volume reaches 50 mL.
4. DTT solution needs to be freshly prepared and stored in the dark before use. The volume of the DTT solution added into the sample should be 1/40 of the total sample volume. Therefore, in this protocol, an aliquot of 1.25 μL 200 mM DTT is added to 50 μL of protein sample.
5. Similar to the DTT solution, the IAA solution must also be freshly prepared and put in the dark before use due to its high light-sensitivity. The volume of the IAA should be four times the volume of the previously added DTT solution. Therefore, an aliquot of 5 μL is used.
6. This step is to make sure that the sample has a mild alkaline condition for trypsin activity. It can be performed by taking a trace amount (< 1 μL) from the sample using a pipette and applying it to a pH paper strip.
7. Before use, the trypsin enzyme usually needs to be resuspended using the resuspension buffer that comes with the enzyme. The amount of trypsin required for digestion should be 1/25 of the weight of the proteins in the sample. Therefore, an aliquot of 0.4 μg is used to cleave 10 μg of proteins. Depending on the concentration after resuspension, a correlated volume of trypsin solution should be added. For example, an aliquot of 0.4 μL of trypsin solution will be added if the concentration equals 1 μg/μL.
8. Centrifuging the sample before loading it onto the LC system is critical to remove any undissolved particles.
9. Complex biological samples, such as human serum, contain more types of glycoproteins than standards; therefore, a shallower and longer elution gradient has been developed for better separation performance.

Acknowledgments

This work is supported by NIH (1R01GM130091-04), Robert A. Welch Foundation (No. D-0005), and The CH Foundation.

References

1. Mellquist J, Kasturi L, Spitalnik S et al (1998) The amino acid following an asn-X-Ser/Thr sequon is an important determinant of N-linked core glycosylation efficiency. Biochemistry 37(19):6833–6837
2. Aebi M, Bernasconi R, Clerc S et al (2010) N-glycan structures: recognition and processing in the ER. Trends Biochem Sci 35(2): 74–82
3. Steen PV, Rudd PM, Dwek RA et al (1998) Concepts and principles of O-linked glycosylation. Crit Rev Biochem Mol Biol 33(3): 151–208
4. Wopereis S, Lefeber DJ, Morava E et al (2006) Mechanisms in protein O-glycan biosynthesis and clinical and molecular aspects of protein O-glycan biosynthesis defects: a review. Clin Chem 52(4):574–600
5. Vainauskas S, Guntz H, McLeod E et al (2021) A broad-specificity O-glycoprotease that enables improved analysis of glycoproteins and glycopeptides containing intact complex O-glycans. Anal Chem 94(2):1060–1069
6. Bergstrom KS, Xia L (2013) Mucin-type O-glycans and their roles in intestinal homeostasis. Glycobiology 23(9):1026–1037
7. Zhu R, Huang Y, Zhao J et al (2020) Isomeric separation of N-glycopeptides derived from glycoproteins by porous graphitic carbon (PGC) LC-MS/MS. Anal Chem 92(14): 9556–9565
8. Gutierrez-Reyes CD, Jiang P, Atashi M et al (2022) Advances in mass spectrometry-based glycoproteomics: an update covering the period 2017–2021. Electrophoresis 43(1–2): 370–387
9. Huang Y, Nie Y, Boyes B et al (2016) Resolving isomeric glycopeptide glycoforms with hydrophilic interaction chromatography (HILIC). J Biomol Tech 27(3):98
10. Ji ES, Lee HK, Park GW et al (2019) Isomer separation of sialylated O-and N-linked glycopeptides using reversed-phase LC–MS/MS at high temperature. J Chromatogr B 1110:101–107
11. Ding Z, Wang N, Ji N et al (2022) Proteomics technologies for cancer liquid biopsies. Mol Cancer 21(1):1–11
12. Park GW, Lee JW, Lee HK et al (2020) Classification of mucin-type O-glycopeptides using higher-energy collisional dissociation in mass spectrometry. Anal Chem 92(14):9772–9781
13. Riley NM, Malaker SA, Driessen MD et al (2020) Optimal dissociation methods differ for N-and O-glycopeptides. J Proteome Res 19(8):3286–3301
14. Bern M, Kil YJ, Becker C (2012) Byonic: advanced peptide and protein identification software. Curr Protoc Bioinformatics 40(1): 13.20. 11–13.20. 14
15. Damerell D, Ceroni A, Maass K et al (2012) The GlycanBuilder and GlycoWorkbench glycoinformatics tools: updates and new developments. Biol Chem 393(11):1357–1362

Part V

Considerations and Alternatives for Glycoprotein Production and Purification

Chapter 18

Solubilization of Oligomeric Cell-Free Synthesized Proteins Using SMA Copolymers

Jessica Ullrich, Lisa Haueis, Carsten Ohlhoff, Anne Zemella, Stefan Kubick, and Marlitt Stech

Abstract

Although membrane proteins are abundant in nature, their investigation is limited due to bottlenecks in heterologous overexpression and consequently restricted accessibility for downstream applications. In this chapter, we address these challenges by presenting a fast and straightforward synthesis platform based on eukaryotic cell-free protein synthesis (CFPS) and an efficient solubilization strategy using styrene-maleic acid (SMA) copolymers. We demonstrate CFPS of TWIK-1, a dimeric ion channel, based on *Sf*21 (*Spodoptera frugiperda*) insect lysate showing homooligomerization and N-glycosylation enabled by endoplasmic reticulum-derived microsomes. Furthermore, we employ SMA copolymers for protein solubilization, which preserves the native-like microsomal environment. This approach not only retains the solubilized protein's suitability for downstream applications but also maintains the oligomerization and glycosylation of TWIK-1 post-solubilization. We validate the solubilization procedure using autoradiography, particle size analysis, and biomolecular fluorescence assay and confirm the very efficient, structurally intact solubilization of cell-free synthesized TWIK-1.

Key words Cell-free protein synthesis (CFPS), K_{2P} potassium channels, TWIK-1, Oligomerization, SMA copolymers, SMALPs, Solubilization of membrane proteins, Biomolecular fluorescence complementation assay

1 Introduction

Oligomeric membrane proteins (MPs) are abundant in nature. They play crucial roles in pathophysiological conditions but are difficult to analyze by downstream applications due to challenges in heterologous protein expression and purification. MPs are commonly overexpressed using cell-based approaches; however, low protein yields, altered functions, and denaturing solubilization

The original version of this chapter was previously published without open access. A correction to this chapter is available at https://doi.org/10.1007/978-1-0716-3666-4_22

Jessica Ullrich and Lisa Haueis have equally contributed to this chapter.

Steven B. Bradfute (ed.), *Recombinant Glycoproteins: Methods and Protocols*, Methods in Molecular Biology, vol. 2762, https://doi.org/10.1007/978-1-0716-3666-4_18, © The Author(s) 2024, Corrected Publication 2025

procedures significantly limit their study. To extract MPs, conventional detergents mimic the phospholipid bilayer environment which often results in degradation, conformational changes, and loss of function [1, 2]. Although lipid nanodiscs containing phospholipids and membrane scaffold proteins (MSPs) provide a more accurate technique, their synthetic lipid composition and instability may impact functionality and require individual validation for every protein. An emerging alternative approach uses polymer-based nanodiscs. Among this group are styrene-maleic acid (SMA) copolymers, which enable the direct solubilization of MPs from their native membrane environment and are compatible with several structure-determination studies [3, 4]. SMA copolymers interact with the hydrophobic core of the phospholipid bilayer and the hydrophilic surrounding, resulting in the formation of water-soluble, disc-shaped lipid nanostructures that embed MPs with the same physicochemical properties as the native membrane [5]. By varying the styrene-to-maleic acid ratio, polymers with different hydrophobicity can be produced, making this approach feasible for maintaining the structural conformation and thus oligomerization of MPs [1, 6]. This technique has been reported for functional solubilization of several MP classes including transporters, ion channels, G-protein coupled receptors (GPCRs), and protein complexes from bacterial, insect, yeast, plant, and mammalian cells [7–10].

Due to their low protein yields and cytotoxicity, heterologous overexpression of MPs is still limiting their investigation. Cell-free protein synthesis (CFPS) based on eukaryotic translationally active lysates has become a promising alternative platform [11–13]. In particular, lysates based on insect cells (*Spodoptera frugiperda, Sf*21) have demonstrated successful protein synthesis of various functional MPs, including transporters, ion channels, GPCRs, and protein complexes due to the presence of endogenous, endoplasmic reticulum-derived microsomes [14–17]. These membranous structures play a crucial role in facilitating MP translocation and post-translational modifications, such as disulfide bridging and core glycosylation [18].

This chapter outlines the practicability for the combination of eukaryotic cell-free synthesis and the solubilization of assembled MPs in a native-like environment using SMA copolymers. The dimeric ion channel TWIK-1 was selected as a model protein due to its pharmacological relevance in brain diseases and successful synthesis with post-translation modifications of TWIK-1 in cell-free *Sf*21 lysate. Furthermore, we present a highly efficient method for solubilizing glycosylated and dimeric protein based on SMA copolymers qualified by autoradiography, particle size analysis, and biomolecular fragment complementation assay. In summary, this chapter demonstrates the feasibility of the cell-free synthesis of an ion channel and its conformationally stable solubilization.

2 Materials

All components should be prepared with ultrapure water on ice and stored at −20 °C if not stated otherwise.

2.1 Cell-Free Synthesis Based on *Sf21* Lysate

1. Ice bucket.
2. Reaction tubes (1.5 mL).
3. TT-mixture (10×): 300 mM Hepes-KOH (pH 7.6), 1 mM of each of the 20 canonical amino acids, 2.5 mM spermidine, 39 mM $Mg(OAc)_2$, and 1350 mM KOAc. Stored at −80 °C.
4. Energy components (5×): 100 mM creatine phosphate, 1.5 mM GTP, 1.5 mM CTP, 1.5 mM UTP, 8.75 mM ATP, and 0.5 mM $m^7G(ppp)G$ cap analog. Stored at −80 °C.
5. *Sf*21 lysate. Stored at −80 °C (*see* **Note 1**).
6. ^{14}C-leucine (200 dpm/pmoL, 1000 μM, f.c. 30–50 μM).
7. T7-RNA polymerase (final concentration (f.c.): 1 U/μL).
8. Polyguanylic acid (polyG, f.c.: 15 μM).
9. Protein-encoding plasmid containing the gene of interest (*see* **Note 2**).
10. Thermomixer with lid.
11. Cooling centrifuge (4 °C) with rotor suitable for reaction tubes.
12. Phosphate-buffered saline (PBS) without divalent cations.

2.1.1 Quantitative Protein Analysis

1. Ice bucket.
2. Glas tubes (10 mL).
3. Trichloroacetic acid (TCA, f.c.: 5%, 10%(v/v) supplemented with casein hydrolysate (f.c.: 2%(w/v)).
4. Water bath with heating function up to 80 °C.
5. Vacuum filtration system.
6. Glass fiber filters.
7. Acetone.
8. Scintillation vials.
9. Scintillation cocktail.
10. Orbital shaker.
11. Scintillation counter.

2.1.2 Qualitative Protein Analysis

1. Sodium dodecyl sulfate polyacrylamide gel electrophoresis (SDS-PAGE) running buffer: 1× lithium dodecyl sulfate sample (LDS)-buffer.

2. For deglycosylation assay: glycoprotein denaturing buffer (f.c. 0.5% SDS, 40 mM DTT), deglycosylation buffer (f.c. 50 mM sodium phosphate), suitable endoglycosidase (PNGase F), thermomixer.
3. SDS-PAGE gels (4–12% Tris-glycine) with running buffer.
4. Gel chamber system.
5. Protein ladder.
6. Coomassie Brilliant Blue stain.
7. Vacuum gel dryer.
8. Radioactively labeled ink.
9. Phosphor screen.
10. Phosphor imager e.g. Amersham Typhoon RGB Imager.

2.2 Solubilization of Cell-Free Expressed Membrane Proteins

1. Styrene-maleic acid copolymers: SMA 1: XIRAN SL25010P20, SMA 2: XIRAN SL30010P20, SMA3: XIRAN SL4005P20.
2. Thermomixer with lid.
3. Amicon® Ultra-0.5 centrifugal filters (MwCO: 10 kDa, 0.5 mL).
4. Cooling centrifuge (4 °C) with rotor suitable for reaction tubes.

2.3 Particle Size Measurement

1. Cuvette suitable for analyzed volume.
2. Particle size analyzer, e.g., Malvern Zetasizer Nano ZS.
3. Software for data analysis.

2.4 Complementation Assay

1. Fusion constructs (*see* **Notes 2** and **10**).
2. Confocal laser scanning microscope with immersion oil, e.g., LSM 510, Carl Zeiss Microscopy GmbH.
3. μ-IBIDI-Slide (18 well, flat).
4. Software for image processing.

3 Methods

3.1 Cell-Free Protein Synthesis

1. Design and prepare the protein-encoding plasmid as described (*see* **Note 2**).
2. Thaw all required components on ice and gently vortex them before use.
3. Execute CFPS in a coupled batch mode. Therefore, transcription and translation are performed in one reaction tube. For synthesis, add 40% *Sf*21 lysate (v/v), polyG (f.c.: 15 μM), TT-mixture (f.c.: 1×) and T7-RNA polymerase (f.c.: 1 U/μL) in a master mixture. Ensure thorough mixing, while avoiding bubble formation (*see* **Note 1**). Supplement protein-encoding

plasmid (125 ng/μL) and energy components (f.c.: 1×) and fill up the volume with double-distilled water.

4. If quantification is required, add ^{14}C-leucine (f.c.: 30 μM), which will be statistically incorporated into the target protein and allow analysis of the *de novo* synthesized protein. Reaction mixtures ranging from 15 to 150 μL can be used.
5. Place the translation mixture (TM) in a thermomixer and shake gently at 600 rpm for 3 h at 27 °C (*see* **Note 3**).
6. Centrifuge the TM at 4 °C and 16,000 × g for 10 min to separate the soluble (SN1) from the vesicular fraction (VF1). For schematic representation see Fig. 2a (*see* **Note 4**).
7. Transfer the SN1 to a fresh reaction tube and resuspend the pellet, the vesicular fraction (VF1), in PBS at a volume equivalent to that of the transferred supernatant (*see* **Note 5**).

3.1.1 Quantitative Analysis

1. After synthesis, sample 3 × 3 μL of every fraction and pipette it in an appropriate glass tube.
2. Add 3 mL trichloro acetic acid (TCA) (f.c.: 10%) supplemented with casein hydrolysate (f.c.: 2%) and incubate for 15 min at 80 °C. Chill the samples on ice for at least 30 min or overnight at 4 °C.
3. Apply the sample to a vacuum filtration system (retention: 1.6 μm) to remove the non-incorporated ^{14}C- leucine. Wash the filters twice with 5% TCA and dry with acetone.
4. Transfer the dried filters into scintillation vials and supplement with 3 mL scintillation cocktail. Shake the samples for 60 min at 300 rpm before measuring in a scintillation counter.
5. The resulting counts per minute are converted into micrograms per milliliter based on the number of leucines, the molecular weight, and the specific radioactivity:

$$\text{Specific radioactivity}\left[\frac{\text{dpm}}{\text{pmoL}}\right] = \frac{\text{stock concentration of C} - \text{leucine}\,[\mu\text{M}] * A_{\text{spec}\left[\frac{\text{dpm}}{\text{pmoL}}\right]}\,14}{\text{total concentration of leucine}\,[\mu\text{M}]}$$

$$\text{Protein Yield}\left[\frac{\mu\text{g}}{\text{mL}}\right] = \frac{\text{scintillation counts}\left[\frac{\text{dpm}}{\text{mL}}\right] * \text{molecular weight}\left[\frac{\mu\text{g}}{\text{pmoL}}\right]}{\text{specific radioactivity}\left[\frac{\text{dpm}}{\text{pmoL}}\right] * \text{number of leucines}}$$

3.1.2 Qualitative Analysis and Deglycosylation Assay

1. Collect 5 μL of the desired sample and incubate with non-reducing sample buffer of the same volume (2×) to maintain the structural conformation including disulfide bridges.
2. To perform a deglycosylation assay, sample 5 μL of VF1 and add glycoprotein denaturing buffer (f.c.: 1×). Subsequently, add reaction buffer (f.c.: 1×) along with PNGase F (f.c.: 50,000 U/mL) and incubate for 60 min at 37 °C. PNGase F is an enzyme that cleaves N-linked glycans from the core protein, resulting in a molecular weight shift if proteins were successfully glycosylated.
3. Load the samples on a 4–12% SDS-PAGE gel. Use a ladder that covers the desired molecular weight range. Let the gel run for 55 min, 160 V at room temperature (*see* **Note 6**).
4. Place the gel in water bath and stain with Coomassie Brilliant Blue for 20 min to confirm that the proteins have entered the gel. Subsequently, wash the gel twice with water to remove excess buffer. Short heating up in a microwave may improve the washing process.
5. Dry the gel on cardboard using a vacuum system for 70 min at 70 °C. Label the marker bands with radioactive ink and expose the dried gel to a storage phosphor screen. After at least two days of incubation, use a phosphor imager to read out the results.
6. Figure 1a illustrates an exemplary autoradiogram of TWIK-1 synthesized in a cell-free system and fractionated in TM, SN1,

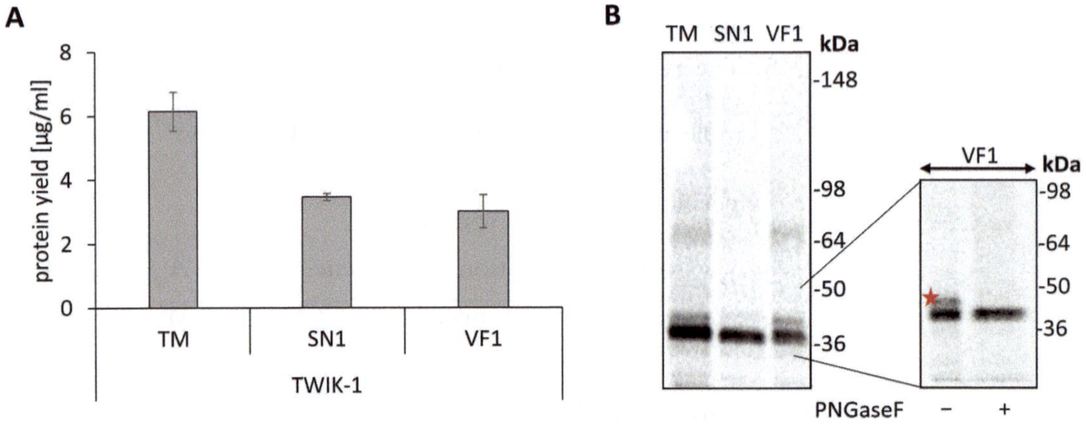

Fig. 1 Cell-free protein synthesis of the membrane protein TWIK-1: TWIK-1 is synthesized in eukaryotic *Sf*21 lysate and labeled with ^{14}C leucine. The translation mixture (TM) is fractionated after synthesis by centrifugation in the supernatant (SN1) and the vesicular fraction (VF1). (**a**) Quantitative analysis determined by TCA-precipitation and liquid scintillation counting. Standard deviations are calculated from triplicate analysis. (**b**) Qualitative analysis of synthesized TWIK-1 by SDS-PAGE (4–12% Tris-Glycine) and autoradiography. To validate the successful glycosylation of TWIK-1, VF1 is subjected to denaturing PNGase F digestion. The glycosylated band is marked with an asterisk

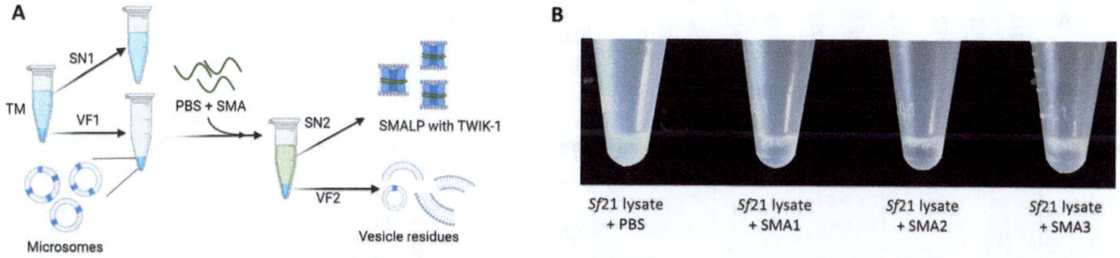

Fig. 2 Fractionation of cell-free translation mixture and solubilization of eukaryotic *Sf*21 lysate: (**a**) Scheme of the fractionation procedure of the cell-free translation mixture (TM) resulting in the supernatant (SN1) and the vesicular fraction (VF1). VF1 contains the synthesized membrane proteins, which are solubilized by incubation with different styrene-maleic acid (SMA) copolymers. VF1 is further fractionated by centrifugation into supernatant 2 (SN2) and vesicular fraction 2 (VF2). (**b**) The solubilization of the endoplasmic reticulum-derived microsomes is evaluated using different SMA copolymers

and VF1 (schematic illustration, *see* Fig. 2a). All fractions show a clear single monomeric band (~39 kDa), while TM and VF1 exhibit a monomeric double band and a high molecular weight band at 78 kDa (Fig. 1a). Previous findings indicate that microsomal structures within the vesicular fraction facilitate protein maturation, including post-translational modifications (PTMs) such as disulfide bridging and core glycosylation. Regarding protein maturation, one N-glycosylation site and homodimeric assembly by disulfide bridging has been previously described for TWIK-1 [19]. Accordingly, PNGase F treatment reveals a molecular weight shift of the monomeric double band to a single band upon cleavage of the glycan and thus proves successful N-glycosylation of the cell-free synthesized protein (Fig. 1b). The high band of the potential high molecular weight dimer disappears, if denaturing conditions (glycoprotein denaturing buffer) are applied, suggesting that the band represents a disulfide-bridged dimer.

3.2 Solubilization of Vesicular Fraction Using SMA Copolymers

1. SMA copolymers are utilized to solubilize the cell-free synthesized proteins. The hydrophobicity of SMA copolymers varies based on the styrene to maleic acid ratio, resulting in different solubilization efficiencies. Therefore, it is recommended to validate SMA copolymers with three different styrene-to-maleic ratios: 3:1 (SMA1), 2.3:1 (SMA2), and 1.2:1 (SMA3). Due to the high viscosity of the SMA solutions, it is recommended to prepare a pre-solution of SMA copolymers by diluting them with PBS to concentrations of 10.5% (SMA1), 11.2% (SMA2), and 11.4% (SMA3). As the SMAs are light-sensitive, it is important to protect all samples from light [20].
2. Split VF1 into three different preparations of equal volume and add the SMA copolymers to a final concentration of 1%. Incubate the mixtures for 2 h at 22 °C with gentle shaking at

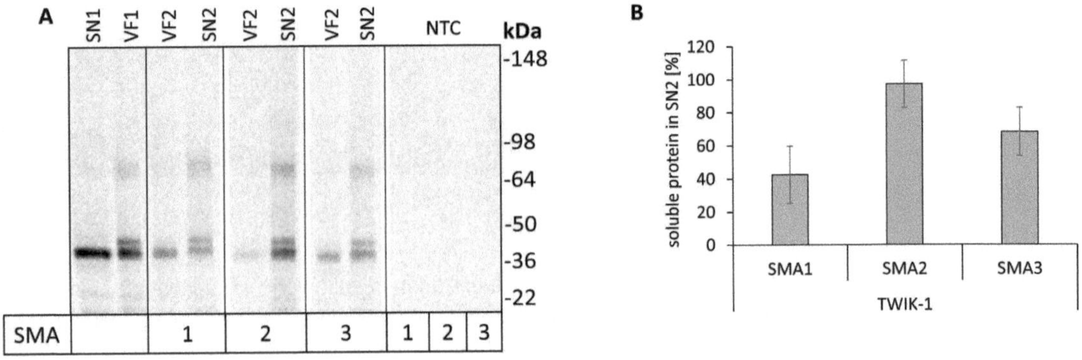

Fig. 3 Solubilization of TWIK-1 by three styrene-maleic acid copolymers: TWIK-1 is synthesized in an insect-based *Sf*21 lysate and ^{14}C-labeled. After synthesis, the translation mixture (TM) is fractionated by centrifugation into the supernatant (SN1) and the vesicular fraction (VF1). Solubilization is performed using various SMA copolymers, followed by fractionation into SN2 containing the solubilized target protein, as well as VF2 containing remaining TWIK-1. (**a**) Autoradiograph depicting the qualitative analysis of the fractions by SDS-PAGE (4–12% Tris-Glycine). As a negative control, a no template control supplemented with the SMAs was used. (**b**) Quantitative analysis of the soluble protein based on trichloro acetic acid precipitation and liquid scintillation counting. Percentage values are calculated based on the ratio of TWIK-1 in VF1 (100%) to TWIK-1 in SN2. Standard deviations are calculated from triplicate analysis

500 rpm (*see* Fig. 2a for schematic illustration of the fractionation) (*see* **Note 7**).

3. The exemplary change of the lysate turbidity resulting from the breakdown of the membrane structures (*see* **Note 8**), is pictured in Fig. 2b. After 2 h of SMA treatment, the eukaryotic cell-free lysate becomes clear for all tested SMAs. This indicates the successful formation of SMA lipid nanoparticles (SMALPs).
4. Centrifuge the samples at 4 °C, 16,000 × g for 10 min to obtain supernatant 2 (SN2) containing the SMALPs with the solubilized membrane protein.
5. Resuspend the remaining pellet in the same volume as SN2 in PBS to obtain VF2.
6. Sample 3 × 3 μL of each fraction (TM, SN1, VF1, SN2, VF2) for TCA analysis and 1 × 4 μL of each fraction for SDS-PAGE and autoradiography.

3.2.1 Ultrafiltration

1. Concentrate the target protein by filling up the SN2 with PBS to 500 μL and transfer the solution to Amicon® Ultra-0.5 centrifugal filters. Centrifuge for 15 min at 14,000 × g at 4 ° C to remove free SMA copolymers.
2. Turn the filter device and collect the retentate with the SMALP in a fresh reaction tube by centrifugation for 5 min at 3000 × g.
3. Figure 3a shows an exemplary autoradiogram of solubilized fractions obtained after cell-free synthesis. Consistent with

our previous findings, the SN1 (~39 kDa) exhibits a single band (corresponding to unglycosylated TWIK-1), while the VF1 shows a monomeric double band (corresponding to unglycosylated and glycosylated TWIK-1) and an additional band with higher molecular weight (~78 kDa), corresponding to TWIK-1 dimer. To solubilize the VF1, three different SMA copolymers with different styrene-to-maleic acid ratios are used. The SN2 exhibits double bands, previously identified as N-glycosylation and a high molecular dimeric band, whereas only a weak monomeric single band is detectable in the VF2. Thus the protein maturation, including oligomerization and glycosylation, are maintained upon solubilization of TWIK-1 using SMAs. However, proteins lacking post-translational modifications remain in VF2 and the faint band might be associated with insolubilized and incompletely folded proteins. Figure 3b shows the quantitive analysis demonstrating varying solubilization efficiencies for the three SMA copolymers (Fig. 3b). Specifically, SMA2 is able to solubilize approximately 97% of the translocated cell-free synthesized protein, whereas SMA3 solubilized 68% and SMA1 only achieves a solubilization efficiency of 42%. These findings are supported by the autoradiograph of the VF2 and SN2 (Fig. 3a).

3.2.2 Particle Size Measurement of SMA Lipid Particles

1. To analyze the particle size distribution of the vesicular fraction (VF1) and the supernatant 2 (SN2) after fractionation and successful solubilization of cell-free synthesized TWIK-1, aliquot 25 μL of each fraction into a quartz glass cuvette and perform particle size analysis using the dynamic light scattering (DLS) technique. To maintain consistency, ensure that the samples are equilibrated to 25 °C and analyzed in triplicates (refer to *see* **Note 9**).
2. Our findings, depicted in Fig. 4, demonstrate the successful solubilization of VF1 based on SMA2, which proved to be most efficient (Fig. 4b). The detected particle size in VF1 ranges from 90 nm to 5000 nm due to large and inhomogeneous membrane structures (Fig. 4a). However, treatment with SMA2 cuts the membranous structure and forms native nanodiscs, resulting in a shift of the particle size in SN2 ranging from 50 nm to 500 nm. To exclude that the shift in particle size is solely due to the addition of SMA2, the copolymer is analyzed as control (Fig. 4b). For SMA2 various species in particle size are detected, ranging from 1 nm to 4 nm, 7 nm to 90 nm, 400 nm to 1200 nm, and around 4500 nm. This indicates that the particles of SN2 in the range of 4–40 nm are free SMA copolymers, that did not contain any membranous structures.

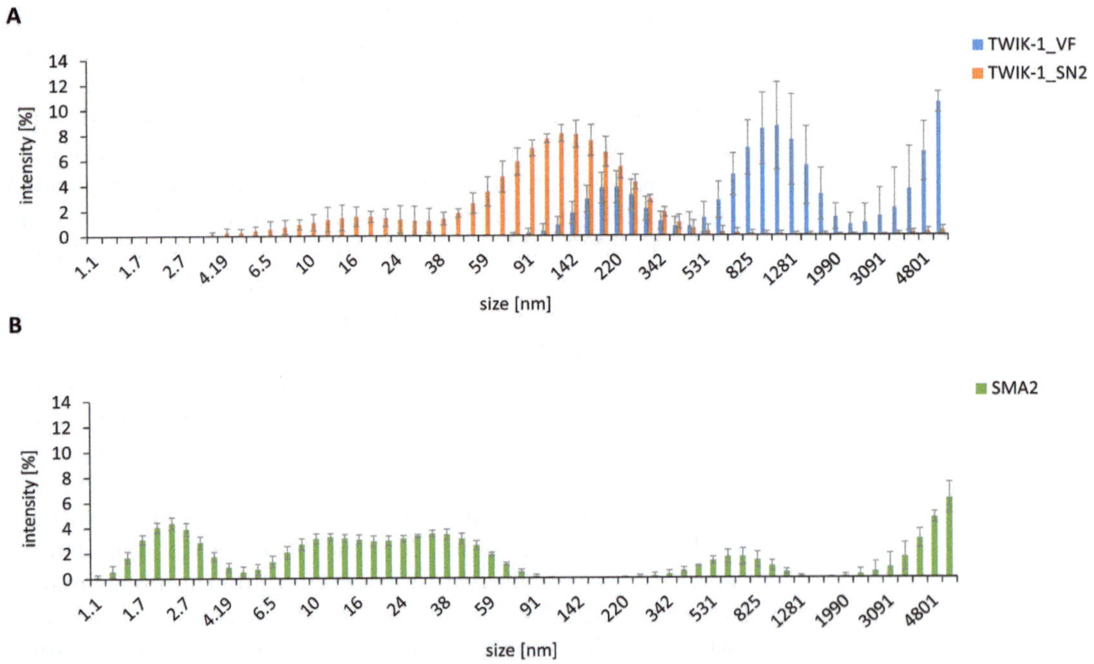

Fig. 4 Analysis of the particle size of two different fractions of cell-free synthesized TWIK-1 solubilized by styrene-maleic acid copolymer 2: (**a**) After cell-free synthesis based on *Sf*21 lysate, the vesicular fraction (VF1) containing membrane-embedded TWIK-1 was analyzed. To obtain supernatant 2 (SN2), VF1 was solubilized with styrene-maleic acid copolymer 2 (SMA2). Both fractions (**a**) and SMA2 (**b**) alone are subjected to particle size measurement. Standard deviations are calculated from triplicate analysis

3.3 Complementation Assay for Fluorescence Detection of Protein-Protein Interactions

1. To detect protein-protein interactions of the solubilized proteins and overcome the issue of SDS instability for protein assemblies, a biomolecular fluorescence complementation assay (BiFC) combined with confocal microscopy is recommended as an alternative approach. This assay is based on fusing non-fluorescent fragments (bait and prey) of a fluorescent protein to a gene of interest. If the proteins come into spatial proximity (< 7 nm), the fragments complement and emit a fluorescence signal. It is important to ensure that the fragment fusion does not interrupt the protein interactions or alter function. Therefore, intensive structure analysis should be performed. If structural information are not available, fuse the protein of interest at the N- or C-terminus, respectively, and validate the fluorescence complementation of all combinations, *see* **Note 10**.
2. To avoid unspecific fluorescence induced by self-assembly, it is recommended to express the proteins individually to address different microsomes. Furthermore, coexpress the terminally fused constructs and select the optimal combination for fluorescence readout.

3. After selecting the desired combination, coexpress the plasmids according to Subheading 3.1, fractionate, solubilize them according to Subheading 3.2, and transfer samples to a μ-IBIDI-Slide. We recommend analyzing both SN1 and VF1 to confirm the successful translocation of the proteins, enabling their oligomerization. After solubilization, the proteins will remain in SN2.
4. For confocal imaging use a plan-achromat objective with at least 40× magnification. For venus protein in microsomal structures, a 63×/1.4 oil objective and excitation of the sample with 488 nm argon laser is recommended. The emitted light may be captured with a band-pass filter in the range of 505 nm–550 nm. The microscope setting regarding pinhole, focus, and laser intensity must be adjusted based on the individual sample. However, ensure the same conditions are applied to the negative controls.
5. Pictures can be processed using ImageJ or Zeiss LSM Imaging Software, e.g., Zen2009.
6. We assessed the solubilization performance of the interacting proteins for TWIK-1 additionally by a split complementation assay (Fig. 5). Coexpression of split venus constructs is carried out, followed by confocal microscopy. If the split fragments are in proximity (< 7 nm), the yellow fluorescent protein is complemented, resulting in fluorescence [21]. In the VF1, strong fluorescence is detectable, indicating successful dimerization and complementation of the split-fragments. As expected, no fluorescence is observed in the SN1, consistent with the previously observed absence of the oligomeric bands (Fig. 1b). After

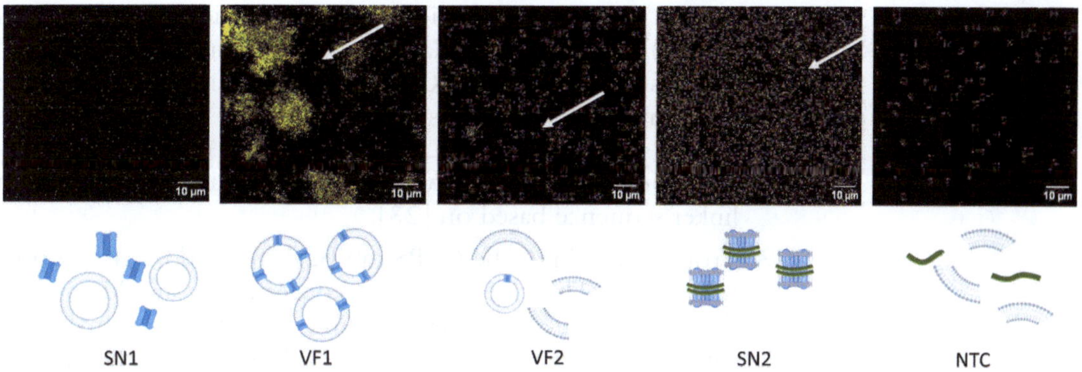

Fig. 5 Biomolecular fluorescence complementation assay of TWIK-1 solubilized by styrene-maleic acid copolymer 2: TWIK-1 venus complements are coexpressed in eukaryotic *Sf*21 lysate. Translation mixture (TM) is fractionated after synthesis by centrifugation in the supernatant (SN1) and the vesicular fraction (VF1). After solubilization with SMA2, sample is fractionated in vesicular fraction 2 (VF2) and supernatant 2 (SN2) and analyzed using confocal microscopy. No template control (NTC) treated with SMA2 is additionally analyzed. For better detection, the contrast of all images was adjusted equally

applying the SMA solubilization procedure, weak fluorescence is detectable in VF2, while SN2 exhibits isolated fluorescent spots with reduced size compared to the structures found in VF1. Weak fluorescent structures are still visible in VF2 due to non-solubilized protein remaining in the microsomes. The solubilized negative control (NTC) shows no significant background fluorescence, consistent with the autoradiographs presented (Fig. 3a) (*see* **Note 11**).

4 Notes

1. Preparation of *Sf*21 lysate [22]. Cells were cultivated in insect cell medium supplemented with fetal bovine serum at 27 °C and harvested at 1 × 10^6cells/mL. Cells were centrifuged (200 × *g*, 4 °C) and washed with homogenization buffer (40 mM Hepes-KOH (pH 7.9), 100 mM KOAc, 10 mM glutathione, and 10 mM glutathione disulfide. Cells were disrupted by passing through a gauge needle and centrifuged at 6500 × *g*, 20 min, 4 °C. Supernatant was chromatographed by a sephadex G-25 column, previously equilibrated with elution buffer (40 mM Hepes-KOH (pH 7.9), 100 mM KOAc). Elution fractions were pooled, and endogenous RNA was digested by micrococcal S7 nuclease (f.c. 10 U/mL) and subsequently stopped by EGTA (f.c. 6.7 mM) supplementation. The lysate was further supplemented with creatinine kinase (100 μg/mL) and shock frozen in liquid nitrogen. Lysate is stored at −80 °C.
2. Coding sequence of TWIK-1 (Uniprot: O00180) was C-terminally fused with a His- and Flag-Tag. Regulatory sequences (5′untranslated region: internal ribosome entry site (IRES) from the intergenic region of the Cricket paralysis virus (CrPV) sequence, T7 RNA polymerase promoter sequence, 3′ untranslated region: T7 RNA polymerase terminator sequence and a multiple cloning site) were cloned with the protein-coding sequence into a pUC vector (BioCat). Venus split complementation fragments were fused N-terminally along with a linker sequence based on [23].
3. Troubleshooting: The CFPS can be optimized regarding protein yields. Therefore, plasmid concertation should be validated in a range of 20–150 ng/μL. The optimal temperature for *Sf*21 cell-free synthesis is 27 °C. However, as the optimal reaction conditions are strongly protein-dependent, we suggest performing a temperature series between 18 °C and 33 °C. Furthermore, a continuous exchange cell-free reaction can be performed, where the reaction mixture is separated by a semi-permeable membrane from a feeding mixture. Fresh components can reach the reaction through the membrane, while

inhibitory byproducts are removed, resulting in a longer reaction time. For further description see [24]. The addition of a N-terminal melittin signal sequence has been demonstrated to improve CFPS and should be considered for higher protein yield and efficient translocation.

4. Pause Point: Samples are stable in the translation mixture and can be stored at −80 °C for several days or at 4 °C overnight.
5. Critical step: SMA copolymers have the potential to chelate with divalent cations (e.g., Mg^{2+} and Ca^{2+}), making them insoluble which could result in disruption of the formed native nanodiscs [25, 26]. Accordingly, use PBS without divalent cations for resuspension.
6. Troubleshooting: To facilitate the entry of MPs into the SDS-Gel, heating or the addition of reducing agents such as dithiothreitol or urea may be required. However, caution should be exercised as this can disrupt protein conformation or complexes and can promote protein aggregation. When investigating membrane proteins, it is important to consider their slightly altered migration behavior due to altered detergent binding [27].
7. Critical step: The membrane protein solubilization with SMA copolymers should be optimized depending on the proteins of interests and their membranous surrounding regarding incubation time, incubation temperature, and total SMA concentration in the sample. To achieve proper solubilization results it is important to protect the samples from light, as the SMA copolymers are light sensitive [20]. Proteins with a size exceeding ~400 kDa pose a challenge for solubilization as their dimensions are too large to accommodate within disc-shaped nanoparticles. Hence, special consideration should be given to solubilize such proteins.
8. Critical step: The pH dependence of SMA copolymers is attributed to the charged ions of the hydrophobic styrene unit and hydrophilic carboxyl group. At low pH values, SMA protonation occurs and results in decreased charge repulsion at low pH and thus aggregate formation. To achieve optimal membrane protein solubilization, the pH should be maintained within a range of 7–8. If acidic conditions are required, available functionalized SMAs could even be applied with a greater tolerance towards pH, salts and cations [4].
9. Critical Step: The particle size was determined based on dynamic light scattering (DLS) which measures the Brownian motion and relates this to the size of the particles in the sample. Thus, the temperature of the sample needs to be stable and should be set before measurement (25 °C). Moreover, when transferring the samples to the quartz glass cuvette it is

necessary to avoid the bubble formation in the cuvette as they would interfere with the measurement. In addition, to ensure that the particle size analyzer is working correctly, particles of a known size should be measured before the samples with unknown particle sizes.

10. Critical Step: Ensure that the investigated protein enables the fusion of the fragments without altering the functionality. For venus complementation assay it is recommended to validate suitable negative controls, as the fluorescent fragments tend to self-assemble. However, this can be sufficiently controlled by a similar expression ratio of the counterparts and single expression. Therefore, validate background fluorescence and self-complementation of the complements by synthesizing proteins in different membrane batches and mix these post syntheses ensuring translocation into different microsomes and thus no oligomerization. Furthermore, validate based on previous reports if the addition of any fusion protein could cause the disruption of the protein-protein interaction. Such validation could be performed in a real-time cycler for practical considerations. However, consider the maturation time of the fluorescent protein and let the sample rest for at least 2 h at room temperature before microscopy. If different fluorescent proteins are chosen, the maturation time might even be exceeded.
11. Troubleshooting: If no signal is detectable but expected, ensure that both proteins are synthesized in a sufficient amount (autoradiography or western-blot analysis is recommended). If necessary, adjust the plasmid concentrations. As fluorescent proteins are pH dependent, ensure a suitable pH value in the sample or vary the split-fluorescence protein. Several biomolecular complementation assays based on fluorescent proteins are available. Furthermore, perform the synthesis in the dark to avoid bleaching of the sample.

Acknowledgments

The authors thank Dana Wenzel (Fraunhofer IZI-BB) for *Sf*21 lysate preparation. This work was supported by the Federal Ministry of Education and Research (BMBF, Nos. 13GW0408C and 16LW0189) and the Federal Ministry for Economic Affairs and Climate Action (BMWK, No. ZF4086508CR6). Funding to pay the open-access publication charges for this article was provided by the Fraunhofer-Gesellschaft zur Förderung der angewandten Forschung e.V. Schematic illustrations were created using BioRender.com.

References

1. Breyton C, Tribet C, Olive J et al (1997) Dimer to monomer conversion of the cytochrome b6 f complex. Causes and consequences. J Biol Chem 272:21892–21900. https://doi.org/10.1074/jbc.272.35.21892
2. Seddon AM, Curnow P, Booth PJ (2004) Membrane proteins, lipids and detergents: not just a soap opera. Biochim Biophys Acta 1666:105–117. https://doi.org/10.1016/j.bbamem.2004.04.011
3. Workman CE, Cawthon B, Brady NG et al (2022) Effects of esterified styrene-maleic acid copolymer degradation on integral membrane protein extraction. Biomacromolecules 23: 4749–4755. https://doi.org/10.1021/acs.biomac.2c00928
4. Dörr JM, Scheidelaar S, Koorengevel MC et al (2016) The styrene-maleic acid copolymer: a versatile tool in membrane research. Eur Biophys J 45:3–21. https://doi.org/10.1007/s00249-015-1093-y
5. Scheidelaar S, Koorengevel MC, Pardo JD et al (2015) Molecular model for the solubilization of membranes into nanodisks by styrene maleic acid copolymers. Biophys J 108:279–290. https://doi.org/10.1016/j.bpj.2014.11.3464
6. Morrison KA, Akram A, Mathews A et al (2016) Membrane protein extraction and purification using styrene-maleic acid (SMA) copolymer: effect of variations in polymer structure. Biochem J 473:4349–4360. https://doi.org/10.1042/BCJ20160723
7. Gulati S, Jamshad M, Knowles TJ et al (2014) Detergent-free purification of ABC (ATP--binding-cassette) transporters. Biochem J 461:269–278. https://doi.org/10.1042/BJ20131477
8. Karlova MG, Voskoboynikova N, Gluhov GS et al (2019) Detergent-free solubilization of human Kv channels expressed in mammalian cells. Chem Phys Lipids 219:50–57. https://doi.org/10.1016/j.chemphyslip.2019.01.013
9. Tedesco D, Maj M, Malarczyk P et al (2021) Application of the SMALP technology to the isolation of GPCRs from low-yielding cell lines. Biochim Biophys Acta Biomembr 1863: 183641. https://doi.org/10.1016/j.bbamem.2021.183641
10. Komar J, Alvira S, Schulze RJ et al (2016) Membrane protein insertion and assembly by the bacterial holo-translocon SecYEG-SecDF-YajC-YidC. Biochem J 473:3341–3354. https://doi.org/10.1042/BCJ20160545
11. Harris NJ, Pellowe GA, Booth PJ (2020) Cell-free expression tools to study co-translational folding of alpha helical membrane transporters. Sci Rep 10:9125. https://doi.org/10.1038/s41598-020-66097-4
12. Nishiguchi R, Tanaka T, Hayashida J et al (2022) Evaluation of cell-free synthesized human channel proteins for in vitro channel research. Membranes (Basel) 13. https://doi.org/10.3390/membranes13010048
13. Spice AJ, Aw R, Bracewell DG et al (2020) Synthesis and assembly of Hepatitis B virus-like particles in a pichia pastoris cell-free system. Front Bioeng Biotechnol 8:72. https://doi.org/10.3389/fbioe.2020.00072
14. Haueis L, Stech M, Schneider E et al (2023) Rapid one-step capturing of native, cell-free synthesized and membrane-embedded GLP-1R. Int J Mol Sci 24. https://doi.org/10.3390/ijms24032808
15. Ullrich J, Ohlhoff C, Dondapati SK et al (2023) Evaluation of the ion channel assembly in a eukaryotic cell-free system focusing on two-pore domain potassium channels K2P. IJMS 24:6299. https://doi.org/10.3390/ijms24076299
16. Dondapati SK, Lübberding H, Zemella A et al (2019) Functional reconstitution of membrane proteins derived from eukaryotic cell-free systems. Front Pharmacol 10:917. https://doi.org/10.3389/fphar.2019.00917
17. Ullrich J, Göhmann PJ, Zemella A et al (2022) Oligomerization of the heteromeric γ-aminobutyric acid receptor GABAB in a eukaryotic cell-free system. Sci Rep 12:20742. https://doi.org/10.1038/s41598-022-24885-0
18. Zemella A, Thoring L, Hoffmeister C et al (2018) Cell-free protein synthesis as a novel tool for directed glycoengineering of active erythropoietin. Sci Rep 8:8514. https://doi.org/10.1038/s41598-018-26936-x
19. Lesage F, Guillemare E, Fink M et al (1996) TWIK-1, a ubiquitous human weakly inward rectifying K+ channel with a novel structure. EMBO J 15:1004–1011
20. Jamshad M, Grimard V, Idini I et al (2015) Structural analysis of a nanoparticle containing a lipid bilayer used for detergent-free extraction of membrane proteins. Nano Res 8:774–789. https://doi.org/10.1007/s12274-014-0560-6
21. Hwang EM, Kim E, Yarishkin O et al (2014) A disulphide-linked heterodimer of TWIK-1 and TREK-1 mediates passive conductance in

astrocytes. Nat Commun 5:3227. https://doi.org/10.1038/ncomms4227

22. Kubick S, Schacherl J, Fleischer-Notter H et al In Vitro translation in an insect-based cell-free system, pp 209–217. https://doi.org/10.1007/978-3-642-59337-6_25
23. Ohashi K, Mizuno K (2014) A novel pair of split venus fragments to detect protein-protein interactions by in vitro and in vivo bimolecular fluorescence complementation assays. Methods Mol Biol 1174:247–262. https://doi.org/10.1007/978-1-4939-0944-5_17
24. Quast RB, Sonnabend A, Stech M et al (2016) High-yield cell-free synthesis of human EGFR by IRES-mediated protein translation in a continuous exchange cell-free reaction format. Sci Rep 6:30399. https://doi.org/10.1038/srep30399
25. Pollock NL, Lee SC, Patel JH et al (2018) Structure and function of membrane proteins encapsulated in a polymer-bound lipid bilayer. Biochim Biophys Acta Biomembr 1860:809–817. https://doi.org/10.1016/j.bbamem.2017.08.012
26. Lee SC, Knowles TJ, Postis VLG et al (2016) A method for detergent-free isolation of membrane proteins in their local lipid environment. Nat Protoc 11:1149–1162. https://doi.org/10.1038/nprot.2016.070
27. Rath A, Glibowicka M, Nadeau VG et al (2009) Detergent binding explains anomalous SDS-PAGE migration of membrane proteins. Proc Natl Acad Sci U S A 106:1760–1765. https://doi.org/10.1073/pnas.0813167106

Chapter 19

Cell-Free Systems for the Production of Glycoproteins

Erik J. Bidstrup, Yong Hyun Kwon, Keehun Kim, Chandra Kanth Bandi, Rochelle Aw, Michael C. Jewett, and Matthew P. DeLisa

Abstract

Cell-free protein synthesis (CFPS), whereby cell lysates are used to produce proteins from a genetic template, has matured as an attractive alternative to standard biomanufacturing modalities due to its high volumetric productivity contained within a distributable platform. Initially, cell-free lysates produced from *Escherichia coli*, which are both simple to produce and cost-effective for the production of a wide variety of proteins, were unable to produce glycosylated proteins as *E. coli* lacks native glycosylation machinery. With many important therapeutic proteins possessing asparagine-linked glycans that are critical for structure and function, this gap in CFPS production capabilities was addressed with the development of cell-free expression of glycoproteins (glycoCFE), which uses the supplementation of extracted lipid-linked oligosaccharides and purified oligosaccharyltransferases to enable glycoprotein production in the CFPS reaction environment. In this chapter, we highlight the basic methods for the preparation of reagents for glycoCFE and the protocol for expression and glycosylation of a model protein using a more productive, yet simplified, glycoCFE setup. Beyond this initial protocol, we also highlight how this protocol can be extended to a wide range of alternative glycan structures, oligosaccharyltransferases, and acceptor proteins as well as to a one-pot cell-free glycoprotein synthesis reaction.

Key words Cell-free glycoprotein synthesis, Glycan, Glycoengineering, Glycoprotein expression, *N*-linked glycosylation, Lipid-linked oligosaccharide, Oligosaccharyltransferase, Synthetic glycobiology

1 Introduction

Glycosylation of asparagine residues, a process known as *N*-linked glycosylation, is one of the most common post-translational modifications and occurs in all domains of life. The significance of this process stems from the many ways that the attached glycan can modulate protein structure and function, including by altering the protein's binding, signaling, folding, and immune recognition [1]. Moreover, because many clinically approved therapeutic proteins are glycosylated [2], efficient conjugation of structurally defined glycans has become a cornerstone of biotherapeutic manufacturing. Maintaining the authenticity of the attached glycan

Steven B. Bradfute (ed.), *Recombinant Glycoproteins: Methods and Protocols*, Methods in Molecular Biology, vol. 2762,
https://doi.org/10.1007/978-1-0716-3666-4_19,

is critical, as minor changes to the glycan structure have been shown to significantly alter the immunogenicity and pharmacokinetics of therapeutics [3, 4]. However, maintaining strict control of the *N*-glycan profile is challenging because, unlike biosynthesis of DNA, RNA, and proteins, the biosynthesis and installation of *N*-glycans is a non-template-driven enzymatic process. As a result, *N*-glycan composition is defined by the availability of glycosyltransferase enzymes and associated carbohydrate building blocks at the time of synthesis and transfer. From the standpoint of recombinant glycoprotein expression, the complexity of glycan construction and conjugation has favored the use of eukaryotic cell-based hosts, such as Chinese hamster ovary (CHO) or murine myeloma NS0 cells, that natively execute *N*-linked glycosylation reactions [5]. However, these eukaryotic hosts suffer from several drawbacks including high cost of goods, slow development timelines, strict cell viability constraints, and the need for centralized manufacturing facilities.

In light of these limitations, cell-free protein synthesis (CFPS) systems have gained increasing attention as an alternative option for producing complex human protein targets, including glycoproteins, using cell extracts supplemented with energy sources, salts, co-factors, and DNA encoding the protein of interest [6]. Historically, these extracts have been generated from yeast [7], mammalian [8], insect [9], and wheat germ cells [10]. However, these extracts must typically be supplemented with microsomes, which can lead to satisfactory glycosylation efficiency but limits the overall glycoprotein yield due to the restricted capacity of microsome compartments [7]. Moreover, microsomal glycosylation is effectively a black box with little to no opportunity to control or engineer the glycosylation reactions for achieving desired *N*-glycan structures. To address this gap, prokaryotic extracts from *Escherichia coli* have been explored for their ability to support glycoprotein expression [11, 12]. While *E. coli* and its cell-free extracts are well known to lack native glycosylation machinery [13], early work demonstrated that commercially available cell-free expression platforms based on *E. coli* (e.g., S30 lysate or PURE system) could furnish *N*-linked glycoproteins by supplementing the reactions with solvent-extracted lipid-linked oligosaccharides (LLOs) bearing the heptasaccharide *N*-glycan from *Campylobacter jejuni* (*Cj*LLOs) and purified oligosaccharyltransferase (OST) enzyme, namely PglB from *C. jejuni* (*Cj*PglB), to catalyze the *en bloc* transfer of the glycan to the acceptor protein of interest [11]. In these cell-free expression systems for glycoproteins (glycoCFE), the open reaction environment allowed for the direct modulation of the LLO substrate and OST catalyst concentrations while the use of lysates that lacked native *N*-glycosylation pathways resulted in a homogenous glycosylation profile. More recently, glycoCFE was further expanded to include a wider range of glycans including several additional bacterial structures (e.g., *E. coli* O9 glycan primer-adaptor) as well as a eukaryotic glycan comprising the trimannosyl core from complex-type human *N*-glycans [12]. The glycoCFE system has also been adapted for flow-based glycosylation using a microfluidic platform that integrated cell-free protein synthesis with protein glycosylation via an immobilized OST, a configuration that promoted higher glycosylation efficiencies than comparable reactions run semi-continuously in a test tube [14].

Building off the glycoCFE concept, a more integrated transcription-translation-glycosylation system known as cell-free glycoprotein synthesis (CFGpS) was created [12]. CFGpS leverages *E. coli* strains that express the requisite glycosylation components (e.g., LLOs, OST) to generate lysates already enriched with a target glycan attached to an LLO and an OST. When supplemented with plasmid DNA encoding a recombinant acceptor protein, these lysates were observed to efficiently express and glycosylate the acceptor protein in a single-pot reaction [12, 15]. Lysates enriched with glycosylation components avoid the need for labor-intensive, solvent-based extraction of LLOs, and membrane purification of OSTs. Combining different CFGpS lysates has enabled rapid prototyping of glycan synthesis pathways and OSTs to explore a wide range of expression and glycosylation conditions for a target glycoprotein, including for *O*-linked glycans [16–19], and multiple glycans on a single protein [20]. Beyond the advantages for glycoprotein expression development, CFGpS reactions have also been used as a distributed biomanufacturing platform for producing effective conjugate vaccines from lyophilized, detoxified, and heat-stable reactions [21, 22]. Compared to CFGpS, an advantage of glycoCFE is that it allows for a more diverse set of glycans to be transferred to proteins than has yet been demonstrated with glycan- and OST-enriched extracts, while retaining the speed and high-throughput capability of protein expression using CFPS. Additionally, expression of both LLOs and OSTs in a lysate-generating strain can occasionally lead to undesirable effects on strain growth and lysate productivity in certain combinations, favoring the use of unenriched lysates for specific glycoprotein targets. While much attention has been spent on making CFGpS methods more accessible [23], there is an opportunity to create an end-user-oriented protocol published for glycoCFE, whereby CFPS-derived proteins are decorated with glycans using exogenous LLOs and OSTs. In this work, we will highlight the core method for glycoCFE with modest refinements to improve accessibility for non-specialized labs. This protocol may be conveniently modified as detailed in the associated notes to work for different glycans, OSTs, and acceptor proteins and may even be modified into a CFGpS-type reaction. Specifically, we will provide detailed protocols with associated pictures and standard results for the preparation of S12 lysate from *E. coli* strain BL21 Star (DE3), the solvent extraction of

*Cj*LLOs from *E. coli* strain CLM24 carrying a plasmid that encodes the glycan biosynthesis pathway, the isolation of active *Cj*PglB OST from recombinant *E. coli* expression, and the setup, implementation, and product characterization of the resulting glycoCFE reactions (Fig. 1).

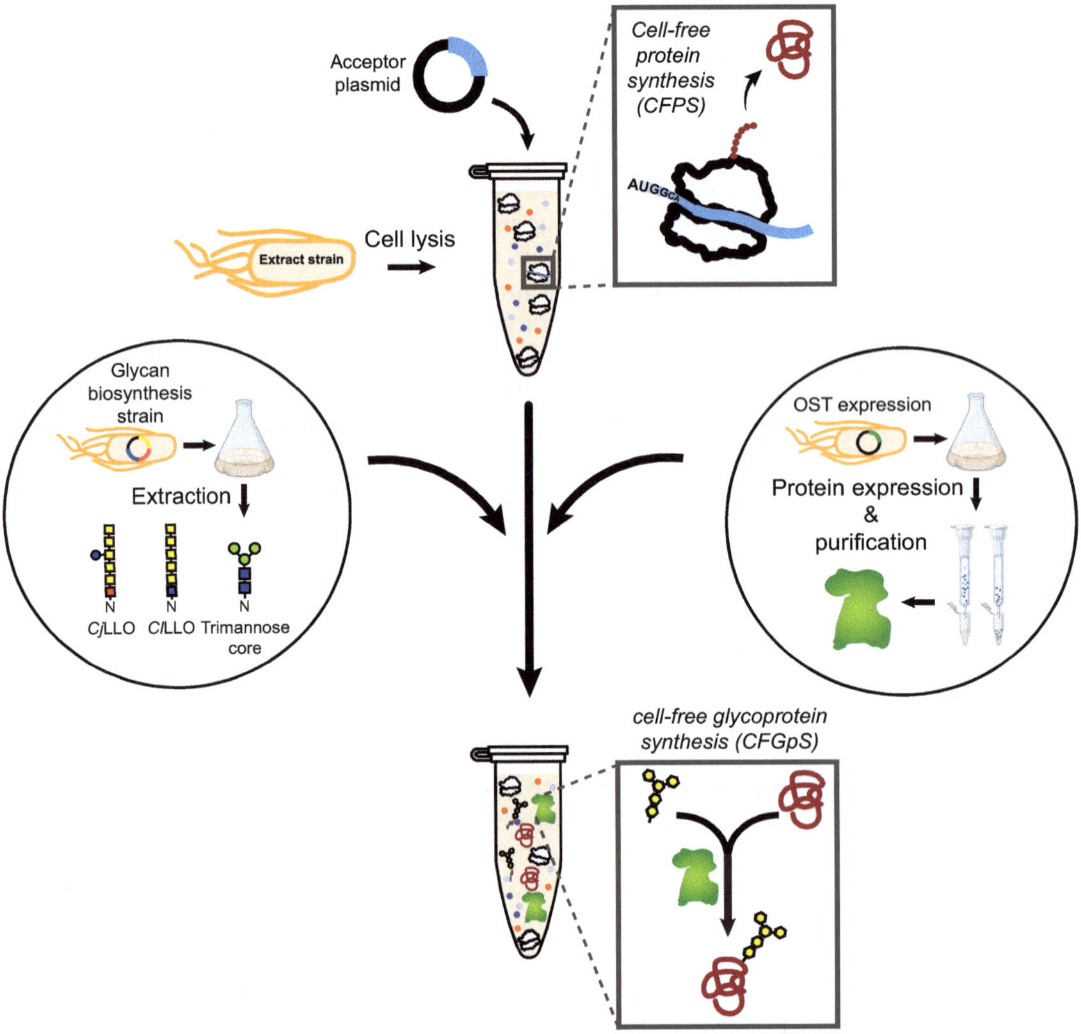

Fig. 1 Overview of glycoCFE protocol. Initially, S12 CFPS lysates are prepared from source strains. In parallel, solvent-extracted LLOs and membrane-purified OSTs are prepared from their source strains. Then, CFPS reactions with requisite chemical additives are activated with acceptor protein DNA to generate the acceptor protein and supplemented with the solvent-extracted LLOs and membrane-purified OSTs. The resulting glycoprotein products can then be characterized using Western blot analysis to confirm protein expression and glycan transfer

2 Materials

2.1 Bacterial Strains and Plasmids

See Table 1.

2.2 Buffers, Media, and Chemicals

1. 2xYTP: 5 g NaCl, 16 g tryptone, 10 g yeast extract, 7 g K_2HPO_4, and 3 g KH_2PO_4, pH to 7.2 via 5N KOH, in 750 mL distilled water.
2. S30 buffer: 10 mM Tris-acetate (pH 8.2), 14 mM magnesium acetate, 60 mM potassium acetate, and 2 mM dithiothreitol.
3. Terrific broth (TB): 12 g of tryptone, 24 g of yeast extract, and 4 mL of glycerol in 900 mL distilled water and 100 mL of potassium phosphate buffer.
4. Potassium phosphate buffer: 0.17 M KH_2PO_4, 0.72 M K_2HPO_4, and filter sterile.
5. 1× in vitro glycosylation (IVG) buffer: 50 mM HEPES, 0.1% (w/v) n-dodecyl-β-D-maltoside (DDM), pH 7.4, and filter sterile.

Table 1
***E. coli* strains and plasmids used in this protocol**

Strains and plasmids	Description	References
Strains		
CLM24	*E. coli* strain W3110 with genomic knockout of *waaL* ligase, which prevents transfer of glycan onto the lipid A-core and strongly expresses the glycan of interest	[24]
BL21 Star (DE3)	Strain carrying T7 DNA polymerase from λDE3 lysogen that facilitates high-level protein expression in CFPS owing to deficiency in both Lon and OmpT proteases as well as RNAse E activity	[25]
BL21(DE3)	Strain carrying T7 DNA polymerase from λDE3 lysogen that facilitates high-level protein expression	[25]
Plasmids		
pMW07-pglΔB	Plasmid encoding protein glycosylation locus (*pgl*) of *C. jejuni* with PglB deletion	[26]
pSF-*Cj*PglB	Plasmid encoding *C. jejuni* PglB with C-terminal FLAG epitope tag	[26]
pSN18	Plasmid encoding *C. jejuni* PglB with C-terminal decahistidine (10xHis) tag	[27]
pJL1-sfGFP$^{DQNAT\text{-}6xHis}$	Plasmid with constitutive T7 RNA polymerase promoter optimized for CFPS expression encoding superfolder GFP (sfGFP) with engineered glycosylation tag (DQNAT) and polyhistidine (6xHis) tag at C-terminus	[12]

6. Chloroform.
7. Methanol.
8. Glucose.
9. Luria-Bertani (LB) broth.
10. Buffer A: 50 mM HEPES, 250 mM NaCl, pH 7.4, 20% (v/v) glycerol, and filter sterile.
11. Buffer B: 1% (w/v) DDM supplemented in Buffer A.
12. Buffer C: Buffer B with 15 mM imidazole supplemented.
13. Buffer D: Buffer B with 250 mM imidazole supplemented.
14. Ni-NTA agarose resin.
15. Protease inhibitor cocktail.
16. Gravity column.
17. PD-10 desalting column.

2.3 CFPS Reagents

1. 15× salt solution: 1.95 M potassium glutamate, 150 mM ammonium glutamate, variable mM magnesium glutamate (concentration is optimized for each batch of lysate).
2. 6.52× reaction mixture: 13.04 mM of each amino acid, 13.04 mM pyroglutamatic acid, 2.15 mM NAD, 1.76 mM coenzyme A, 9.78 mM spermidine, 6.52 mM putrescine, 26.08 mM sodium oxalate, 215.16 mM PEP, 371.64 mM HEPES.
3. 15× master mix: 18 mM ATP, 12.75 mM GTP, 12.75 mM UTP, 12.75 mM CTP, 510 mM folinic acid, 2.55 M tRNA.

2.4 Equipment and Atypical Lab Materials

1. Water bath sonicator.
2. Benchtop centrifuge suitable for spinning 1.5 mL tubes.
3. Floor centrifuge.
4. Ultracentrifuge.
5. Homogenizer A: Avestin EmulsiFlex-B15.
6. Homogenizer B: Avestin EmulsiFLex-C5.
7. 2.5 L culture flask.
8. 100 mM glass Petri dish.
9. 2.8 L Tunair cell culture flask.
10. Shaking incubator for fermentation.
11. Refrigerated shaking incubator.
12. Swinging bucket rotor.
13. Fixed angle rotor that fits a 50 mL centrifuge tube.
14. Fixed angle rotor that fits a 500 mL centrifuge bottle.

3 Methods

3.1 S12 Lysate Preparation (Fig. 2a)

3.1.1 Media Preparation and Inoculation of Overnight Culture (see **Note 1**)

1. For a single batch of lysate, prepare sterilized 750 mL 2xYTP media pH 7.2, 250 mL 7.2% (w/v) glucose, and 50 mL LB.
2. Inoculate BL21 Star (DE3) overnight in 50 mL LB in a 250 mL culture flask.

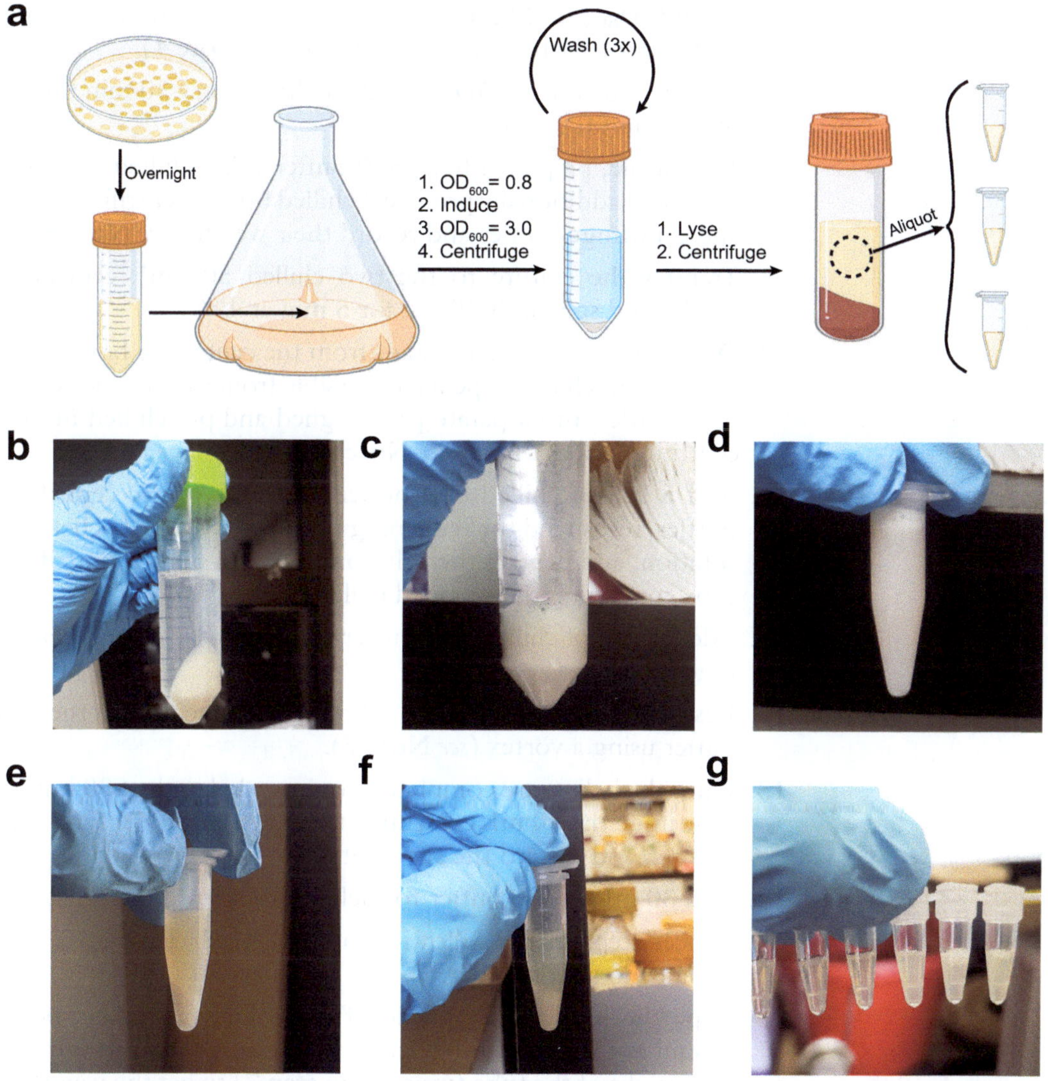

Fig. 2 S12 lysate production process. (**a**) Schematic of the CFPS lysate production process. (**b**) Bacterial pellet following triplicate (3×) wash steps (Subheading 3.1.2, **steps 12–15**). (**c**) Bacteria resuspended in S30 buffer solution prior to lysis (Subheading 3.1.3, **step 1**). (**d**) Solution following bacterial lysis by homogenization (Subheading 3.1.3, **step 2**). (**e**) Bacterial lysate following centrifugation at 12,000 × g (Subheading 3.1.3, **step 3**). (**f**) Bacterial lysate following centrifugation at 10,000 × g (Subheading 3.1.3, **step 5**). (**g**) Final CFPS lysates prior to flash freeze

3.1.2 Lysate Strain Growth and Harvest

1. Mix the 750 mL 2xYTP media pH 7.2 and 250 mL 7.2% glucose to make 1 L 2xYTP + 1.8% (w/v) glucose media (*see* **Note 2**).
2. Check the optical density at 600 nM (OD_{600}) of the overnight culture and inoculate into the 1 L 2xYTP + 1.8% (w/v) glucose media to achieve a starting OD_{600} of approximately 0.08 (*see* **Note 3**).
3. Incubate the 1 L subculture at 37 °C with shaking at 250 rpm.
4. Take the OD_{600} of the culture every 40 min until it reaches an OD_{600} of 0.8.
5. At an OD_{600} of 0.8, induce the culture with 1 mM isopropyl β-D-1-thiogalactopyranoside (IPTG) (*see* **Note 4**).
6. Continue taking an OD_{600} of the culture every 30 min until it reaches an OD_{600} of 2.8–3.0.
7. Meanwhile, prepare a fresh S30 buffer using chilled sterilized water and additionally prepare 2 chilled 50 mL centrifuge tubes and centrifuge bottles and record their weight (*see* **Note 5**).
8. Transfer the culture to the two chilled 500 mL centrifuge bottles and spin at 8000 × g for 5 min at 4 °C.
9. Discard the media supernatant from the centrifuge bottles and scoop as much of the pellet as possible from each of the centrifuge bottles into separate pre-weighed and pre-chilled 50 mL centrifuge tubes on ice (*see* **Note 6**).
10. Add 5 mL of chilled S30 buffer into each of the centrifuge bottles, resuspend the remaining portions of the cell pellet into solution, and deposit the solution into each 50 mL centrifuge tube containing the scooped pellets.
11. Add 25 mL of chilled S30 buffer to each of the 50 mL centrifuge tubes.
12. Resuspend the pellet in each 50 mL centrifuge tube in the S30 buffer using a vortex (*see* **Note 7**).
13. Once the pellets are completely resuspended in the S30 buffer, centrifuge the 50 mL centrifuge tubes at 10,000 × g for 3 min at 4 °C, and discard the supernatant.
14. Add 30 mL of S30 buffer to each 50 mL centrifuge tube.
15. Repeat steps in the following order 12, 13, 14, 12,13, and advance to **step 16** (Fig. 2b).
16. Wipe the inside and outside of each centrifuge tube, being careful not to disturb the pellet; weigh each tube and subtract the weight of the tube (weighed in **step** 7) to get the weight of the pellet (*see* **Note 8**).

3.1.3 Resuspension, Lysis, and Centrifugation

1. Resuspend each pellet in 1 mL S30 buffer per 1 g of cell pellet by vortexing (*see* **Note 7**, Fig. 2c).
2. Combine the resuspended cells and lyse using a single pass-through homogenizer A set to 21,000 psi (*see* **Note 9**, Fig. 2d).
3. Pellet un-lysed cells and debris via centrifugation in 1.5-mL centrifuge tubes at 12,000 × g for 10 min at 4 °C, using a pre-chilled centrifuge (Fig. 2e).
4. Transfer the supernatant to fresh 1.5-mL tubes and discard the pellets (*see* **Note 10**).
5. Spin down the supernatant once more via centrifugation at 10,000 × g for 10 min at 4 °C (Fig. 2f).
6. Collect and combine the supernatants, gently pipet mix the final lysate solution to obtain a homogenous sample, aliquot in appropriate volumes, flash freeze the aliquots in liquid nitrogen, and store long term at −80 °C (*see* **Note 11**, Fig. 2g).

3.1.4 CFPS Reaction and Optimization of Magnesium in Lysate

For each batch of CFPS lysates, it is advised that the concentration of magnesium glutamate used for protein synthesis be optimized as follows:

1. Prepare salt solution variations with the following range of magnesium glutamate concentrations: 30 mM, 60 mM, 90 mM, 120 mM, 150 mM, and 180 mM.
2. For a general CFPS reaction, see the recipe in Table 2. To identify the optimal magnesium glutamate concentration for a given batch of CFPS lysate, prepare 6 CFPS reactions comprising a common DNA construct and each of the 6 different variations of the Salt Solution of Subheading 3.1.4, **step 1** (*see* **Note 12**).
3. Set up each reaction in 2-mL microcentrifuge tubes and centrifuge at 1000 × g for 1 min (*see* **Note 13**).

Table 2
Reaction protocol for CFPS of target acceptor protein

Component	Volume (μL)
15× salt solution	1.0
15× master mix	1.0
6.52× reagent mix	2.3
Plasmid DNA (200 ng/μL)	1.0
CFPS lysate	4.5
Nuclease-free water	5.2
Total	15.0

4. Place the 2 mL microcentrifuge tubes in a heat block or water bath set at 30 °C and let the reaction proceed for 16 h. Once complete, reactions can be stored at −20 °C.
5. Evaluate the protein synthesis yield to determine the optimal magnesium glutamate concentration for the batch of CFPS lysate (*see* **Note 14**).

3.2 Solvent Extraction of LLOs from E. coli *(Fig. 3a)*

Chloroform must be handled in the fume hood and appropriate PPE must be worn at all times. Please coordinate proper disposal procedures for chloroform and other organic wastes. Chloroform can be contained in disposable polypropylene tubes

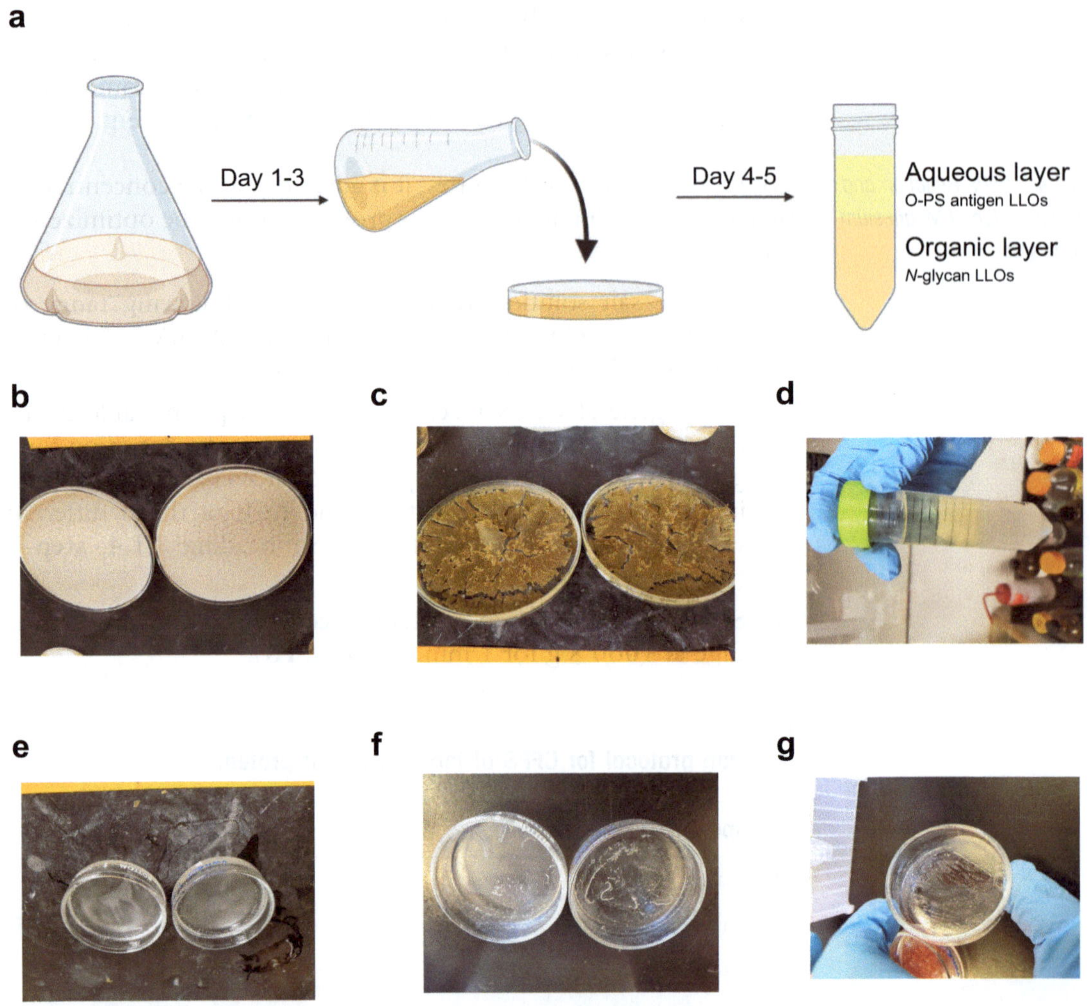

Fig. 3 Solvent extraction of LLOs from *E. coli*. (**a**) Schematic of the LLO extraction process from *E. coli* cells. (**b**) Cell resuspended in methanol and prior to drying (Subheading 3.2.2, **step 7**). (**c**) Dried cell pellets (Subheading 3.2.3, **step 1**). (**d**) Two separated layers at the end of the extraction process (Subheading 3.2.3, **step 15**). (**e**) Organic layer collected and prior to drying (Subheading 3.2.3, **step 17**). (**f**) Dried LLOs in the glass plate (Subheading 3.2.4, **step 1**). (**g**) LLOs resuspended in 1× IVG buffer (Subheading 3.2.4, **step 1**)

only for a short time. Never store chloroform or chloroform waste in the tube for a long period of time as it can melt the tube.

3.2.1 Media Preparation and Inoculation of Overnight (Day 1)

1. Prepare 1 L of TB and potassium phosphate buffer (*see* **Note 15**).
2. Transfer the media into a 2.8-L baffled culture flask and autoclave to sterility.
3. Let the media cool at room temperature overnight.
4. Inoculate the overnight culture by growing CLM24 transformed with pMW07-pglΔB and 34 mg/mL chloramphenicol in 50 mL of LB at 37 °C with shaking at 250 rpm (*see* **Note 16**).

3.2.2 Cell Growth and Harvest (Days 2–3)

1. Add 100 mL of potassium phosphate buffer to chilled 900 mL of media and supplement with appropriate antibiotics.
2. Subculture 10 mL of the overnight culture into the 1 L of TB (1:100 volumetric ratio).
3. Incubate cells in a 37 °C shaker at 250 rpm until the OD_{600} of the culture reaches 0.7–0.8.
4. Induce with L-arabinose to a final concentration of 0.2% (w/v) and adjust the temperature of the shaking incubator to 30 °C; let the culture grow overnight for about 16 h.
5. The next day, harvest cells by centrifugation at 8000 × g for 25 min at 4 °C.
6. Discard the supernatant and fully resuspend cell pellets in 10 mL of methanol per gram of pellet.
7. Pour the resuspended cell solution into the glass Petri dishes and dry in the hood overnight (Fig. 3b).

3.2.3 LLO Extraction (Day 4)

1. Scrape the dried cell pellet material into a 50 mL conical tube. Pulverize the dried pellet with a spatula or a glass rod to aid in the extraction process (Fig. 3c).
2. Add 12 mL of chloroform:methanol (CM) 2:1 and sonicate in a water bath for 10 min.
3. Centrifuge at 3000 × g for 10 min at 4 °C in a centrifuge with a swinging bucket rotor to ensure a flat phase separation.
4. Decant the supernatant and repeat Subheading 3.2.3, **steps 2** and **3**.
5. Add 20 mL of H_2O to the tube and vortex for 10 s to mix.
6. Sonicate the tube in a water bath for 10 min or until the solution is homogeneous (*see* **Note 17**).
7. Centrifuge at 3000 × g for 10 min at 4 °C in a centrifuge with a swinging bucket rotor.
8. Decant and discard the supernatant.

9. Add 18 mL of chloroform:methanol:water (CMW) 10:10:3 (v/v/v) to the tube with the pellet and vortex for 10 s.
10. Sonicate in a water bath for 10 min or until homogeneous.
11. Add 8 mL of methanol and vortex for 3 s.
12. Centrifuge at 3000 × g for 10 min at 4 °C in a centrifuge with a swinging bucket rotor.
13. Transfer the supernatant to a fresh 50 mL conical tube and discard the pellet.
14. Add 8 mL of chloroform and 2 mL of H_2O.
15. Centrifuge at 3000 × g for 10 min at 4 °C in a swinging bucket rotor (Fig. 3d).
16. Aspirate and discard the top layer using a 5 mL pipette (the aqueous layer primarily composed of water and methanol) (*see* **Note 18**).
17. Transfer the bottom organic layer to clean glass Petri dishes and dry it overnight in the hood (Fig. 3e).

3.2.4 LLO Recovery (Day 5)

1. Resuspend the dried LLO on the plate (Fig. 3f) in 1.0 mL of 1× IVG buffer (Fig. 3g) (*see* **Note 19**).
2. The resuspensions should be distributed in 50-μL aliquots and then stored at −20 °C for up to a year. Glycan extraction quality is sufficient if approximately 100% glycosylation efficiency is observed via western blotting during the IVG reaction described in Subheading 3.4.

3.3 Expression, Membrane Solubilization, and Purification of CjPglB (Fig. 4a)

Samples must be kept on ice at all times unless noted otherwise. It is recommended to pre-chill equipment to 4 °C prior to the experiment.

3.3.1 Media Preparation and Inoculation of Overnight Culture (Day 1)

1. Prepare 1 L of TB and potassium phosphate buffer as described in the materials.
2. Transfer the media to a 2.8-L baffled culture flask and autoclave to sterility.
3. Let the media cool at room temperature overnight.
4. Inoculate the overnight culture by growing BL21(DE3) carrying plasmid pSN18 in 25 mL of LB supplemented with 50 μg/mL ampicillin (Amp) at 37 °C with shaking at 250 rpm.

3.3.2 Cell Growth and Protein Induction (Days 2–3)

1. Add 100 mL of potassium phosphate buffer to 900 mL of media and supplement with the appropriate antibiotics.
2. Subculture the 10 mL of the overnight cells into the prepared 1 L TB in a 1:100 volumetric ratio.

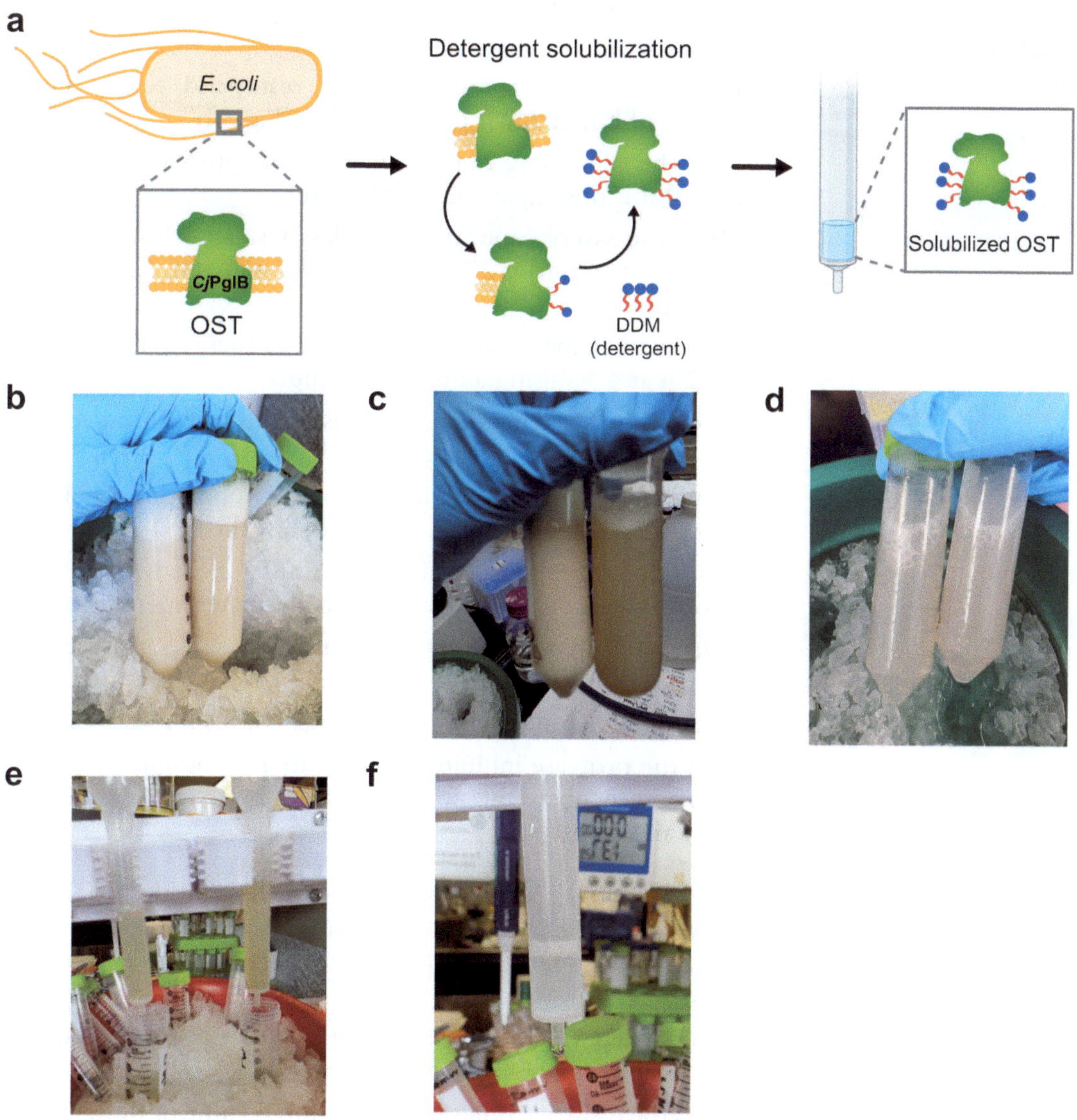

Fig. 4 Membrane isolation and purification of PglB. (**a**) Schematic of *Cj*PglB purification from *E. coli* cells. (**b**) Cells resuspended in buffer A (Subheading 3.3.3, **step 2**). (**c**) Lysed cells on the left and un-lysed cells on the right (Subheading 3.3.3, **step 3**). (**d**) Detergent-solubilized membrane protein (Subheading 3.3.3, **step 7**). (**e**) Sample loaded on the gravity column (Subheading 3.3.4, **step 1**). (**f**) Elution of the protein (Subheading 3.3.4, **step 4**)

3. Incubate cells in a refrigerated shaking incubator set to 37 °C and 250 rpm until the OD_{600} of the culture reaches 0.7–0.8.
4. Induce OST expression with L-arabinose to a final concentration of 0.2% (w/v) and adjust the temperature of the refrigerated shaking incubator to 16 °C; let the culture grow overnight for about 16 h.

3.3.3 Isolation and Extraction of Membrane Protein (Day 3)

1. The next day, harvest cells by centrifugation at 8000 × g for 25 min at 4 °C.
2. Discard the supernatant and fully resuspend cell pellets in 10 mL of buffer A per gram of pellet (Fig. 4b).
3. Lyse the cells using homogenizer B for approximately 1 min per 10 mL of sample volume (Fig. 4c).
4. Once the sample appears lysed (less viscous, clear, darker in color), centrifuge at 27,000 × g for 30 min at 4 °C to remove the cell debris.
5. Collect the supernatants and centrifuge them at 100,000 × g for 2 h at 4 °C using an ultracentrifuge.
6. Discard the supernatants and resuspend the resulting cell pellets with 25 mL of buffer B.
7. Incubate the sample rotating at room temperature for 1 h to allow the DDM to extract and solubilize the OST (Fig. 4d).
8. Centrifuge the sample again at 100,000 × g for 1 h at 4 °C to remove the membrane debris and collect the supernatants.
9. During the spin, equilibrate 0.8 mL of Ni-NTA agarose resin slurry with 5 column volumes of buffer B (*see* **Note 20**).
10. Add the pre-equilibrated Ni-NTA agarose resin to the samples.
11. Add the protease inhibitor cocktail to 1× working concentration and incubate the protein with Ni-NTA agarose resin for 16 h at 4 °C.

3.3.4 Protein Purification Using Affinity Column (Day 4)

1. Pour samples into a clean gravity column and let the sample run down completely, being sure to collect the flow through. Rinse the tube with flow through to ensure collection of all of the resin (Fig. 4e).
2. Repeat **step 15** once again with the flow through, collecting this fraction as the flow-through fraction.
3. Wash the resin five times with a column volume of Buffer C and collect the wash fraction. The first wash fraction should be incubated with the resin for at least 1 min.
4. After the washes, elute the protein by adding 1 column volume of Buffer D to the resin, incubating for 1 min, and subsequently eluting. Repeat this process a total of six times (Fig. 4f).
5. Remove the imidazole by buffer exchange with Buffer B using PD-10 desalting columns. If necessary, concentrate the sample using a 30 K MWCO protein concentrator to a volume of 3.6 mL. Expected A280 protein yield for a 1 L culture should be approximately 7 mg.

Table 3
Reaction protocol for IVG using unpurified CFPS product

Component	Volume (μL)
10× IVG buffer	5
1 M $MnCl_2$	1
sfGFP$^{DQNAT\text{-}6xHis}$ CFPS reaction sample	0.5
Extracted *Cj*LLOs	10
Membrane-purified *Cj*PglB	15
RNAase-free water	18.5
Total volume	50

3.4 IVG Utilizing CFPS Reaction Sample

1. Carry out a CFPS reaction via Subheading 3.1.4, **steps 2–4** with DNA construct encoding a target protein of interest. Utilize the salt solution with optimal magnesium glutamate concentration for the given batch of CFPS lysate.
2. To carry out an IVG reaction on the resulting CFPS reaction sample, resulting in glycosylated target proteins, utilize the recipe in Table 3 (*see* **Note 21**).
3. Set up each reaction in 0.2 mL PCR tubes.
4. Place the tubes in a heat block or water bath set at 30 °C and let the reaction proceed for 16 h.
5. Carry out Western blot analysis of the produced sfGFP$^{DQNAT\text{-}6xHis}$ glycoprotein with an anti-His antibody to detect the protein and hR6 serum to detect the glycan; alternatively, soybean agglutinin (SBA), a commercially available lectin specific for *N*-acetylgalactosamine, can be used for detection of the *C. jejuni* heptasaccharide glycan (*see* **Note 22**).
6. A representative image of target protein expression and glycosylation via CFPS and IVG is provided (Fig. 5).

4 Notes

1. When producing lysates for CFGpS reactions in Subheading 3.1.1, **step 2**, the strain of interest should be CLM24 containing a plasmid incorporating a gene cluster for the glycan biosynthesis pathway of interest and a plasmid encoding the OST of interest.
2. Use 250 mL baffled flasks or larger. Picking a colony to inoculate with is preferred over direct inoculation from glycerol stocks. The strain of interest should ideally be a B strain with a DE3 locus to induce T7 polymerase expression in

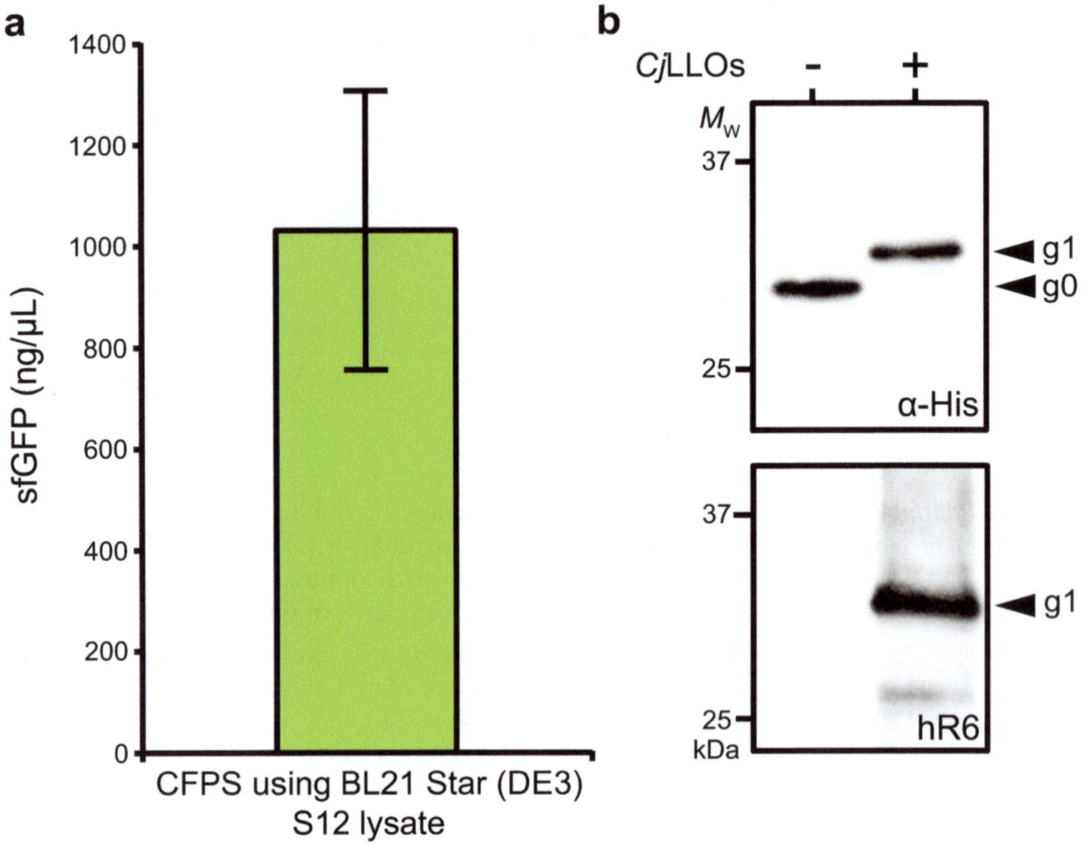

Fig. 5 Characterization of glycoproteins produced via glycoCFE. (**a**) Quantification of sfGFP$^{DQNAT\text{-}6xHis}$ production in CFPS using BL21 Star (DE3) S12 lysate. Mean of three technical replicates of CFPS reactions from the same lysate with error reported as standard deviation. (**b**) Representative Western blot of sfGFP$^{DQNAT\text{-}6xHis}$ produced in CFPS reaction subjected to IVG reaction with (+) and without (−) *Cj*LLOs. Anti-His antibody (α-His) was used to detect His-tagged sfGFP$^{DQNAT\text{-}6xHis}$ and anti-glycan serum (hR6) was used to detect the *C. jejuni* heptasaccharide glycan produced from plasmid pMW07-pglΔB

Subheading 3.1.2, **step 6**. When producing lysates for CFGpS reactions, do not add glucose to the media if your glycan biosynthesis pathway and OST plasmids are induced by L-arabinose, as is the case for the plasmids pMW07-pglΔB and pSF-*Cj*PglB. It should be avoided as it will suppress the expression of the induced proteins. Instead, bring the media to the 1-L working volume with autoclaved water.

3. Set aside 5 mL of 2× YTP + 1.8% (w/v) glucose media prior to inoculation for blanking. Pre-chill the harvesting centrifuge to 4 °C. 2.5 L Tunair cell culture flasks are preferred for optimal lysate preparation but can be substituted for more typical culture flasks. When producing lysates for CFGpS reactions, the appropriate antibiotics should be used in Subheading 3.1.1, **step 2** and Subheading 3.1.2, **step 2**.

4. If the strain of interest does not require induction, skip **steps 4** and **5**, continue culture growth until OD_{600} of 3.0, and advance to **step 6**. When producing lysates for CFGpS reactions, the induction in Subheading 3.1.2, **step 5** should induce expression of both the OST of interest and the LLO synthesis pathway and the temperature should be lowered to 30 °C post induction. Typically, this is done with 0.02% (w/v) L-arabinose for the plasmids noted.
5. We have observed that the time interval between cell harvest and lysis shows a negative correlation with lysate quality. As such, it is advised to minimize the lag between cell harvest and lysis by pre-assembling all materials and reagents required and keeping them on hand prior to cell harvest.
6. Wiping the inside of the centrifuge bottle prior to scraping the pellet from the bottle will help minimize media contamination in your lysates. Make sure not to disturb the pellet while you wipe.
7. Set the vortex to about 3/4 of its maximum power setting, with 15 s of vortexing followed by 15 s on ice to prevent the sample from warming.
8. The protocol can be paused at this step by flash freezing the pellet and storing at −80 °C. When deciding to pause, the weight of the pellet should be recorded and saved.
9. For preparation of lysates from BL21(DE3), sonication or homogenization may be used [28]. However, for preparation of CFGpS lysates from CLM24, homogenization is highly preferred for improved glycosylation efficiency [15].
10. It is crucial that no portion of the pellet is collected alongside the supernatant. When transferring the supernatant, it is advised to leave behind a small volume of the supernatant so as not to disrupt the pellet. For a 1 mL sample, it is advised to take roughly 800 μL supernatant, as not to disturb the pellet. When producing lysates for CFGpS reactions, between Subheading 3.1.3, **steps 4** and **5**, incubate the supernatant at 37 ° C, shaking at 250 rpm, for 1 h in a step typically referred to as a run-off reaction. All lysates made from *E. coli* K strains should include the run-off reaction.
11. If the lysed cells were divided into aliquots in **steps 3–5**, first combine the centrifuged supernatants prior to gently mixing.
12. We typically utilize $sfGFP^{DQNAT\text{-}6xHis}$ during the manganese optimization process as the $sfGFP^{DQNAT\text{-}6xHis}$ yield is easily quantified via fluorescence readings and it has robust expression in CFPS reactions. When carrying out CFGpS reactions, include 0.3 μL of commercial T7 RNA polymerase in the mixture. Additionally, incubate at 30 °C for 5 min in Subheading 3.1.4, **step 4**.

13. We have observed that the higher surface volume of the reaction leads to an increase in protein synthesis yield. The use of 2 mL microcentrifuge tubes and the corresponding centrifugation step is intended to increase the surface volume of the reaction.
14. Protein synthesis yield in CFPS cannot be quantified reliably with Coomassie staining for most proteins due to background from *E. coli* proteins present in the lysates. The DNA construct should ideally be of a fluorescent protein, for which the protein synthesis yield can be determined via its fluorescence. Alternatively, the synthesis yield of proteins with fused tags can be determined by quantitative Western blot analysis.
15. Potassium phosphate buffer should be separately prepared from the above materials and added to the media after the autoclave.
16. MC4100 *gmd*::kan Δ*waaL* should be used for $Man_3GlcNAc_2$ glycan extraction since the *gmd* knockout increases the intracellular level of GDP-mannose, an important precursor for the production of the $Man_3GlcNAc_2$ glycan.
17. Vortexing the sample occasionally can speed up the homogenization process.
18. O-antigen polysaccharides (O-PS) from gram-negative bacteria accumulate in the aqueous layer rather than the organic layer. Therefore, the aqueous fraction must be saved for extraction of O-PS LLOs rather than the organic fraction. Each fraction should be approximately 20 mL in volume.
19. Glycans can be further purified using graphite carbon columns in an HPLC and quantified using MALDI-TOF mass spectrometry.
20. The amount of Ni-NTA agarose resin should be adjusted proportionally depending on the culture size, mass of the cell pellet, and the final yield of the protein.
21. When producing lysates for CFGpS reactions for Subheading 3.3.4, substitute the recipe with that in Table 4. The amount of CFPS reaction samples to utilize in the IVG reaction will differ

Table 4
Reaction protocol for one-pot CFGpS of a target glycoprotein

Component	Volume (μL)
CFPS reaction sample	15
10× IVG buffer (500 mM HEPES pH 7.4, 1% (w/v) DDM)	2
133 mM $MnCl_2$	3

depending on the target protein yield from the CFPS reaction. Adjust the recipe accordingly to optimize IVG reactions for the target protein of interest, desired glycosylation efficiency, and desired glycoprotein yield. The IVG reaction can also be scaled to glycosylate the entire volume of CFPS product.

22. If the yield of the target protein in the CFPS reaction is low, it may be necessary to scale the reactions and purify the target proteins from the CFPS/IVG reaction prior to gel electrophoresis. Additionally, unpurified and purified products may be characterized via LC-MS analysis for further confirmation of protein production, glycan attachment, and glycosylation efficiency. It is expected that approximately 100% glycosylation efficiency will be achieved for the system described here.

Conflicts of Interest

The authors declare the following competing financial interest(s): M.P.D. has a financial interest in Gauntlet, Inc., Glycobia, Inc., SwiftScale Biologics, Inc., Versatope, Inc., Gauntlet Bio, and UbiquiTx, Inc. M.C.J. has a financial interest in SwiftScale Biologics, Gauntlet Bio, Pearl Bio, Inc., Design Pharmaceutics, and Stemloop Inc. M.P.D.s and M.C.J.s interests are reviewed and managed by Cornell University and Northwestern University and Stanford University in accordance with their competing interest policies. All other authors declare no competing interests.

Funding

This work was supported by Defense Threat Reduction Agency (HDTRA1-20-10004 to M.P.D. and M.C.J.), Army Contracting Command (W52P1J-21-9-3023 to M.C.J.), the National Science Foundation (CBET-1936823 to M.P.D. and CBET-1936789 to M.P.D. and M.C.J.), and the Defense Advanced Research Projects Agency (DARPA contract W911NF-23-2-0039 to M.P.D. and M. C.J.). E.J.B was supported by an NIH/NIGMS Chemical Biology Interface Training Grant (T32GM138826) and an NSF Graduate Research Fellowship (DGE-2139899).

References

1. Varki A (2017) Biological roles of glycans. Glycobiology 27:3–49
2. Walsh G, Walsh E (2022) Biopharmaceutical benchmarks 2022. Nat Biotechnol 40:1722–1760
3. Kaneko Y (2006) Anti-inflammatory activity of immunoglobulin G resulting from Fc sialylation. Science 313:670–673
4. Lin C-W, Tsai M-H, Li S-T et al (2015) A common glycan structure on immunoglobulin

G for enhancement of effector functions. Proc Natl Acad Sci U S A 112:10611–10616
5. Durocher Y, Butler M (2009) Expression systems for therapeutic glycoprotein production. Curr Opin Biotechnol 20:700–707
6. Hershewe J, Kightlinger W, Jewett MC (2020) Cell-free systems for accelerating glycoprotein expression and biomanufacturing. J Ind Microbiol Biotechnol 47:977–991
7. Rothblatt JA, Meyer DI (1986) Secretion in yeast: reconstitution of the translocation and glycosylation of α-factor and invertase in a homologous cell-free system. Cell 44:619–628
8. Gurramkonda C, Rao A, Borhani S et al (2018) Improving the recombinant human erythropoietin glycosylation using microsome supplementation in CHO cell-free system. Biotechnol Bioeng 115:1253–1264
9. Tarui H, Imanishi S, Hara T (2000) A novel cell-free translation/glycosylation system prepared from insect cells. J Biosci Bioeng 90:508–514
10. Brödel AK, Sonnabend A, Roberts LO et al (2013) IRES-mediated translation of membrane proteins and glycoproteins in eukaryotic cell-free systems. PLoS One 8:e82234
11. Guarino C, DeLisa MP (2012) A prokaryote-based cell-free translation system that efficiently synthesizes glycoproteins. Glycobiology 22:596–601
12. Jaroentomeechai T, Stark JC, Natarajan A et al (2018) Single-pot glycoprotein biosynthesis using a cell-free transcription-translation system enriched with glycosylation machinery. Nat Commun 9:2686
13. Zemella A, Thoring L, Hoffmeister C et al (2015) Cell-free protein synthesis: pros and cons of prokaryotic and eukaryotic systems. Chembiochem 16:2420–2431
14. Aquino AK, Manzer ZA, Daniel S et al (2021) Glycosylation-on-a-Chip: a flow-based microfluidic system for cell-free glycoprotein biosynthesis. Front Mol Biosci 8:782905
15. Hershewe JM, Warfel KF, Iyer SM et al (2021) Improving cell-free glycoprotein synthesis by characterizing and enriching native membrane vesicles. Nat Commun 12:2363
16. Kightlinger W, Duncker KE, Ramesh A et al (2019) A cell-free biosynthesis platform for modular construction of protein glycosylation pathways. Nat Commun 10:5404
17. Natarajan A, Jaroentomeechai T, Cabrera-Sánchez M et al (2020) Engineering orthogonal human *O*-linked glycoprotein biosynthesis in bacteria. Nat Chem Biol 16:1062–1070
18. Kightlinger W, Lin L, Rosztoczy M et al (2018) Design of glycosylation sites by rapid synthesis and analysis of glycosyltransferases. Nat Chem Biol 14:627–635
19. Techner J-M, Kightlinger W, Lin L et al (2020) High-throughput synthesis and analysis of intact glycoproteins using SAMDI-MS. Anal Chem 92:1963–1971
20. Lin L, Kightlinger W, Prabhu SK et al (2020) Sequential glycosylation of proteins with substrate-specific N-glycosyltransferases. ACS Cent Sci 6:144–154
21. Stark JC, Jaroentomeechai T, Moeller TD et al (2021) On-demand biomanufacturing of protective conjugate vaccines. Sci Adv 7:eabe9444
22. Warfel KF, Williams A, Wong DA et al (2023) A low-cost, thermostable, cell-free protein synthesis platform for on-demand production of conjugate vaccines. ACS Synth Biol 12:95–107
23. Stark JC, Jaroentomeechai T, Warfel KF et al (2023) Rapid biosynthesis of glycoprotein therapeutics and vaccines from freeze-dried bacterial cell lysates. Nat Protoc:1–25
24. Feldman MF, Wacker M, Hernandez M et al (2005) Engineering *N*-linked protein glycosylation with diverse O antigen lipopolysaccharide structures in Escherichia coli. Proc Natl Acad Sci U S A 102:3016–3021
25. Lopez PJ, Marchand I, Joyce SA et al (1999) The C-terminal half of RNase E, which organizes the *Escherichia coli* degradosome, participates in mRNA degradation but not rRNA processing in vivo. Mol Microbiol 33:188–199
26. Ollis AA, Zhang S, Fisher AC et al (2014) Engineered oligosaccharyltransferases with greatly relaxed acceptor-site specificity. Nat Chem Biol 10:816–822
27. Kowarik M, Numao S et al (2006) *N*-linked glycosylation of folded proteins by the bacterial oligosaccharyltransferase. Science 314(5802): 1148–1150
28. Kwon Y-C, Jewett MC (2015) High-throughput preparation methods of crude extract for robust cell-free protein synthesis. Sci Rep 5:8663

Chapter 20

Considerations for Glycoprotein Production

Elizabeth C. Clarke

Abstract

This chapter is intended to provide insights for researchers aiming to choose an appropriate expression system for the production of recombinant glycoproteins. Producing glycoproteins is complex, as glycosylation patterns are determined by the availability and abundance of specific enzymes rather than a direct genetic blueprint. Furthermore, the cell systems often employed for protein production are evolutionarily distinct, leading to significantly different glycosylation when utilized for glycoprotein production. The selection of an appropriate production system depends on the intended applications and desired characteristics of the protein. Whether the goal is to produce glycoproteins mimicking native conditions or to intentionally alter glycan structures for specific purposes, such as enhancing immunogenicity in vaccines, understanding glycosylation present in the different systems and in different growth conditions is essential. This chapter will cover *Escherichia coli*, baculovirus/insect cell systems, *Pichia pastoris*, as well as different mammalian cell culture systems including Chinese hamster ovary (CHO) cells, human endothelial kidney (HEK) cell lines, and baby hamster kidney (BHK) cells.

Key words Glycosylation, Expression system, Glycans, Recombinant protein production, Glycoprotein, Complex glycans, N-linked glycosylation, O-linked glycosylation, *Escherichia coli*, Baculovirus, *Pichia pastoris*, Chinese hamster ovary cells, Human endothelial kidney

1 Introduction

Glycans provide an additional layer of complexity, information, and structure to proteins, and have roles in cell adhesion, signaling, immune recognition, and growth and repair. Glycosylated proteins are essential for multicellularity [1], and the glycans add stability, binding sites, and targeting to a diverse array of proteins [2]. It is therefore necessary for methods of protein production to consider what kind of glycosylation a recombinantly produced protein will have, if any. Selection of an appropriate expression system will be dependent on the intended applications and desired features of the recombinant protein, but this chapter lays out some of the primary considerations for choosing an expression system.

Steven B. Bradfute (ed.), *Recombinant Glycoproteins: Methods and Protocols*, Methods in Molecular Biology, vol. 2762, https://doi.org/10.1007/978-1-0716-3666-4_20,

The considerations for producing glycoproteins can be more complicated than for producing nucleic acids and proteins, as the combinations of specific glycans imparted in different patterns are almost limitless for a glycoprotein with multiple sites of N-linked glycosylation. Unlike the discrete programming of nucleic acids and amino acids, glycosylation is not controlled by specific interactions with "blueprint" molecules, but by availability and abundance of specific glycosylating enzymes, such as enzymes that catalyze the formation of glycosidic linkages (i.e., glycotransferases) and enzymes that cleave glycan structures during glycosylation (i.e., glycosyl hydrolases). Further complexity may be added by the separate system of O-linked glycosylation. The balance between the enzymes in these systems therefore influences the outcome of the glycosylation pattern on the produced glycoprotein, and the expressed enzyme and identity in different production systems varies greatly.

This chapter will consider the production of glycoproteins from the perspective that the aim is to be producing glycoproteins as close to native conditions as possible. There are many reasons to wish to alter the glycans present on a glycoprotein for separate purposes, including preventing shielding by glycans, increasing binding to a specific lectin, or altering the stability of the protein. In contrast to the aim of producing human-like glycoproteins, there are numerous applications when more foreign or immunogenic glycans might be desirable. Consider production of glycoconjugates as bacterial vaccines, while not strictly recombinant glycoproteins, which utilizes the foreign sugar as a target for the immune system. Relatedly, there may be instances where, for glycoproteins that are to be used as vaccine antigens, less human-like glycosylation may be advantageous at targeting the protein to immune cells that are valuable for vaccine function [3].

This chapter will discuss N-linked glycans in terms of the specific monosaccharide residues imparted onto the protein. N-linked glycans are classed in four broad categories: high mannose, hybrid, complex, or truncated. The high mannose glycans contain five or more mannose sugars attached to two core N-acetyl glucosamines (GlcNAcs); hybrid glycans have one GlcNAc antenna and one antenna with core mannose; and complex glycans have at least one GlcNAc attached to each of the core mannose branches. Truncated glycans include paucimannose, where only one to four mannose sugars attached, and may have a fucosylated core GlcNAc; and hypermannosylated glycans are ones with many additional mannose sugars added. The fucosylation, present or absence of fucose on the core GlcNAc, does not affect the classification into the four categories, but glycans can be further classified as fucosylated, difucosylated, or afucosylated. Other sugars which may influence the character of the glycoproteins include galactose and sialic acid residues. There are ten common monosaccharide building

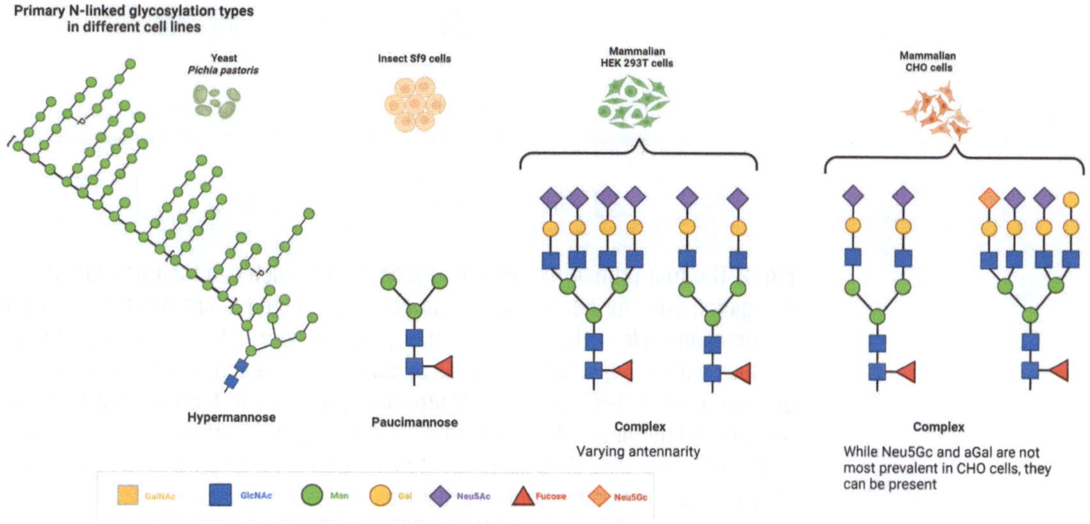

Fig. 1 N-glycan types produced by popular protein production cell lines. Pichia pastoris produce primarily hypermannose glycans with up to hundreds of additional mannose residues; Sf9 insect cell lines produce paucimannose type glycans; HEK293 human cell lines produce primarily complex glycans with terminal sialyation that differ in their number of antennae; CHO cells generally produce similar glycans to humans, but can carry the immunogenic αGal sugars and the terminal Neu5Gc. Branching and specific residue levels in mammalian cells may vary with mechanical and chemical environment

blocks to mammalian glycans, including glucose (Glc), fucose (Fuc), galactose (Gal), xylose (Xyl), mannose (Man), N-acetylglucosamine (GlcNAc), N-acetylgalactosamine (GalNAc), glucuronic acid (GlcA), iduronic acid (IdoA), and sialic acid (Neu5Ac) (Fig. 1).

The other primary type of glycosylation to consider is O-glycosylation, sometimes called mucin glycosylation, where the glycan is bound to the oxygen atom of serine (S) or threonine (T) amino acid, primarily in the cis-Golgi. Unlike N-linked glycosylation, O-linked glycosylation does not have a conserved amino acid motif to guide the enzymatic addition of the glycans. Most types of O-glycosylation are carried out in the secretory pathway in the Golgi and generally occur once the protein has folded [4].

Proteins with O-linked glycans often have clustered O-linked sites, as in mucins, the heavily glycosylated proteins secreted by epithelial cells. The clustering of O-linked sites makes it more difficult to assign functions to individual sites than with N-linked sites. O-linked glycans can be classified by the core that initially binds to the S/T that begins elongation. The first step in the synthesis of O-linked glycans requires catalysis by N-acetylgalactosaminyltransferase (GALNT) enzymes in the presence of UDP-N-acetylgalactosamine (UDP-GalNAc) as the carbohydrate donor [5]. The type of O-linked glycans can be classified by

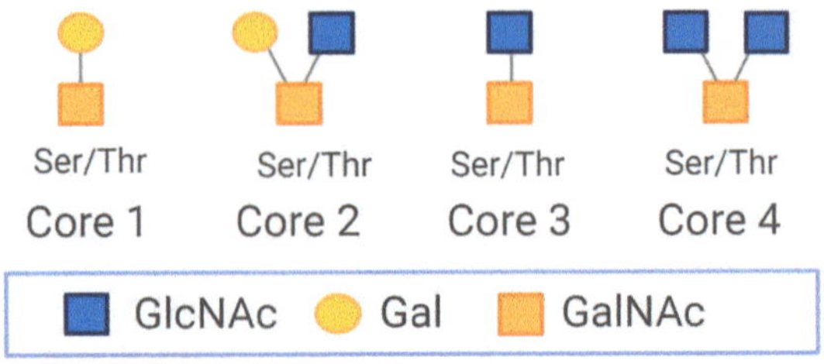

Fig. 2 The four primary cores of O-glycosylation which can then be modified by elongation with further monosaccharides. Core 1 and 2 are most common in mammalian cells. CHO cells primarily produce core 1 O-glycans and cannot produce core 3 O-glycans. HEK cells primarily produce core 1 and 2 also but can produce core 3 O-glycans, which are most prevalent in the epithelia from the gastrointestinal and respiratory tracts in humans. The Sf9 insect cell line may produce GalNAc only Tn O-glycans. *Pichia pastoris* do not produce a typical core O-glycan, but may add mannose residues to Ser/Thr residues

core, and there are 8 known mucin-like cores that begin with the GalNAc. Of these, core 1 and 2 are the most abundant, core 1 is composed of a single galactose and is attached to the base GalNAc. core 2 is formed by the addition of β1-6 N-acetylglucosamine to the N-acetylgalactosamine of core 1 [5] (Fig. 2). Subsequent elongation and termination of O-linked glycans is carried out by several glycosyltransferases. The relative expression and subcellular distribution of the various glycosyltransferases determine the outcome of O-glycan biosynthesis. Termination of O-linked glycans usually includes GalNAc, Galactose, N-acetylglucosamine, fucose, and sialic acid and up to 20 sugars can be added to an individual O-linked site. Each core can be extended by up to 20 sugar residues to give linear or branched chains. By far the most common modification of the core structures most abundant in mammals is the addition of sialic acid to result in mono-, di-, or trisialylation of core 1 and 2. Finally, there are antigenic O-glycans with incomplete glycosylation. T antigen (tumor-associated) antigen is a core 1 Gal-β(1 $\rightarrow$ 3)GalNAc that results from removal of sialic acids from the core 1, or from a lack of glycotransferases to achieve the elongation. The Tn antigen is just the GalNAc without any further extension.

There are several commonly used expression systems for protein production, and this chapter will discuss the merits and considerations most relevant for each for the production of glycoproteins, *Escherichia coli*, baculovirus/insect cell systems, *Pichia pastoris*, and different mammalian cell culture systems are described, as well as more recent cell-free production systems.

2 Expression Systems

2.1 Bacterial Expression Systems

Bacterial expression systems are easy to culture with a rapid life cycle and are amendable to genetic manipulation. The *Escherichia coli* system was the first culture system used to produce a biopharmaceutical, human insulin produced by Eli Lilly in 1982, to replace porcine insulin [6]. As well as the lack of glycosylation, the lack of capacity of *E. coli* for protein folding and other post-translational modifications such as pegylation [7] and forming disulfide is not sufficient for many recombinant proteins. While there are some difficulties in producing soluble proteins that are overexpressed in the *E. coli* system, the primary difficulties in producing human-relevant proteins in *E. coli* is the lack of human post-translational modifications, including glycosylation [8]. Finally, many of the advantages in cost and yield in the industrialization of *E. coli*-produced proteins occur because the large amounts of proteins are produced in *E. coli* inclusion bodies and chemically refolded. The compatibility of this system, with the highest yields and lowest costs associated, with a glycosylated protein is currently unfeasible.

Bacterial systems were initially thought to possess no N-linked glycosylation machinery, but N-linked glycosylation systems were initially discovered in epsilonproteobacterium *Campylobacter jejuni* [9], making it clear that some members of the bacteria domain have N-linked glycosylation pathways, albeit with significant differences compared to eukaryotic glycosylation pathways [10]. In the bacterial system, glycosylation occurs post-translationally on accessible residues in flexible areas, and is a block transfer where the whole sugar is added to the glycosylated residue in its final state, compared to eukaryotic systems where N-linked glycosylation occurs co-translationally and can influence protein folding and the nascent glycan is modified sequentially in the endoplasmic reticulum [8]. The *C. jejuni* system is the best characterized of bacterial glycosylation pathways [11], although other bacteria have glycosylation machinery, either naturally or through recombinant engineering [12]. One significant difference between eukaryotic glycosylation and bacterial is the consensus motif, where eukaryotes use Asparagine_X_Serine/Threonine (where X is any amino acid except Proline, and bacteria require a longer sequence that requires Aspartate or Glutamate at -2 position Aspartate/Glutamate_X1_-Asparagine_X2_Serine/Threonine (where X1 and X2 are any amino acids except Proline) [13]. The specific consensus region may also vary between different bacterial species that have the ability to impart glycans [14]. The changed acceptor site means that any sequences from mammalian systems that would be desirable to express and glycosylate in the modified bacterial system would have to have sequence modification to achieve this.

In *C. jejuni*, a heptasaccharide is formed on the cytoplasmic side of the inner membrane on undecaprenyl pyrophosphate (Und-PP) compared to the eukaryotic dolichyl pyrophosphate (Dol-PP) formed in the endoplasmic reticulum (ER) membrane [15]. The lipid-linked oligosaccharide is then translocated into the periplasmic space by ATP-dependent flippase, PglK, and transferred onto asparagine residues on the protein by bacterial oligo saccharyltransferase PglB13. The resulting glycoprotein has a distinctive heptasaccharide added, the α-d-GalNAc(1–4)α-d-GalNAc (1–4)α-d-GalNAc(β-d-Glc(1–3))(1–4)α-d-GalNAc(1–4)α-d-GalNAc(1–3)α-d-di-N-acetylbacillosamine (diNAcBac) -β1 sugar. The actual heptasaccharide sugar added to the protein in these bacterial systems does not resemble the sugars expressed in humans and other mammals [15]. Additionally, the bacterial N-glycan is highly immunogenic, specifically being recognized by the human macrophage galactose-type lectin (MGL) [16], which may pose problems for various in vivo applications.

Differences in the glycan added to the protein are perhaps the major limitation in expressing most glycosylated proteins in modified bacterial systems, and would seem to limit the *C. jejuni* enzymes-based systems to producing glycoconjugate vaccines where bacterial glycans are desirable, enzymes for biomedical use where the glycosylation is required for function, or bacterial proteins [15]. However, there are various efforts to work around the central issue of the glycans themselves. The first involves initial production in the bacterial expression systems, followed by in vitro modification of the product. Using the *C. jejuni* system expressed in *E. coli*, the GlcNAc-Asn linkage is established, but the bacterial glycans are then trimmed and remodeled by enzymatic transglycosylation [17] to produce eukaryotic N-glycoproteins.

The in vitro enzymatic modification of *E. coli* glycosylated proteins may then require further purification to obtain the correct glycoforms and can remove many of the advantages otherwise presented by the bacterial system, namely speed of production and cost [18]. The second set of methods to try to solve the problem of bacterial glycans involves the in vivo construction of a eukaryotic glycosylation pathway in *E. coli* that generates human-like N-glycans. The transfer of four eukaryotic glycosyltransferases, including the yeast uridine diphosphate N-acetylglucosamine transferases Alg13 and Alg14 and the mannosyltransferases Alg1 and Alg2 into *E. coli*, along with the *C . jejuni* PglB, allowed for the production of proteins with simple eukaryotic glycans [19]. However, while the achievement of eukaryotic glycosylation in *E. coli* is impressive, the yields of glycosylated proteins were less than 1% of the total protein produced, and so an efficient in vivo system in bacteria remains an elusive challenge despite much speculation on the subject [20]. A protocol for updated bacterial systems is described in Chap. 10 of this book.

2.2 Yeast Expression Systems

The yeast protein production systems are widely used as they can produce high yields of proteins and are easy to scale to higher volume culture. Unlike bacteria, yeasts are capable of performing many human post-translational modification reactions. However, the glycosylation in *Pichia pastoris* differs significantly from mammalian and specifically human glycosylation as yeast produce hypermannosylated glycans. *P. pastoris* are the most widely used yeast production system, and have been shown to produce proteins at up to 14.8 g/L [21]. Recent work with tetanus toxin fragment has shown that the alteration of specific glycosylation sites on glycoproteins produced by *P. pastoris* dramatically alters the yield of the expressed protein, dependent on the site of the glycan and the number of glycan sites [22].

The initial stages of the eukaryotic glycosylation in all systems are very similar. The glycan itself is initially generated on the cytoplasmic side of the ER membrane as Man5GlcNAc2-P-dolichol which is then moved to the ER membrane by the flipase enzyme, to be extended to Glc3Man9GlcNAc2 and transferred as a unit on the N-X-S-T motif of the nascent peptide by the oligosaccharyltransferase enzyme complex. The glycan is then trimmed, with three glucose residues and one mannose residue removed to generate Man8GlcNAc2. After this point, mammalian and yeast systems diverge. Mammalian cells trim initially to Man5GlcNAc2, and then continue to trim and extend the glycan to produce a huge variety of different glycans, but primarily sialylated complex glycans. In contrast, in yeast systems the Man8GlcNAc2 structure is not trimmed, and more mannose sugar moieties may be added, resulting in hypermannosylated glycoproteins. *P. pastoris* produce relatively shorter hyper-mannosylated compared to other yeast species such as *Saccharomyces cerevisiae* [23].

The hypermannosylated glycoproteins are highly immunogenic in humans, perhaps unsurprisingly as many yeasts are associated with human disease. Some pathogenic yeast (*Candida albicans*) add an additional poly-β1,2-Man extension to their N-glycans, making them distinct from non-pathogenic yeasts and contributing to virulence.

As with bacteria, there have been extensive efforts to modify *P. pastoris* glycosylation machinery to result in a more human-like glycosylation. Efforts to humanize the glycosylation process in Pichia have focused on eliminating the enzyme responsible hyperglycosylation, the enzyme OCH1 (α-1,6 mannosyltransferase). One method of humanizing the glycoforms of *P. pastoris* for therapeutically relevant glycoproteins has been achieved with glycoengineered *P. pastoris* strains using the GlycoSwitch® [24]. The Glycoswitch general strategy involves the sequential introduction of glycosylation enzymes after the OCH1 knockout [24]. The conversion of any wild-type strain into a strain that modifies its glycoproteins with Gal2GlcNAc2Man3GlcNAc2 N-glycans requires the introduction of five GlycoSwitch vectors. The efforts to humanize the glycosylation pathways in *P. pastoris* are reviewed in [25].

2.3 Insect Cells and Baculovirus Systems

Insect cell systems have been used successfully for the production of glycosylated proteins and are considered particularly versatile due to the ability to use the cells with the baculovirus expression system. Notably, insect cell production has been used to produce Flublok, a trivalent Influenza A hemagglutinin vaccine, which was initially FDA-approved in 2013 (and approved as a quadrivalent vaccine in 2016) [26]. The baculovirus-insect expression systems work through viral infection of insect cell lines to drive the expression of virally encoded proteins of interest. The baculoviruses, arthropod-specific viruses in the Baculoviridae family, infect a variety of insect cell lines derived from lepidopteran (moth and butterfly) family members. The most widely used of these are the *Spodoptera frugiperda* (Sf) cells, including Sf9 and Sf21 cell lines [27]. The insect cell line for production choice itself is important, and is not limited to just Sf species. Cell lines stemming from *Trichoplusia ni*, including BTI-TN-5B1-4, commercially available as "High-Five™"cells [28], are claimed to give higher expression of certain proteins than Sf cell lines.

One of the key roles of glycosylation plays is chaperoning of the glycoproteins in the endoplasmic reticulum. After the initial OST transfer of the glycan, the protein carries a tri-glucosylated residue, which is cleaved by α-glucosidases in the ER to result in a mono-glucosylated glycoprotein. Misfolded proteins are recognized by UDP-Glc:glycoprotein glucosyltransferase and reglucosylated. Mono-glucosylated N-glycans can serve as a recognition signal for calnexin and calreticulin, and for some glycoproteins this process is essential for their correct folding [29–31]. This pathway, including the α-glucosidases and UDP-Glc:glycoprotein glucosyltransferase, exists in insects, which can limit the frequency of misfolded glycoproteins when compared to yeast and bacterial systems [32].

In fact, the early stages of glycosylation in insect cells and mammalian cells are essentially identical. The presence of oligomannosidic N-glycans with five to nine mannose residues (Man5, Man6, Man7, Man8, and Man9) on proteins produced from insect cells is common [33]. However, the high mannose type glycans may not make up the majority of the glycans imparted in insect systems [34]. Insect cells tend to produce clipped paucimannose structures (Man1–3GlcNac2Fuc0–1). The paucimannose glycans imparted in insect cells may initially seem to be a simple form of glycan, but the biosynthesis of these glycans occurs after the addition of modifications like the addition of glucose and galactose associated with complex glycans. The biosynthesis of paucimannosidic glycans requires the prior action of N-acetylglucosaminyltransferase I (GlcNAc-TI), the enzyme that acts GlcNAc residues for further elongation, in contrast to the high mannose glycans that also have terminal mannose residues and are more correctly classed as "simple" glycans. For the paucimannose glycans, the GlcNAc-TI activity generates the necessary "GO" signal for core fucosylation [35]. The removal of the GlCNAc residues necessary for further elongation, and added by GlcNAc-TI, is mediated by the fused lobes hexosaminidase (Fdl).

Efforts to create a more mammalian type of glycosylation in proteins produced in insect cells may involve mutations in the fused lobe gene or overexpression of mammalian glycosyltransferase enzymes that cap the GlcNAc residues [33]. In general, mammalian cells do not produce high levels of paucimannose glycans, but do carry homologues of the fused lobes gene. In neutrophils, paucimannose glycans have been observed modulating the activity of human neutrophil elastase (HNE), an important N-glycosylated serine protease in the innate immune system, in activated neutrophils [36]. There is evidence that paucimannose glycans found in insect cell production are differentially immunogenic to mammalian glycoproteins in humans [37].

New insect cell lines with desirable features are still being developed, with an *Ascalapha odorata* cell line A038 shown to have similar glycosylation to Sf cell lines and the advantages of High Five cells for protein/cell mass, as well as being free of the alphanodavirus that was recently reported to contaminate cultures of High Five cells [38]. *Mamestra brassicae*-derived MB0503 cells are able to impart different fucosylation patterns by transferring fucose into α1–3 and α1–6 linkage to the first N-acetylglucosamine, whereas primarily α1–6-fucosyl linkages are produced in Sf9 cells [39]. Another lepidopteran cell line, *Estigmene acrea*, have also been used for protein production a baculovirus vector and glycosylation analysis has shown N-glycans from the fowl plague virus hemagglutinin mostly produced of glycan cores elongated by at least one terminal N-acetylglucosamine residue compared to Sf9 cells which produce predominantly trimannosyl core oligosaccharides[26].

The advantages of this system include the technical ease of cloning protein sequences into baculovirus vectors, particularly with the use of commercially available vectors, where the desired gene from the pFastBac vector recombines with the parent bacmid in DH10Bac *E. coli* competent cells to form an expression bacmid. The bacmid is then transfected into insect cells for the production of recombinant baculovirus particles [40].

Insect cells can be grown on serum-free media, as opposed to some mammalian cells, and particularly compared to mammalian cell culture on the lab scale. For large-scale culture, conditions have been developed which meet the temperature and gas requirements of insect cells [41]. However, baculovirus itself must be eliminated from the culture during the purification process, as the virus itself can be immunogenic in mammals. This can be a problem for virus-like particles produced in insect cells [42].

It is not clear how the baculovirus infection itself influences glycosylation, particularly of complex glycans, in the system. Using the human placental-secreted alkaline phosphatase (SEAP) protein, significant differences were observed comparing different baculoviruses and culture conditions. *Autographa californica nucleopolyhedrovirus* (AcMNPV), the most widely used baculovirus, produced 1% hybrid and no complex oligosaccharides attached to SEAP, in both *T. ni* (Tn-4h) and Sf9 cells. However, using *Trichoplusia ni nucleopolyhedrovirus* (TnSNPV) with *T. ni* (Tn-4h) cells, around 25% of the glycans contained terminal N-acetylglucosamine and/or galactose residues. The same study demonstrated the effect of the addition of 10% Fetal Bovine serum to culture, which resulted in 50% hybrid or complex glycans on SEAP compared to the 1% hybrid glycans attached in serum-free conditions [43].

In insect glycoproteins the proximal GlcNAc residue can be fucosylated at position 6, as is well known from mammalian glycoproteins, as well as in position 3 [33]. Due to the a1,3-fucosylated core GlcNAc, the peptide-N4-(N-acetyl-b-glucosaminyl)asparagine amidase (PNGase) used to cleave the glycan from the peptide should be the almond PNGase A, rather than the more widely used PNGase F from *Flavobacterium meningosepticum*. This difference in the fucosylation of the core GlcNAc is important as the a1,3 fucosylated GlcNAc may be highly immunogenic in humans, as shown by the IgE responses to this moiety in honey-bee venom allergies [44, 45]. It is difficult to know the extent of a1,3 fucosylation in Sf9 and Sf21 cells, as frequently mass spectrometry analysis techniques use PNGase F as the enzyme for release of the glycans for analysis, which will leave the a1,3-fucosylated core intact [44], but there have been reports of the linkage in Sf species cell lines [46].

This also relates to the specific cell line used, as *M. brassicae*-derived MB0503 cells are able to impart different fucosylation patterns by transferring fucose into α1–3 and α1–6 linkage to the first N-acetylglucosamine, compared to the only α1–6-fucosyl linkages are produced in Sf9 cells [44].

While the baculovirus system has advantages in terms of ease of use, stable transfection of insect cells presents an interesting alternative. The differences in culture conditions and cell types mean there is not a single type of glycosylation imparted on proteins using the baculovirus-insect cell system, but the marked alterations between different variations in culture conditions, cell lines, and viruses used should be investigated further to provide a toolbox of culture conditions to be tailored for specific uses.

2.4 Cell-Free Systems

Cell-free systems are an alternative method for producing proteins but have not been used extensively for glycosylated proteins. Cell-free systems use components derived from cells, using cell extracts or protein components generated synthetically for protein biosynthesis [47]. Optimized cell-free systems based on *E. coli* components can rapidly synthesize grams of protein per liter, are highly scalable, and the components can be freeze-dried for shelf stability, meaning they are an appealing system for producing recombinant proteins rapidly and in large amounts [48].

A current limitation of using cell-free systems for protein production are the challenges associated with producing glycosylated proteins. Many of the naturally occurring enzymes involved in many glycosylation modifications are membrane bound, and require the membrane architecture for function, making extracting them in a functional form in cell extracts challenging. Methods have been developed to create membrane proteins that associate with cell-derived vesicles or using synthetic nanodisks to provide structure for membrane protein function [49, 50]. However, progress to topologically alter membrane-bound glycotransferases into water-soluble versions expressed at higher levels in the cytoplasm is being made [51]. Rationally redesigned glycotransfereases provide an effective biosynthetic route to large quantities of these enzymes that can be used in cell-free systems for glycoprotein production [51]. Current cell-free systems have not been used to produce large quantities of glycoprotein; however, cell-free systems have the potential to produce glycans in a more tightly regulated manner compared to control of glycosylation components at precise ratios and in precise sequence in vivo so efforts to improve glycosylation in cell-free systems may result in significant benefits for production in the future [48].

2.5 Mammalian Cell Systems

A large number of mammalian-derived cell lines have been used to produce proteins, including Chinese hamster ovary (CHO) cells, human embryonic kidney (HEK) cells, and baby hamster kidney (BHK) cells. Additionally, there are specialized murine myeloma cell lines used to produce specific monoclonal antibodies. However, there is an assumption that using a mammalian cell line produces more "native" glycosylation on a recombinant glycoprotein, which ignores the huge diversity in the glycosylation patterns and levels imparted by different cell lines, and even the same cell lines grown in different growth conditions. Depending on the downstream use of the glycoprotein, the presence of many of the glycans that are immunogenic in humans may not pose an issue, for example, as small animal models do not have circulating antibodies against Neu5GC or α-Gal.

There are likely to be differences in producing a protein in cell culture conditions compared to the native cell the glycoprotein is produced in, particularly in viral antigens. For example, the gp120 from HIV is heavily glycosylated, and in native peripheral blood mononuclear cells that the virus infects, the N-glycans are primarily α2,6-linked sialic acids, but HEK293T-derived gp120 primarily carried α2,3-linked sialic acids [52]. The differences for antigens may be significant in interactions with lectins or binding to immune cells, as well as interactions with neutralizing antibodies, which are able to bind and recognize specific glycans [53].

2.5.1 Chinese Hamster Ovary Cells

Chinese hamster ovary (CHO) cells are the most established mammalian cell line used in biopharmaceuticals [54]. CHO cells can be cultured at very high cell densities in serum-free media at high volumes. For clinical-grade protein production, CHO cells have the longest track record, with up to 70% of therapeutic proteins produced in CHO cells [55]. CHO cells are frequently used for expression after stable transfections with high expression levels. For transient expression, the CHO cell line is more difficult to transfect compared to transfected HEK cells, although modified CHO lines for transfection exist [56].

N-glycans of human proteins possess both α2,6- and α2,3-linked terminal sialic acid (SA). Recombinant glycoproteins produced in CHO only have α2,3-linkage due to the absence of α2,6-sialyltransferase (St6gal1) expression [57]. However, more human-like glycosylation can be restored with transient sialyltransferase expression [58] or use of sialyltransferase expressing modified CHO cell lines [59].

As well as lacking certain desirable glycans, CHO cells also produce potentially immunogenic glycans for therapeutic use, specifically α-gal [60] and Neu5Gc [61]. Human N-glycans are lacking the terminal Gal 1-3Gal (α-Gal) modification, and humans, unlike most other mammals, do not produce Neu5Gc. Humans have Neu5Gc-specific antibodies, and CHO cell-produced monoclonal antibody cetuximab (Erbitux) was shown to carry Neu5Gc. Anti-Neu5Gc antibodies from healthy humans interact with cetuximab in a Neu5Gc and generate immune complexes in vitro [62].

It was initially thought that CHO cells lacked the biosynthetic machinery to synthesize glycoproteins with α-Gal antigens, unlike murine myeloma cell lines, and this was one of the advantages of the cell line. However, it has since been shown that CHO cells carry N-acetyllactosaminide 3-α-galactosyltransferase-1, which is responsible for the synthesis of the α-Gal epitope, and glycoproteins carrying this epitope are produced in CHO cells under certain conditions [60]. In an industrial setting the presence of the glycan can be monitored and clones of the cell line with absent expression selected for, but the possibility of the residue occurring should be considered in smaller production settings [55].

As CHO cells are the most widely used in biomanufacturing, there are numerous CHO cell lines that have been genetically modified for specific purposes. As well as modifications to glycosylation pathways that result in more human-like glycosylation, there are CHO cell lines specifically for the production of recombinant antibodies that have modifications to give desirable antibody features [63–65], such as afucosylation on the core GlcNAc through knockouts of the α1,6-fucosyltransferase (encoded by FUT8 gene) [66]. Interestingly, modifications to one aspect of the glycosylation machinery of the cell may also alter the abundance of non-altered glycans, as sialyltransferases and glucosyltransferases were decreased in FUT8 knockout CHO cells [66].

2.5.2 Human Endothelial Kidney (HEK) 293 Cells

The other primary cell line used in protein production is human endothelial kidney (HEK) 293 cells. The HEK293s are easy to grow with a rapid reproduction time, can be grown in serum-free conditions, and are efficient at producing proteins [67]. HEK293s are used commercially to produce a handful of recombinant proteins, including Dulaglutide for diabetes and a glycosylated recombinant factor VIII Fc fusion protein for hemophilia B treatment, and are both FDA-licensed treatments [68, 69]. HEK cells are human embryonic kidney cells that have been transformed by exposing cells to sheared fragments of adenovirus type 5 DNA [70]. Recombinant Hek293 cells constitutively expressing the Epstein Barr virus nuclear antigen (EBNA) allow for high levels of plasmid amplification, and recombinant HEK293 cells expressing the SV40 large T antigen are highly amenable to transfection and are capable of producing high titers of viral gene vectors for use in gene therapy [71]. HEK293 variants that have enabled the re-engineering of the cells toward enhanced use for manufacture-scale production of recombinant biopharmaceuticals including suspension adaptation and development of defined serum-free media are available [72].

While CHO cells dominate biomanufacturing, HEK cells are used extensively in academic laboratories, due to their ease of transfection and amenability to a number of transfection methods, and are often used in studies where small amounts of protein are sufficient, before moving to a stable cell line expression of CHO cells [68]. It is therefore important to verify that results obtained with glycoproteins produced in one cell line translate when switching to another cell line given the differences seen in glycosylation between the two [73]. When compared directly in the production of recombinant human coagulation factor VII (rFVII), HEK293-produced rFVII glycans have more structural variation and more terminal sialyation compared to CHO cell-produced rFVII glycans, along with a higher level of N-acetyl galactosamine [74]. Unlike CHO cells, HEK293 cells do not produce Neu5GC or α-gal, the primary immunogenic glycans CHO cells may produce under specific conditions. HEK293 cells do tend to produce a heterogeneous population of glycans, with individual N-linked sites on the Ebola glycoprotein produced in HEK293T cells carrying as many as 15 different glycans [34, 75].

2.5.3 Murine Hybridoma Cell Lines

Hybridoma cells are specialized antibody production cell lines that result from the fusion of B cells specific to an antigen and a myeloma cancer cell line. Since producing antibodies is necessary in biotherapeutics and research settings, and antibodies are glycoproteins with glycosylation sites that significantly impact function [65], the systems used to produce them should be carefully selected. A variety of hybridoma cell lines that express specific antibodies, such as BCF2 hybridoma from a Balb/c mouse that produces an IgG1 monoclonal antibody against toxin 2 of the Mexican bark scorpion (*Centruroides noxius Hoffmann*) [76, 77]. Since the antibodies are murine rather than human, they are generally not used to produce biotherapeutic antibodies for humans as the Fc portion of the protein is not always completely compatible with the human Fc receptors. However, even aside from effector function, murine antibodies should not be used as biotherapeutics as murine myeloma glycosylation differs significantly from glycans found in humans as they impart the highly immunogenic Galα(1,3)Gal epitope and N-glycolylneuraminic acid (Neu5Gc), sometimes referred to as NGNA, residues, without modification [78].

3 Additional Considerations

3.1 Culture Conditions in Mammalian Cell Culture

In large-scale production of glycoproteins, most systems do not use serum in the media used to grow cells. Both HEK293F cells and CHO cells have been adapted to grow in serum-free conditions, but as with the insect cell culture, serum-free systems produce different glycoforms compared to cells grown with serum in the media. In both serum-free and DMEM-containing fetal bovine serum media, there is heterogeneity in the glycans imparted onto the antibody structures, which is slightly decreased in the serum free media [77]. More significantly, the type of glycans differs depending on the media conditions [77, 79].

There are two mechanisms for the effects of the media on glycosylation: either the media itself contains proteins that alter the soluble protein directly, or metabolism of the cell changes and glycosylates the produced proteins differently. In the first case, β-galactosidase activity was three times higher in SFM cultures compared with DMEM/FBS cultures, which explains a decrease in terminal galactose residues on glycoforms of monoclonal antibodies produced in SFM conditions, where agalactosylated G0 glycans made up 58% of SFM antibody glycans compared to only 28% in DMEM/FBS [77]. In the second case, the process of adaptation to serum-free conditions results in changes to the glycosylation. The process of adaptation involves gradually reducing serum concentration in the cell culture medium. For intermediate levels of serum, a decrease of fucosylation, but increase in sialylation were observed, whereas sialyation recovered and fucosylation was lost when in the final lowest concentration of serum and SFM. The differences may be directly related to the serum level, or may be related to the growth and density of the cells achieved at the different stages of adaptations, or may relate the changes from adherent to suspension cell culture [80].

Aside from the serum, the mechanical and chemical control of cell culture can also influence glycosylation. A decreasing dissolved oxygen level modified the glycosylation pattern of a murine mAb in a murine hybridoma cell line, CC9C10, where the culture produced antibodies at dissolved oxygen (DO) concentrations of 10, 50, and 100% of air saturation. The dominant glycans produced were core fucosylated asialo chains with a shift toward decreased galactosylation as the DO concentrations were reduced [81]. In CHO cells, the cell surface polysialylation of glycoproteins may decrease with elevated partial pressure of CO_2 [82]. Finally, shear stress increases minimize site occupancy in glycoproteins produced in suspension CHO cells [83].

3.2 O-Linked Glycosylation

If the protein of interest has O-linked glycosylation in nature, the expression system should also be considered carefully, as eliminating O-glycans can have unclear effects on the protein. O-glycans are added in a truly post-translational manner, and so changes in O-glycans do not necessarily result in changes in broad structure, although they may still alter folding [4, 5].

While the role of O-glycans is understudied, there is evidence of the importance of O-linked glycans in viral lifecycles which should be considered if producing a viral antigen that would have O-glycans in nature. In herpes simplex virus type 1 (HSV1), the glycoprotein gC-1 contains numerous O-linked glycans [53]. Two basic amino acids were mutated to alanine and increased the number of O-glycans in the O-glycan rich region considerably, and resulted in a virus with impaired binding to cells expressing chondroitin sulfate (herpes auxiliary receptor) [84]. Herpes simplex virus type 1 glycoproteins have also been demonstrated to require elongated O-glycans for the propagation of the virus [85]. In Ebola glycoprotein, the loss of the heavily O-glycosylated mucin-like domain reduced the antibody shielding effects that were observed on the virion [86]. Effects such as shielding of antigenic sites if the glycoprotein is to be used as a vaccine or target for antibodies should be carefully considered, as removal of the shielding may be beneficial or detrimental depending on the usage.

Yeast species do not impart human-like O-glycans onto yeast proteins, but as with N-linked modifications, there are now recombinant yeast strains modified to produce mucin-like glycosylation

[87]. Some yeasts do impart yeast O-glycosylation onto proteins, including *P. pastoris*. The initial synthesis of O-linked glycans in *P. pastoris* is the transfer of mannose from dolichol-phosphomannose in the yeast secretory pathway by members of the protein-O-mannosyltransferase (PMT) family, and these cores may be extended resulting in heavily mannosylated O-glycans [88]. β-Man- and phosphate-containing O-linked glycans synthesized by *P. pastoris* have been found [89]. As with the hypermannosylated N-glycans, the mannosylated O-glycans are immunogenic in mammals [90].

O-glycosylation has been studied in three baculovirus/insect expression cell lines, Sf-9, Mb, and Tn. All three insect cell lines express the GalNAcα-O-Ser/Thr antigen that reflects high UDP-GalNAc:polypeptide N-acetylgalactosaminyltransferase activity and carry a Tn antigen (GalNAc only). Only some O-linked GalNAc residues are further processed by the addition of β1,3-linked Gal residues to form T-antigen or core 1 [91].

A comparison of the pseudorabies virus envelope protein gp50 expressed in Sf9, Vero (a mammalian cell line derived from African Green Monkey endothelial cells), or CHO cells showed that only the mammalian cell lines produced core 1 disaccharide Galβ1-3GalNAc, and those had most of the T-antigen substituted by one or two sialic acid residues. Most of the gp50 produced by insect cells contained only O-linked GalNAc (Tn-antigen) [92]. The Sf9 cell line had no O-glycans substituted by sialic acid residues. This was further demonstrated in Ebola glycoprotein production in Sf9 cells, where only GalNAc Tn antigen was observed [34]. The Tn antigen may be immunogenic in mammals, and further work should investigate the impact of the Tn antigen on the immunogenicity of insect-produced mucin glycosylated proteins. However, contrary to this work, Gaunitz et al. [93] found O-glycans containing hexuronic acid (HexA) prevalent in both Sf9 and High Five cells. Sulfate (Hi-5 and Sf9) and phosphocholine (Sf9) O-glycan substitutions were also detected, and have not been seen previously in other species.

Mammalian cell lines may also express altered O-glycans compared to native protein expression in mammals, particularly when grown in serum-free media. Many truncated glycans found in tumor cells, and considered tumor antigens, occur in specific growth conditions and in some cell lines derived from cancers, including the Tn antigen and the STn antigen (sialyated Tn antigen) [94]. CHO cells also have some differences in O-glycosylation compared to human cells. Specifically, CHO cells produce mostly mono- or di-sialyated core 1. CHO-K1 cells, in general, are not able to elongate or branch core 1 O-glycans because they lack the necessary glycosyltransferase activity. In addition, core 3 O-glycans are also absent in CHO-K1 cells, unlike in most human cell lines, where they are not abundant, but are present. As with N-linked glycosylation, there are cell lines with specific mutations introduced to alter the character of the O-glycans imparted, including recent work engineering sulfated o-glycans in HEK293 [95], and developing more murine-like glycans in CHO cells [96]. However, there have been generally limited studies examining the O-glycoproteome differences between mammalian cell lines [97]. There have been recent improvements in protocols to analyze O-glycosylation, and particularly map the sites of O-glycans, and future studies may be able to better untangle differences in O-glycans between cell lines [98].

3.3 Purification of Glycoproteins

Proteins are often purified on a lab scale through the use of affinity tags on the proteins. There are a diverse range of available affinity tags, reviewed in [99], and some of these may have an impact on the folding, thermal stability, and glycosylation of the produced proteins. Polyhistidine tags, a sequence of multiple histidine residues that can chelate with Nickel for purification, alter the thermal stability of the proteins they are added to [100]. The use of serum in cell culture media may additionally interfere with downstream purification of the glycoprotein, as polyhistidine-tagged proteins that are purified with nickel column purification have been shown to be contaminated with a protein was identified by mass spectrometry as bovine serotransferase, a component of the fetal bovine serum (FBS) [34]. The contamination is specific to nickel purification, and Anti-His Affinity Resin can be used for purification to prevent bovine serotransferase protein contamination [75].

3.4 Protein Sequence Modifications

Aside from the use of tags, other modifications as part of production may also impact glycosylation. Many industrially produced proteins that are glycoproteins in their native form have sequence modifications to remove glycosylation motifs, where the glycosylation is not essential for the desired therapeutic function of the protein [101]. Eukaryotic systems, including *P. pastoris,* baculovirus/insect cell, and mammalian cell culture systems promote good protein folding and include forms of glycosylation [54, 102]. For many recombinant proteins, and particularly viral antigens, the structure of the protein may be stabilized or have domains added to facilitate multimer formation, removal of transmembrane domains for solubilization of proteins, or expression of different multimers from native forms [103]. For example, it has been shown that membrane-truncated soluble HIV-1 envelope glycoprotein (gp140) does not have the same glycosylation patterns, and particularly the high mannose glycans present on the membrane-anchored gp140 in trimeric form [104, 105].

4 Conclusions

Choosing a production system for glycoprotein expression requires the consideration of multiple factors, including considering the type of glycosylation specific to the cells in that system, the culture conditions of the system, and the downstream purification methods for the glycoprotein. Unlike in protein synthesis, the glycosylation is stochastic, and in higher organism production systems there is a great deal of heterogeneity in the glycans imparted. The aim of producing recombinant glycoproteins with human-like glycosylation has resulted in the production of recombinant humanized glycosylation systems in bacteria, yeast, insect cells, and even other mammalian cell lines. For glycoproteins intended for use as antigens, the immunogenicity of glycans from the different production systems should be considered, as they may alter the function and binding of the antigen.

The same general considerations for protein production apply to glycoproteins, including technical ease, the costs associated, the yield of the system, and the purification method. However, some of the methods popularly used to reduce cost, such as serum-free media cultures, should also be considered in terms of their impact of the glycans imparted. Changes such as removal of transmembrane domains, addition of large purification tags, and addition of multimerization domains should also be considered as potentially altering the type of glycosylation and site occupancy in the produced glycoproteins. Finally, the end purpose of the production should be considered, as glycoproteins that are intended for development into therapeutic proteins will have more stringent glycosylation requirements than glycoproteins produced for research purposes.

References

1. Lauc G, Krištić J, Zoldoš V (2014) Glycans – the third revolution in evolution. Front Genet 5. https://doi.org/10.3389/fgene.2014.00145
2. Gagneux P, Aebi M, Varki A (2015) Evolution of glycan diversity. In: Varki A, Cummings RD, Esko JD, Stanley P, Hart GW, Aebi M, Darvill AG, Kinoshita T, Packer NH, Prestegard JH, Schnaar RL, Seeberger PH (eds) Essentials of glycobiology, 3rd edn. Cold Spring Harbor Laboratory Press, Cold Spring Harbor
3. Warfield KL, Posten NA, Swenson DL et al (2007) Filovirus-like particles produced in insect cells: immunogenicity and protection in rodents. J Infect Dis 196:S421–S429. https://doi.org/10.1086/520612
4. Vasudevan D, Haltiwanger RS (2014) Novel roles for O-linked glycans in protein folding. Glycoconj J 31:417–426. https://doi.org/10.1007/s10719-014-9556-4
5. Brockhausen I, Stanley P (2015) O-GalNAc glycans. In: Varki A, Cummings RD, Esko JD, Stanley P, Hart GW, Aebi M, Darvill AG, Kinoshita T, Packer NH, Prestegard JH, Schnaar RL, Seeberger PH (eds) Essentials of glycobiology, 3rd edn. Cold Spring Harbor Laboratory Press, Cold Spring Harbor
6. Keefer LM, Piron MA, De Meyts P (1981) Human insulin prepared by recombinant DNA techniques and native human insulin interact identically with insulin receptors. Proc Natl Acad Sci U S A 78:1391–1395. https://doi.org/10.1073/pnas.78.3.1391

7. Rosendahl MS, Doherty DH, Smith DJ et al (2005) A long-acting, highly potent interferon alpha-2 conjugate created using site-specific PEGylation. Bioconjug Chem 16: 200–207. https://doi.org/10.1021/bc049713n
8. Sahdev S, Khattar SK, Saini KS (2008) Production of active eukaryotic proteins through bacterial expression systems: a review of the existing biotechnology strategies. Mol Cell Biochem 307:249–264. https://doi.org/10.1007/s11010-007-9603-6
9. Szymanski CM, Yao R, Ewing CP et al (1999) Evidence for a system of general protein glycosylation in campylobacter jejuni. Mol Microbiol 32:1022–1030. https://doi.org/10.1046/j.1365-2958.1999.01415.x
10. Szymanski CM, Logan SM, Linton D et al (2003) Campylobacter – a tale of two protein glycosylation systems. Trends Microbiol 11: 233–238. https://doi.org/10.1016/S0966-842X(03)00079-9
11. Nothaft H, Szymanski CM (2013) Bacterial protein N-glycosylation: new perspectives and applications *. J Biol Chem 288:6912–6920. https://doi.org/10.1074/jbc.R112.417857
12. Wacker M, Linton D, Hitchen PG et al (2002) N-linked glycosylation in campylobacter jejuni and its functional transfer into E. coli. Science 298:1790–1793. https://doi.org/10.1126/science.298.5599.1790
13. Kowarik M, Numao S, Feldman MF et al (2006) N-linked glycosylation of folded proteins by the bacterial oligosaccharyltransferase. Science 314:1148–1150. https://doi.org/10.1126/science.1134351
14. Schwarz F, Lizak C, Fan Y-Y et al (2011) Relaxed acceptor site specificity of bacterial oligosaccharyltransferase in vivo. Glycobiology 21:45–54. https://doi.org/10.1093/glycob/cwq130
15. Nothaft H, Szymanski CM (2010) Protein glycosylation in bacteria: sweeter than ever. Nat Rev Microbiol 8:765–778. https://doi.org/10.1038/nrmicro2383
16. Van Sorge NM, Bleumink NMC, Van Vliet SJ et al (2009) N-glycosylated proteins and distinct lipooligosaccharide glycoforms of campylobacter jejuni target the human C-type lectin receptor MGL. Cell Microbiol 11: 1768–1781. https://doi.org/10.1111/j.1462-5822.2009.01370.x
17. Schwarz F, Huang W, Li C et al (2010) A combined method for producing homogeneous glycoproteins with eukaryotic N-glycosylation. Nat Chem Biol 6:264–266. https://doi.org/10.1038/nchembio.314
18. Li C, Wang L-X (2018) Chemoenzymatic methods for the synthesis of glycoproteins. Chem Rev 118:8359–8413. https://doi.org/10.1021/acs.chemrev.8b00238
19. Valderrama-Rincon JD, Fisher AC, Merritt JH et al (2012) An engineered eukaryotic protein glycosylation pathway in Escherichia coli. Nat Chem Biol 8:434–436. https://doi.org/10.1038/nchembio.921
20. Pratama F, Linton D, Dixon N (2021) Genetic and process engineering strategies for enhanced recombinant N-glycoprotein production in bacteria. Microb Cell Factories 20:198. https://doi.org/10.1186/s12934-021-01689-x
21. Werten MWT, van den Bosch TJ, Wind RD et al (1999) High-yield secretion of recombinant gelatins by Pichia pastoris. Yeast 15: 1087–1096. https://doi.org/10.1002/(SICI)1097-0061(199908)15:11<1087::AID-YEA436>3.0.CO;2-F
22. Wang N, Wang KY, Xu F et al (2020) The effect of N-glycosylation on the expression of the tetanus toxin fragment C in Pichia pastoris. Protein Expr Purif 166:105503. https://doi.org/10.1016/j.pep.2019.105503
23. Gemmill TR, Trimble RB (1999) Overview of N- and O-linked oligosaccharide structures found in various yeast species. Biochim Biophys Acta 1426:227–237. https://doi.org/10.1016/s0304-4165(98)00126-3
24. Jacobs PP, Geysens S, Vervecken W et al (2009) Engineering complex-type N-glycosylation in Pichia pastoris using GlycoSwitch technology. Nat Protoc 4:58–70. https://doi.org/10.1038/nprot.2008.213
25. Hamilton SR, Gerngross TU (2007) Glycosylation engineering in yeast: the advent of fully humanized yeast. Curr Opin Biotechnol 18:387–392. https://doi.org/10.1016/j.copbio.2007.09.001
26. Research C for BE and Flublok quadrivalent (2022) FDA
27. Kost TA, Condreay JP, Jarvis DL (2005) Baculovirus as versatile vectors for protein expression in insect and mammalian cells. Nat Biotechnol 23:567–575. https://doi.org/10.1038/nbt1095
28. High Five™ Cells in Express Five™ Medium. https://www.thermofisher.com/order/catalog/product/B85502. Accessed 27 Feb 2023
29. Kozlov G, Pocanschi CL, Rosenauer A et al (2010) Structural basis of carbohydrate recognition by calreticulin. J Biol Chem 285: 38612–38620. https://doi.org/10.1074/jbc.M110.168294

30. Lamriben L, Graham JB, Adams BM et al (2016) N-glycan based ER molecular chaperone and protein quality control system: the calnexin binding cycle. Traffic Cph Den 17: 308–326. https://doi.org/10.1111/tra.12358
31. Kozlov G, Gehring K (2020) Calnexin cycle – structural features of the ER chaperone system. FEBS J 287:4322–4340. https://doi.org/10.1111/febs.15330
32. Altmann F, Staudacher E, Wilson IB et al (1999) Insect cells as hosts for the expression of recombinant glycoproteins. Glycoconj J 16:109–123. https://doi.org/10.1023/a:1026488408951
33. Aumiller JJ, Mabashi-Asazuma H, Hillar A et al (2012) A new glycoengineered insect cell line with an inducibly mammalianized protein N-glycosylation pathway. Glycobiology 22:417–428. https://doi.org/10.1093/glycob/cwr160
34. Clarke EC, Collar AL, Ye C (2017) Production and purification of filovirus glycoproteins in insect and mammalian cell lines. Sci Rep 7: 15091. https://doi.org/10.1038/s41598-017-15416-3
35. Schiller B, Hykollari A, Yan S et al (2012) Complicated N-linked glycans in simple organisms. Biol Chem 393:661–673. https://doi.org/10.1515/hsz-2012-0150
36. Loke I, Østergaard O, Heegaard NHH et al (2017) Paucimannose-rich N-glycosylation of spatiotemporally regulated human neutrophil elastase modulates its immune functions. Mol Cell Proteomics 16:1507–1527. https://doi.org/10.1074/mcp.M116.066746
37. Lin S-C, Jan J-T, Dionne B et al (2013) Different immunity elicited by recombinant H5N1 hemagglutinin proteins containing Pauci-mannose, high-mannose, or complex type N-Glycans. PLoS One 8:e66719. https://doi.org/10.1371/journal.pone.0066719
38. Hashimoto Y, Zhang S, Blissard GW (2010) Ao38, a new cell line from eggs of the black witch moth, Ascalapha odorata (Lepidoptera: Noctuidae), is permissive for AcMNPV infection and produces high levels of recombinant proteins. BMC Biotechnol 10:50. https://doi.org/10.1186/1472-6750-10-50
39. Staudacher E, Kubelka V, März L (1992) Distinct N-glycan fucosylation potentials of three lepidopteran cell lines. Eur J Biochem 207: 987–993. https://doi.org/10.1111/j.1432-1033.1992.tb17134.x
40. Bac-to-Bac™ Baculovirus expression system. https://www.thermofisher.com/order/catalog/product/10359016. Accessed 27 Feb 2023
41. Possee RD, Chambers AC, Graves LP et al (2019) Recent developments in the use of Baculovirus expression vectors. Curr Issues Mol Biol 34:215–230. https://doi.org/10.21775/cimb.034.215
42. Tang X-C, Lu H-R, Ross TM (2011) Baculovirus-produced influenza virus-like particles in mammalian cells protect mice from lethal influenza challenge. Viral Immunol 24:311–319. https://doi.org/10.1089/vim.2011.0016
43. Joshi L, Davis TR, Mattu TS et al (2000) Influence of Baculovirus-host cell interactions on complex N-linked glycosylation of a recombinant human protein. Biotechnol Prog 16:650–656. https://doi.org/10.1021/bp000057p
44. Tretter V, Altmann F, Kubelka V et al (1993) Fucose α1,3-linked to the core region of glycoprotein N-Glycans creates an important epitope for IgE from honeybee venom allergic individuals. Int Arch Allergy Immunol 102: 259–266. https://doi.org/10.1159/000236534
45. Prenner C, Mach L, Glössl J et al (1992) The antigenicity of the carbohydrate moiety of an insect glycoprotein, honey-bee (Apis mellifera) venom phospholipase A2. The role of α1,3-fucosylation of the asparagine-bound N-acetylglucosamine. Biochem J 284:377–380. https://doi.org/10.1042/bj2840377
46. Palmberger D, Ashjaei K, Strell S et al (2014) Minimizing fucosylation in insect cell-derived glycoproteins reduces binding to IgE antibodies from the sera of patients with allergy. Biotechnol J 9:1206–1214. https://doi.org/10.1002/biot.201400061
47. Kigawa T, Yabuki T, Yoshida Y et al (1999) Cell-free production and stable-isotope labeling of milligram quantities of proteins. FEBS Lett 442:15–19. https://doi.org/10.1016/S0014-5793(98)01620-2
48. Carlson ED, Gan R, Hodgman CE et al (2012) Cell-free protein synthesis: applications come of age. Biotechnol Adv 30:1185–1194. https://doi.org/10.1016/j.biotechadv.2011.09.016
49. Schoborg JA, Hershewe JM, Stark JC et al (2018) A cell-free platform for rapid synthesis and testing of active oligosaccharyltransferases. Biotechnol Bioeng 115:739–750. https://doi.org/10.1002/bit.26502
50. Hershewe JM, Warfel KF, Iyer SM et al (2021) Improving cell-free glycoprotein synthesis by characterizing and enriching native

membrane vesicles. Nat Commun 12:2363. https://doi.org/10.1038/s41467-021-22329-3

51. Jaroentomeechai T, Kwon YH, Liu Y et al (2022) A universal glycoenzyme biosynthesis pipeline that enables efficient cell-free remodeling of glycans. Nat Commun 13:6325. https://doi.org/10.1038/s41467-022-34029-7
52. Pritchard LK, Harvey DJ, Bonomelli C et al (2015) Cell- and protein-directed glycosylation of native cleaved HIV-1 envelope. J Virol 89:8932–8944. https://doi.org/10.1128/JVI.01190-15
53. Bagdonaite I, Wandall HH (2018) Global aspects of viral glycosylation. Glycobiology 28:443–467. https://doi.org/10.1093/glycob/cwy021
54. Zhu J (2012) Mammalian cell protein expression for biopharmaceutical production. Biotechnol Adv 30:1158–1170. https://doi.org/10.1016/j.biotechadv.2011.08.022
55. Tihanyi B, Nyitray L (2020) Recent advances in CHO cell line development for recombinant protein production. Drug Discov Today Technol 38:25–34. https://doi.org/10.1016/j.ddtec.2021.02.003
56. Daramola O, Stevenson J, Dean G et al (2014) A high-yielding CHO transient system: coexpression of genes encoding EBNA-1 and GS enhances transient protein expression. Biotechnol Prog 30:132–141. https://doi.org/10.1002/btpr.1809
57. Ghaderi D, Zhang M, Hurtado-Ziola N et al (2012) Production platforms for biotherapeutic glycoproteins. Occurrence, impact, and challenges of non-human sialylation. Biotechnol Genet Eng Rev 28:147–175. https://doi.org/10.5661/bger-28-147
58. Fukuta K, Yokomatsu T, Abe R et al (2000) Genetic engineering of CHO cells producing human interferon-γ by transfection of sialyltransferases. Glycoconj J 17:895–901. https://doi.org/10.1023/A:1010977431061
59. Lin N, Mascarenhas J, Sealover NR et al (2015) Chinese hamster ovary (CHO) host cell engineering to increase sialylation of recombinant therapeutic proteins by modulating sialyltransferase expression. Biotechnol Prog 31:334–346. https://doi.org/10.1002/btpr.2038
60. Bosques CJ, Collins BE, Meador JW et al (2010) Chinese hamster ovary cells can produce galactose-α-1,3-galactose antigens on proteins. Nat Biotechnol 28:1153–1156. https://doi.org/10.1038/nbt1110-1153
61. Noguchi A, Mukuria CJ, Suzuki E et al (1995) Immunogenicity of N-Glycolylneuraminic acid-containing carbohydrate chains of recombinant human erythropoietin expressed in Chinese hamster ovary cells. J Biochem 117:59–62. https://doi.org/10.1093/oxfordjournals.jbchem.a124721
62. Ghaderi D, Taylor RE, Padler-Karavani V et al (2010) Implications of the presence of N-glycolylneuraminic acid in recombinant therapeutic glycoproteins. Nat Biotechnol 28:863–867. https://doi.org/10.1038/nbt.1651
63. Fischer S, Handrick R, Otte K (2015) The art of CHO cell engineering: a comprehensive retrospect and future perspectives. Biotechnol Adv 33:1878–1896. https://doi.org/10.1016/j.biotechadv.2015.10.015
64. Davies J, Jiang L, Pan L-Z et al (2001) Expression of GnTIII in a recombinant anti-CD20 CHO production cell line: expression of antibodies with altered glycoforms leads to an increase in ADCC through higher affinity for FCγRIII. Biotechnol Bioeng 74:288–294. https://doi.org/10.1002/bit.1119
65. Ferrara C, Brünker P, Suter T et al (2006) Modulation of therapeutic antibody effector functions by glycosylation engineering: influence of Golgi enzyme localization domain and co-expression of heterologous β1, 4-N-acetylglucosaminyltransferase III and Golgi α-mannosidase II. Biotechnol Bioeng 93:851–861. https://doi.org/10.1002/bit.20777
66. Yang G, Wang Q, Chen L et al (2021) Glycoproteomic characterization of FUT8 Knock-out CHO cells reveals roles of FUT8 in the glycosylation. Front Chem 9
67. Swiech K, Picanço-Castro V, Covas DT (2012) Human cells: new platform for recombinant therapeutic protein production. Protein Expr Purif 84:147–153. https://doi.org/10.1016/j.pep.2012.04.023
68. Dumont J, Euwart D, Mei B et al (2016) Human cell lines for biopharmaceutical manufacturing: history, status, and future perspectives. Crit Rev Biotechnol 36:1110–1122. https://doi.org/10.3109/07388551.2015.1084266
69. Mancuso ME, Mannucci PM (2014) Fc-fusion technology and recombinant FVIII and FIX in the management of the hemophilias. Drug Des Devel Ther 8:365–371. https://doi.org/10.2147/DDDT.S47312
70. Graham FL, Smiley J, Russell WC et al (1977) Characteristics of a human cell line transformed by DNA from human adenovirus

type 5. J Gen Virol 36:59–72. https://doi.org/10.1099/0022-1317-36-1-59

71. Yamaguchi K, Itoh K, Ohnishi N et al (2003) Engineered long terminal repeats of retroviral vectors enhance transgene expression in hepatocytes in vitro and in vivo. Mol Ther 8:796–803. https://doi.org/10.1016/j.ymthe.2003.08.005
72. Pulix M, Lukashchuk V, Smith DC et al (2021) Molecular characterization of HEK293 cells as emerging versatile cell factories. Curr Opin Biotechnol 71:18–24. https://doi.org/10.1016/j.copbio.2021.05.001
73. Croset A, Delafosse L, Gaudry J-P et al (2012) Differences in the glycosylation of recombinant proteins expressed in HEK and CHO cells. J Biotechnol 161:336–348. https://doi.org/10.1016/j.jbiotec.2012.06.038
74. Böhm E, Seyfried BK, Dockal M et al (2015) Differences in N-glycosylation of recombinant human coagulation factor VII derived from BHK, CHO, and HEK293 cells. BMC Biotechnol 15:87. https://doi.org/10.1186/s12896-015-0205-1
75. Collar AL, Clarke EC, Anaya E et al (2017) Comparison of N- and O-linked glycosylation patterns of ebolavirus glycoproteins. Virology 502:39–47. https://doi.org/10.1016/j.virol.2016.12.010
76. Laustsen AH, Solà M, Jappe EC et al (2016) Biotechnological trends in spider and scorpion antivenom development. Toxins 8:226. https://doi.org/10.3390/toxins8080226
77. Serrato JA, Hernández V, Estrada-Mondaca S et al (2007) Differences in the glycosylation profile of a monoclonal antibody produced by hybridomas cultured in serum-supplemented, serum-free or chemically defined media. Biotechnol Appl Biochem 47:113–124. https://doi.org/10.1042/BA20060216
78. Yoo EM, Yu LJ, Wims LA et al (2010) Differences in N-glycan structures foundon recombinant IgA1 and IgA2 producedin murine myeloma and CHO cell lines. mAbs 2:320–334
79. Costa AR, Withers J, Rodrigues ME et al (2013) The impact of cell adaptation to serum-free conditions on the glycosylation profile of a monoclonal antibody produced by Chinese hamster ovary cells. New Biotechnol 30:563–572. https://doi.org/10.1016/j.nbt.2012.12.002
80. LeFloch F, Tessier B, Chenuet S et al (2006) Related effects of cell adaptation to serum-free conditions on murine EPO production and glycosylation by CHO cells. Cytotechnology 52:39–53. https://doi.org/10.1007/s10616-006-9039-y
81. Kunkel JP, Jan DCH, Jamieson JC et al (1998) Dissolved oxygen concentration in serum-free continuous culture affects N-linked glycosylation of a monoclonal antibody. J Biotechnol 62:55–71. https://doi.org/10.1016/S0168-1656(98)00044-3
82. Zanghi JA, Schmelzer AE, Mendoza TP et al (1999) Bicarbonate concentration and osmolality are key determinants in the inhibition of CHO cell polysialylation under elevated pCO2 or pH. Biotechnol Bioeng 65:182–191. https://doi.org/10.1002/(SICI)1097-0290(19991020)65:2<182::AID-BIT8>3.0.CO;2-D
83. Senger RS, Karim MN (2003) Effect of shear stress on intrinsic CHO culture state and glycosylation of recombinant tissue-type plasminogen activator protein. Biotechnol Prog 19:1199–1209. https://doi.org/10.1021/bp025715f
84. Mårdberg K, Nyström K, Tarp MA et al (2004) Basic amino acids as modulators of an O-linked glycosylation signal of the herpes simplex virus type 1 glycoprotein gC: functional roles in viral infectivity. Glycobiology 14:571–581. https://doi.org/10.1093/glycob/cwh075
85. Bagdonaite I, Nordén R, Joshi HJ et al (2016) Global mapping of O-glycosylation of varicella zoster virus, human cytomegalovirus, and Epstein-Barr virus. J Biol Chem 291:12014–12028. https://doi.org/10.1074/jbc.M116.721746
86. Noyori O, Matsuno K, Kajihara M et al (2013) Differential potential for envelope glycoprotein-mediated steric shielding of host cell surface proteins among filoviruses. Virology 446:152–161. https://doi.org/10.1016/j.virol.2013.07.029
87. Amano K, Chiba Y, Kasahara Y et al (2008) Engineering of mucin-type human glycoproteins in yeast cells. Proc Natl Acad Sci 105:3232–3237. https://doi.org/10.1073/pnas.0710412105
88. Nett JH, Cook WJ, Chen M-T et al (2013) Characterization of the Pichia pastoris protein-O-mannosyltransferase gene family. PLoS One 8:e68325. https://doi.org/10.1371/journal.pone.0068325
89. Trimble RB, Lubowski C, Hauer CR III et al (2004) Characterization of N- and O-linked glycosylation of recombinant human bile salt–stimulated lipase secreted by Pichia pastoris. Glycobiology 14:265–274. https://doi.org/10.1093/glycob/cwh036

90. Gustafsson A, Sjöblom M, Strindelius L et al (2011) Pichia pastoris-produced mucin-type fusion proteins with multivalent O-glycan substitution as targeting molecules for mannose-specific receptors of the immune system. Glycobiology 21:1071–1086. https://doi.org/10.1093/glycob/cwr046
91. Lopez M, Tetaert D, Juliant S et al (1999) O-glycosylation potential of lepidopteran insect cell lines. Biochim Biophys Acta 1427: 49–61. https://doi.org/10.1016/S0304-4165(98)00176-7
92. Thomsen DR, Post LE, Elhammer ÅP (1990) Structure of O-glycosidically linked oligosaccharides synthesized by the insect cell line Sf9. J Cell Biochem 43:67–79. https://doi.org/10.1002/jcb.240430107
93. Gaunitz S, Jin C, Nilsson A, Liu J et al (2013) Mucin-type proteins produced in the trichoplusia ni and spodoptera frugiperda insect cell lines carry novel O-glycans with phosphocholine and sulfate substitutions. Glycobiology 23:778–796. https://doi.org/10.1093/glycob/cwt015
94. Cervoni GE, Cheng JJ, Stackhouse KA et al (2020) O-glycan recognition and function in mice and human cancers. Biochem J 477: 1541–1564. https://doi.org/10.1042/BCJ20180103
95. de Haan N, Narimatsu Y, Koed Møller Aasted M et al (2022) In-depth profiling of O-glycan isomers in human cells using C18 nanoliquid chromatography–mass spectrometry and glycogenomics. Anal Chem 94:4343–4351. https://doi.org/10.1021/acs.analchem.1c05068
96. Gupta S, Shah B, Fung CS et al (2023) Engineering protein glycosylation in CHO cells to be highly similar to murine host cells. Front Bioeng Biotechnol 11:1113994
97. Tejwani V, Andersen MR, Nam JH et al (2018) Glycoengineering in CHO cells: advances in systems biology. Biotechnol J 13: 1700234. https://doi.org/10.1002/biot.201700234
98. Yang W, Song A, Ao M et al (2020) Large-scale mapping of site-specific O-GalNAc glycoproteome. Nat Protoc 15:2589–2610. https://doi.org/10.1038/s41596-020-0345-1
99. Kimple ME, Brill AL, Pasker RL (2013) Overview of affinity tags for protein purification. Curr Protoc Protein Sci 73:Unit-9.9. https://doi.org/10.1002/0471140864.ps0909s73
100. Booth WT, Schlachter CR, Pote S et al (2018) Impact of an N-terminal polyhistidine tag on protein thermal stability. ACS Omega 3:760–768. https://doi.org/10.1021/acsomega.7b01598
101. Chen R (2012) Bacterial expression systems for recombinant protein production: E. coli and beyond. Biotechnol Adv 30:1102–1107. https://doi.org/10.1016/j.biotechadv.2011.09.013
102. Harrison RL, Jarvis DL (2006) Protein N-glycosylation in the baculovirus-insect cell expression system and engineering of insect cells to produce "mammalianized" recombinant glycoproteins. Adv Virus Res 68:159–191. https://doi.org/10.1016/S0065-3527(06)68005-6
103. Krammer F, Margine I, Tan GS et al (2012) A carboxy-terminal trimerization domain stabilizes conformational epitopes on the stalk domain of soluble recombinant hemagglutinin substrates. PLoS One 7:e43603. https://doi.org/10.1371/journal.pone.0043603
104. Go EP, Herschhorn A, Gu C et al (2015) Comparative analysis of the glycosylation profiles of membrane-anchored HIV-1 envelope glycoprotein trimers and soluble gp140. J Virol 89:8245–8257. https://doi.org/10.1128/JVI.00628-15
105. Behrens A-J, Crispin M (2017) Structural principles controlling HIV envelope glycosylation. Curr Opin Struct Biol 44:125–133. https://doi.org/10.1016/j.sbi.2017.03.008

Correction to: SARS-CoV-2 S-Protein–Ace2 Binding Analysis Using Surface Plasmon Resonance

Jason Baardsnes and Béatrice Paul-Roc

Correction to:
Chapter 5 in: Steven B. Bradfute (ed.), *Recombinant Glycoproteins: Methods and Protocols*, Methods in Molecular Biology, vol. 2762,
https://doi.org/10.1007/978-1-0716-3666-4_5

The original version of Chapter 5 "SARS-CoV-2S-Protein–Ace2 Binding Analysis Using Surface Plasmon Resonance" was inadvertently published with a mistake in the chapter title. The chapter title has been updated as "SARS-CoV-2 S-Protein–Ace2 Binding Analysis Using Surface Plasmon Resonance"

The correction chapter has been updated with these changes.

The updated version of this chapter can be found at
https://doi.org/10.1007/978-1-0716-3666-4_5

Steven B. Bradfute (ed.), *Recombinant Glycoproteins: Methods and Protocols*, Methods in Molecular Biology, vol. 2762,
https://doi.org/10.1007/978-1-0716-3666-4_21,

Correction to: Solubilization of Oligomeric Cell-Free Synthesized Proteins Using SMA Copolymers

Jessica Ullrich, Lisa Haueis, Carsten Ohlhoff, Anne Zemella, Stefan Kubick, and Marlitt Stech

Correction to:
Chapter 18 in: Steven B. Bradfute (ed.), *Recombinant Glycoproteins: Methods and Protocols,* Methods in Molecular Biology, vol. 2762,
https://doi.org/10.1007/978-1-0716-3666-4_18

The chapter "Solubilization of Oligomeric Cell-Free Synthesized Proteins Using SMA Copolymers", written by Jessica Ullrich, Lisa Haueis, Carsten Ohlhoff, Anne Zemella, Stefan Kubick, and Marlitt Stech, was originally published without open access. The copyright of the chapter changed on 14-Mar-2024 to © The Author(s) 2025, and the chapter is now distributed under the terms of the Creative Commons Attribution 4.0 International License (http://creativecommons.org/licenses/by/4.0/), which permits use, sharing, adaptation, distribution and reproduction in any medium or format, as long as you give appropriate credit to the original author(s) and the source, provide a link to the Creative Commons licence and indicate if changes were made.

The images or other third party material in this chapter are included in the chapter's Creative Commons licence, unless indicated otherwise in a credit line to the material. If material is not included in the chapter's Creative Commons licence and your intended use is not permitted by statutory regulation or exceeds the permitted use, you will need to obtain permission directly from the copyright holder.

The updated version of this chapter can be found at
https://doi.org/10.1007/978-1-0716-3666-4_18

INDEX

Steven B. Bradfute (ed.), *Recombinant Glycoproteins: Methods and Protocols*, Methods in Molecular Biology, vol. 2762, https://doi.org/10.1007/978-1-0716-3666-4,

T

V

MIX
Papier aus verantwortungsvollen Quellen
Paper from responsible sources
FSC® C105338

If you have any concerns about our products,
you can contact us on
ProductSafety@springernature.com

In case Publisher is established outside the EU,
the EU authorized representative is:
Springer Nature Customer Service Center GmbH
Europaplatz 3, 69115 Heidelberg, Germany

Printed by Libri Plureos GmbH
in Hamburg, Germany